CAMBRIDGE TRACTS IN MATHEMATICS

General Editors

B. BOLLOBÁS, W. FULTON, A. KATOK,
F. KIRWAN, P. SARNAK, B. SIMON, B. TOTARO

176   The Monster Group and Majorana Involutions

The Monster Group and Majorana Involutions

A. A. IVANOV

Imperial College of Science, London
and
Institute for System Analysis, Moscow

CAMBRIDGE UNIVERSITY PRESS

Cambridge, New York, Melbourne, Madrid, Cape Town, Singapore, São Paulo, Delhi

Cambridge University Press
The Edinburgh Building, Cambridge CB2 8RU, UK

Published in the United States of America by Cambridge University Press, New York

www.cambridge.org
Information on this title: www.cambridge.org/9780521889940

First published 2009

Printed in the United Kingdom at the University Press, Cambridge

A catalogue record for this publication is available from the British Library

ISBN 978-0-521-88994-0 hardback

To Love and Nina

Contents

Preface   *page* **xi**

1 M_{24} and all that   **1**
1.1   Golay code   1
1.2   Todd module   5
1.3   Anti-heart module   10
1.4   Extraspecial extensions   13
1.5   Parker loop $(\mathcal{L}, \circ)$ 16
1.6   Aut$(\mathcal{L}, \circ)$ 19
1.7   Back to extraspecial extensions   23
1.8   Leech lattice and the monomial subgroup   25
1.9   Hexacode   31
1.10 Centralizer–commutator decompositions   34
1.11 Three bases subgroup   37

2   The Monster amalgam $\mathcal{M}$ **41**
2.1   Defining the amalgam   41
2.2   The options for G_1 43
2.3   Analysing G_{12} 45
2.4 G_2^s/Z_2 and its automorphisms   49
2.5   Assembling G_2 from pieces   52
2.6   Identifying $\{G_1, G_2\}$ 57
2.7   Conway's realization of G_2 59
2.8   Introducing G_3 62
2.9   Complementing in G_3^s 65
2.10 Automorphisms of G_3^s 68
2.11 $L_3(2)$-amalgam   70

	2.12	Constructing G_3	73
	2.13	G_3 contains $L_3(2)$	75
	2.14	Essentials	76
3		**196 883-representation of $\mathcal{M}$**	**80**
	3.1	Representing $\{G_1, G_2\}$	81
	3.2	Incorporating G_3	97
	3.3	Restricting to G_3^s	98
	3.4	Permuting the $\varphi(G_3^s)$-irreducibles	102
	3.5	G_3^φ is isomorphic to G_3	105
4		**2-local geometries**	**107**
	4.1	Singular subgroups	107
	4.2	Tilde geometry	111
	4.3	$2^{10+16}.\Omega_{10}^+(2)$-subgroup	112
	4.4	$2^2 \cdot (^2E_6(2)) : S_3$-subgroup	117
	4.5	Acting on the 196 883-module	119
5		**Griess algebra**	**121**
	5.1	Norton's observation	122
	5.2	3-dimensional S_4-algebras	124
	5.3	Krein algebras	126
	5.4	Elementary induced modules	128
	5.5	$(\Omega_{10}^+(2), \Pi_{155}^{10})$ is a Norton pair	132
	5.6	Allowances for subalgebras	134
	5.7	G_1-invariant algebras on $C_\Pi(Z_1)$	136
	5.8	G_2-invariant algebras on $C_\Pi(Z_2)$	138
	5.9	Producing $A^{(z)}$	145
	5.10	Expanding $A^{(z)}$	146
6		**Automorphisms of Griess algebra**	**149**
	6.1	Trace form	149
	6.2	Some automorphisms	150
	6.3	Involution centralizer	152
	6.4	Explicit version of $A^{(z)}$	154
	6.5	222-triangle geometry	163
	6.6	Finiteness and simplicity of $\varphi(G)$	165
7		**Important subgroups**	**168**
	7.1	Trident groups	169
	7.2	Tri-extraspecial groups	172
	7.3	Parabolics in $2^{11} \cdot M_{24}$	175

7.4	$3 \cdot Fi_{24}$-subgroup	178
7.5	$2 \cdot BM$-subgroup	184
7.6	p-locality	190
7.7	Thompson group	191
7.8	Harada–Norton group	195
8	**Majorana involutions**	**199**
8.1	$196\,883+1=196\,884$	199
8.2	Transposition axial vectors	200
8.3	Spectrum	201
8.4	Multiplicities	205
8.5	Fusion rules	208
8.6	Main definition	209
8.7	Sakuma's theorem	212
8.8	Majorana calculus	214
8.9	Associators	224
9	**The Monster graph**	**228**
9.1	Collinearity graph	228
9.2	Transposition graph	230
9.3	Simple connectedness	232
9.4	Uniqueness systems	233
	Fischer's story	**235**
	References	**245**
	Index	**251**

Preface

The *Monster* is the most amazing among the finite simple groups. The best way to approach it is via an amalgam called the *Monster amalgam.*

Traditionally one of the following three strategies are used in order to construct a finite simple group H:

(I) realize H as the automorphism group of an object Ξ;

(II) define H in terms of generators and relations;

(III) identify H as a subgroup in a 'familiar' group F generated by given elements.

The strategy offered by the *amalgam method* is a symbiosis of the above three. Here the starting point is a carefully chosen generating system $\mathcal{H} = \{H_i \mid i \in I\}$ of subgroups in H. This system is being axiomatized under the name of *amalgam* and for a while lives a life of its own independently of H. In a sense this is almost like (III) although there is no 'global' group F (familiar or non-familiar) in which the generation takes place. Instead one considers the class of all *completions* of $\mathcal{H}$ which are groups containing a quotient of $\mathcal{H}$ as a generating set. The axioms of $\mathcal{H}$ as an abstract amalgam do not guarantee the existence of a completion which contains an isomorphic copy of $\mathcal{H}$. This is a familiar feature of (II): given generators and relations it is impossible to say in general whether the defined group is trivial or not. This analogy goes further through the *universal completion* whose generators are all the elements of $\mathcal{H}$ and relations are all the identities hold in $\mathcal{H}$. The *faithful* completions (whose containing a generating copy of $\mathcal{H}$) are of particular importance. To expose a similarity with (I) we associate with a faithful completion X a combinatorial object $\Xi = \Xi(X, \mathcal{H})$ known as the *coset geometry* on which X induces a flag-transitive action. This construction equips some group theoretical notions with topological meaning: the homomorphisms of faithful completions correspond to local isomorphisms of the coset geometries; if X is the universal completion of $\mathcal{H}$, then $\Xi(X, \mathcal{H})$ is simply connected and vice versa. The ideal outcome is when the group H we are after is the universal completion of its subamalgam $\mathcal{H}$. In the classical situation, this is always the case whenever H is taken to be the universal central cover of a finite simple group of Lie type of rank at least 3 and $\mathcal{H}$ is the amalgam of parabolic subgroups containing a given Borel subgroup.

By the classification of flag-transitive Petersen and tilde geometries accomplished in [Iv99] and [ISh02], the Monster is the universal completion of an amalgam formed by a triple of subgroups

$$G_1 \sim 2^{1+24}_+ . Co_1,$$
$$G_2 \sim 2^{2+11+22} . (M_{24} \times S_3),$$
$$G_3 \sim 2^{3+6+12+18} . (3 \cdot S_6 \times L_3(2)),$$

where $[G_2 : G_1 \cap G_2] = 3$, $[G_3 : G_1 \cap G_3] = [G_3 : G_2 \cap G_3] = 7$. In fact, explicitly or implicitly, this amalgam has played an essential role in proofs of all principal results about the Monster, including discovery, construction, uniqueness, subgroup structure, Y-theory, moonshine theory.

The purpose of this book is to build up the foundation of the theory of the Monster group adopting the amalgam formed by G_1, G_2, and G_3 as the first principle. The strategy is similar to that followed for the fourth Janko group J_4 in [Iv04] and it amounts to accomplishing the following principal steps:

(A) 'cut out' the subset $G_1 \cup G_2 \cup G_3$ from the Monster group and axiomatize the partially defined multiplication to obtain an abstract *Monster amalgam* $\mathcal{M}$;

(B) deduce from the axioms of $\mathcal{M}$ that it exists and is unique up to isomorphism;

(C) by constructing a faithful (196 883-dimensional) representation of $\mathcal{M}$ establish the existence of a faithful completion;

(D) show that a particular subamalgam in $\mathcal{M}$ possesses a unique faithful completion which is the (non-split) extension $2 \cdot BM$ of the group of order 2 by the Baby Monster sporadic simple group BM (this proves that every faithful completion of $\mathcal{M}$ contains $2 \cdot BM$ as a subgroup);

(E) by enumerating the suborbits in a graph on the cosets of the $2 \cdot BM$-subgroup in a faithful completion of $\mathcal{M}$ (known as the *Monster graph*), show that for any such completion the number of cosets is the same (equal to the index of $2 \cdot BM$ in the Monster group);

(F) defining G to be the universal completion of $\mathcal{M}$ conclude that G is the
Monster as we know it, that is a non-abelian simple group, in which G_1
is the centralizer of an involution and that

$$|G| = 2^{46} \cdot 3^{20} \cdot 5^9 \cdot 7^6 \cdot 11^2 \cdot 13^3 \cdot 17 \cdot 19 \cdot 23 \cdot 29 \cdot 31 \cdot 41 \cdot 47 \cdot 59 \cdot 71.$$

In terms of the Monster group G, the Monster graph can be defined as
the graph on the class of $2A$-involutions in which two involutions are
adjacent if and only if their product is again a $2A$-involution. The
centralizer in G of a $2A$-involution is just the above-mentioned subgroup
$2 \cdot BM$. It was known for a long time that the $2A$-involutions in the
Monster form a class of 6-transpositions in the sense that the product of
any two such involutions has order at most 6. At the same time the
$2A$-involutions act on the $196\,884$-dimensional G-module in a very
specific manner, in particular we can establish a G-invariant
correspondence of the $2A$-involutions with a family of so-called *axial
vectors* so that the action of an involution is described by some simple
rules formulated in terms of the axial vector along with the G-invariant
inner and algebra products on this module (the latter product goes under
the name of *Griess algebra*). The subalgebras in the Griess algebra
generated by pairs of axial vectors were calculated by Simon
Norton [N96]: there are nine isomorphism types and the dimension is at
most eight. By a remarkable result recently proved by Shinya Sakuma in
the framework of the Vertex Operator Algebras [Sak07], these nine types
as well as the 6-transposition property are implied by certain properties
of the axial vectors and the corresponding involutions. In this volume we
axiomatize these properties under the names of *Majorana axial vectors*
and *Majorana involutions*. The fact that the Monster is generated by
Majorana involutions will certainly dominate the future studies.

1

M_{24} and all that

This chapter can be considered as a usual warming up with Mathieu and Conway groups, prior to entering the realm of the Monster. It is actually aimed at a specific goal to classify the groups which satisfy the following condition:

$$T \sim 2_+^{1+22}.M_{24}$$

The quotient $O_2(T)/Z(T)$ (considered as a $GF(2)$-module for $T/O_2(T) \cong M_{24}$) has the irreducible Todd module $\mathcal{C}_{11}^*$ as a submodule and the irreducible Golay code module $\mathcal{C}_{11}$ as the corresponding factor module. It turns out that there are exactly two such groups T: one splits over $O_2(T)$ with $O_2(T)/Z(T)$ being the direct sum $\mathcal{C}_{11}^* \oplus \mathcal{C}_{11}$, while the other does not split, and the module $O_2(T)/Z(T)$ is indecomposable. The latter group is a section in the group which is the first member $2_+^{1+24}.Co_1$ of the Monster amalgam.

1.1 Golay code

Let F be a finite field, and let (m, n) be a pair of positive integers with $m \le n$. A linear (m, n)-code over F is a triple $(V_n, \mathcal{P}, \mathcal{C})$ where V_n is an n-dimensional F-space, $\mathcal{P}$ is a basis of V_n, and $\mathcal{C}$ is a m-dimensional subspace in V_n. Although the presence of V_n and $\mathcal{P}$ is always assumed, it is common practice to refer to such a code simply by naming $\mathcal{C}$. It is also assumed (often implicitly) that V_n is endowed with a bilinear form b with respect to which $\mathcal{P}$ is an orthonormal basis

$$b(p, q) = \delta_{pq} \text{ for } p, q \in \mathcal{P}.$$

1

The dual code of C is the orthogonal complement of C in V_n with respect to b, that is

$$\{e \mid e \in V_n, b(e, c) = 0 \text{ for every } c \in C\}.$$

Since b is non-singular, the dual of an (m, n)-code is an $(n - m, n)$-code. Therefore, C is self-dual if and only if it is totally singular of dimension half the dimension of V_n. The weight $wt(c)$ of a codeword $c \in C$ is the number of non-zero components of c with respect to the basis $\mathcal{P}$. The minimal weight of C is defined as

$$m(C) = \min_{c \in C \setminus \{0\}} wt(c).$$

The codes over the field of two elements are known as *binary codes*. In the binary case, the map which sends a subset of $\mathcal{P}$ onto the sum of its elements provides us with an identification of V_n with the power set of $\mathcal{P}$ (the set of all subsets of $\mathcal{P}$). Subject to this identification, the addition is performed by the symmetric difference operator, the weight is just the size and b counts the size of the intersection taken modulo 2, i.e. for $u, v \subseteq \mathcal{P}$ we have

$$u + v = (u \cup v) \setminus (u \cap v);$$
$$wt(u) = |u|;$$
$$b(u, v) = |u \cap v| \bmod 2.$$

A binary code is said to be *even* or *doubly even* if the weights (i.e. sizes) of all the codewords are even or divisible by four, respectively. Notice that a doubly even code is always totally singular with respect to b.

A binary $(12, 24)$-code is called a (binary) *Golay code* if it is doubly even, self-dual of minimal weight 8. Up to isomorphism there exists a unique Golay code which we denote by C_{12}. In view of the above discussion, C_{12} can be defined as a collection of subsets of a 24-set $\mathcal{P}$ such that C_{12} is closed under the symmetric difference, the size of every subset in C_{12} is divisible by four but not four and $|C_{12}| = 2^{12}$. The subsets of $\mathcal{P}$ contained in C_{12} will be called *Golay sets*.

There are various constructions for the Golay code. We are going to review some basic properties of C_{12} and of its remarkable automorphism group M_{24}. The properties themselves are mostly construction-invariant while the proofs are not. We advise the reader to refer to his favorite construction to check the properties (which are mostly well-known anyway) while we will refer to Section 2.2 of [Iv99].

The weight distribution of C_{12} is

$$0^1 \, 8^{759} \, 12^{2576} \, 16^{759} \, 24^1,$$

which means that besides the improper subsets $\emptyset$ and $\mathcal{P}$ the family of Golay sets includes 759 subsets of size 8 (called *octads*), 759 complements of octads, and 2576 subsets of size 12 called *dodecads* (splitting into 1288 complementary pairs). If $\mathcal{B}$ is the set of octads, then $(\mathcal{P}, \mathcal{B})$ is a Steiner system of type $S(5, 8, 24)$ (this means that every 5-subset of $\mathcal{P}$ is in a unique octad). Up to isomorphism $(\mathcal{P}, \mathcal{B})$ is the unique system of its type and C_{12} can be redefined as the closure of $\mathcal{B}$ with respect to the symmetric difference operator in the unique Steiner system of type $S(5, 8, 24)$.

If $(V_{24}, \mathcal{P}, C_{12})$ is the full name of the Golay code, then

$$C_{12}^* := V_{24}/C_{12}$$

is known as the 12-dimensional *Todd module*. We continue to identify V_{24} with the power set of $\mathcal{P}$ and for $v \subseteq \mathcal{P}$ the coset $v + C_{12}$ (which is an element of C_{12}^*) will be denoted by v^*. It is known that for every $v \subseteq \mathcal{P}$ there is a unique integer $t(v) \in \{0, 1, 2, 3, 4\}$ such that $v^* = w^*$ for some $w \subseteq \mathcal{P}$ with $|w| = t(v)$. Furthermore, if $t(v) < 4$, then such w is uniquely determined by v; if $t(v) = 4$, then the collection

$$\mathcal{S}(v) = \{w \mid w \subseteq \mathcal{P}, |w| = 4, v^* = w^*\}$$

forms a *sextet*. The latter means that $\mathcal{S}(v)$ is a partition of $\mathcal{P}$ into six 4-subsets (also known as *tetrads*) such that the union of any two tetrads from $\mathcal{S}(v)$ is an octad. Every tetrad w is in the unique sextet $\mathcal{S}(w)$ and therefore the number of sextets is

$$1771 = \binom{24}{4}/6.$$

The automorphism group of the Golay code (which is the set of permutations of $\mathcal{P}$ preserving C_{12} as a whole) is the sporadic simple Mathieu group M_{24}. The action of M_{24} on $\mathcal{P}$ is 5-fold transitive and it is similar to the action on the cosets of another Mathieu group M_{23}. The stabilizer in M_{24} of a *pair* (a 2-subset of $\mathcal{P}$) is an extension of the simple Mathieu group M_{22} of degree 22 (which is the elementwise stabilizer of the pair) by an outer automorphism of order 2. The stabilizer of a *triple* is an extension of $L_3(4)$ (sometimes called the Mathieu group of degree 21 and denoted by M_{21}) by the symmetric group S_3 of the triple.

The sextet stabilizer $M(\mathcal{S})$ is an extension of a group $K_{\mathcal{S}}$ of order $2^6 \cdot 3$ by the symmetric group S_6 of the set of tetrads in the sextet. The group $K_{\mathcal{S}}$ (which is the kernel of the action of $M(S)$ on the tetrads in the sextet is a semidirect product of an elementary abelian group Q_S of order 2^6 and a group X_S of order 3 acting on Q_S fixed-point freely. If we put

$$Y_S = N_{M(S)}(X_S),$$

then $Y_S \cong 3 \cdot S_6$ is a complement to Q_S in $M(S)$; Y_S does not split over X_S and $C_{Y_S}(X_S) \cong 3 \cdot A_6$ is a perfect central extension of A_6. Furthermore, Y_S is the stabilizer in M_{24} of a 6-subset of $\mathcal{P}$ not contained in an octad (there is a single M_{24}-orbit on the set of such 6-subsets).

Because of the 5-fold transitivity of the action of M_{24} on $\mathcal{P}$, and since $(\mathcal{P}, \mathcal{B})$ is a Steiner system, the action of M_{24} on the octads is transitive. The stabilizer of an octad is the semidirect product of an elementary abelian group $Q_\mathcal{O}$ of order 2^4 (which fixes the octad elementwise) and a group $K_\mathcal{O}$ which acts faithfully as the alternating group A_8 on the elements in the octad and as the linear group $L_4(2)$ on $Q_\mathcal{O}$ (the latter action is by conjugation). Thus, the famous isomorphism $A_8 \cong L_4(2)$ can be seen here. The action of M_{24} on the dodecads is transitive, with the stabilizer of a dodecad being the simple Mathieu group M_{12} acting on the dodecad and on its complement as on the cosets of two non-conjugate subgroups each isomorphic to the smallest simple Mathieu group M_{11}. These two M_{11}-subgroups are permuted by an outer automorphism of M_{12} realized in M_{24} by an element which maps the dodecad onto its complement.

The following lemma is easy to deduce from the description of the stabilizers in M_{24} of elements in $\mathcal{C}_{12}$ and in $\mathcal{C}_{12}^*$.

Lemma 1.1.1 *Let u and v be elements of $\mathcal{C}_{12}$, and let $M(u)$ and $M(v)$ be their respective stabilizers in M_{24}. Then:*

(i) *$M(u)$ does not stabilize non-zero elements of $\mathcal{C}_{12}^*$;*
(ii) *if u and v are octads, then $(u \cap v)^*$ is the only non-zero element of $\mathcal{C}_{12}^*$ stabilized by $M(u) \cap M(v)$.*   □

A presentation $d = u + v$ of a dodecad as the sum (i.e. symmetric difference) of two octads determines the pair $u \cap v$ in the dodecad complementary to d and also a partition of d into two *heptads* (6-subsets) $u \setminus v$ and $v \setminus u$. If $\mathcal{K}$ is the set of all heptads obtained via such presentations of d, then $(d, \mathcal{K})$ is a Steiner system of type $S(5, 6, 12)$ (every 5-subset of d is in a unique heptad). There is a bijection between the pairs of complementary heptads from $\mathcal{K}$ and the set of pairs in $\mathcal{P} \setminus d$ such that if $d = h_1 \cup h_2$ corresponds to $\{p, q\}$, then $h_1 \cup \{p, q\}$ and $h_2 \cup \{p, q\}$ are octads, and d is their symmetric difference.

Lemma 1.1.2 *Let d be a dodecad, $\{p, q\}$ be a pair disjoint from d, and let $d = h_1 \cup h_2$ be the partition of d into heptads which correspond to $\{p, q\}$. Let A be the stabilizer in M_{24} of d and $\{p, q\}$, and let B be the stabilizer in M_{24} of h_1, h_2, and $\{p, q\}$. Then:*

(i) *$A \cong \mathrm{Aut}(S_6)$, while $B \cong S_6$;*
(ii) *$A \setminus B$ contains an involution.*

   Proof. (i) is Lemma 2.11.7 in [Iv99] while (ii) is a well-known property of the automorphism group of S_6. $\square$

Lemma 1.1.3 ([CCNPW]) *The following assertions hold:*

(i) *the outer automorphism group of M_{24} is trivial;*
(ii) *the Schur multiplier of M_{24} is trivial.* $\square$

1.2 Todd module

The 24-dimensional space V_{24} containing C_{12} and identified with the power set of $\mathcal{P}$ carries the structure of the $GF(2)$-permutation module of M_{24} acting on $\mathcal{P}$. With respect to this structure, C_{12} is a 12-dimensional submodule known as the *Golay code module*. Let $V^{(1)}$ and $V^{(23)}$ be the subspaces in V_{24} formed by the improper and even subsets of $\mathcal{P}$, respectively. Then $V^{(1)}$ and $V^{(23)}$ are the M_{24}-submodules contained in C_{12} and containing C_{12}, respectively. Put

$$C_{11} = C_{12}/V^{(1)} \text{ and } C_{11}^* = V^{(23)}/C_{12}.$$

The elements of $V_{24}/V^{(1)}$ are the partitions of $\mathcal{P}$ into pairs of subsets. There are two M_{24}-orbits on $C_{11} \setminus \{0\}$. One of the orbits consists of the partitions involving octads and other one the partitions into pairs of complementary dodecads. Acting on $C_{11}^* \setminus \{0\}$, the group M_{24} also has two orbits, this time indexed by the pairs and the sextets

$$|C_{11}| = 1 + 759 + 1288; \quad |C_{11}^*| = 1 + 276 + 1771.$$

Already from this numerology it follows that both C_{11} and C_{11}^* are irreducible and not isomorphic to each other. The modules C_{11} and C_{11}^* are known as the *irreducible Golay code and Todd modules* of M_{24}, respectively.

   Since C_{12} is totally singular and $V^{(1)}$ is the radical of b, the bilinear form b establishes a duality between C_{12} and C_{12}^* and also between C_{11} and C_{11}^*. Since M_{24} does not stabilize non-zero vectors in C_{12}^*, the latter is indecomposable. Because of the dually, C_{12} is also indecomposable.

Lemma 1.2.1 *The series*

$$0 < V^{(1)} < C_{12} < V^{(23)} < V_{24}$$

is the only composition series of V_{24} considered as the module for M_{24}.

Proof. We have seen that the above series is indeed a composition series. Since both C_{12} and C_{12}^* are indecomposable, in order to prove the uniqueness it is sufficient to show that $V_{24}/V^{(1)}$ does not contain C_{11}^* as a submodule. Such a submodule would contain an M_{24}-orbit X indexed by the pairs from $\mathcal{P}$. On the other hand, by the 5-fold transitivity of M_{24} on $\mathcal{P}$, the stabilizer of a pair stabilizes only one proper partition of $\mathcal{P}$ (which is the partition into the pair and its complement). Therefore, X has no choice but to consist of all such partitions. But then X would generate the whole of $V^{(23)}/V^{(1)}$, which proves that X does not exist.   □

If K is a group and U is a $GF(2)$-module for K, then $H^1(K,U)$ and $H^2(K,U)$ denote the first and the second cohomology groups of U. Each of these groups carries a structure of a $GF(2)$-module, in particular it is elementary abelian. The order of $H^1(K,U)$ is equal to the number of classes of complements to U in the semidirect product $U:K$ of U and K (with respect to the natural action), while the elements of $H^2(K,U)$ are indexed by the isomorphism types of extensions of U by K with the identity element corresponding to the split extension $U:K$. If W is the largest indecomposable extension of U by a trivial module, then $W/U \cong H^1(K,U)$ and all the complements to W in the semidirect product $W:K$ are conjugate and $H^1(K,W)$ is trivial. Dually, if V is the largest indecomposable extension of a trivial module V_0 by U, then $V_0^* \cong H^1(K,U^*)$ (here U^* is the dual module of U)

Lemma 1.2.2 *The following assertions hold:*

(i) $H^1(M_{24}, C_{11})$ *is trivial;*
(ii) $H^2(M_{24}, C_{11})$ *is trivial;*
(iii) $H^1(M_{24}, C_{11}^*)$ *has order 2;*
(iv) $H^2(M_{24}, C_{11}^*)$ *has order 2.*   □

Proof. The first cohomologies were computed in Section 9 in [Gri74]. The second cohomologies calculations are commonly attributied to D.J. Jackson [Jack80] (compare [Th79a]). All the assertions were rechecked by Derek Holt using his computer package for cohomology calculating.   □

In view of the paragraph before the lemma, by (ii) every extension of C_{11} by M_{24} splits; by (i) all the M_{24}-subgroups in the split extension $C_{11}:M_{24}$ are conjugate; by (iv) there exists a unique non-split extension (denoted by

$\mathcal{C}_{11}^* \cdot M_{24}$), while by (iii) the split extension $\mathcal{C}_{11}^* : M_{24}$ contains two classes of complements (these classes are fused in $\mathcal{C}_{12}^* : M_{24}$). Furthermore, every extension by $\mathcal{C}_{11}^*$ of a trivial module is semisimple (which means decomposable), while $\mathcal{C}_{12}$ is the largest indecomposable extension by $\mathcal{C}_{11}$ of a trivial module.

The following result has been recently established by Derek Holt using computer calculations and in a sense it assures success of our construction of the Monster amalgam.

Lemma 1.2.3 *The following assertions hold (where $\otimes$ and $\wedge$ denote the tensor and exterior products of modules):*

(i) $H^1(M_{24}, \mathcal{C}_{11}^* \otimes \mathcal{C}_{11}^*)$ *has order* 2;
(ii) $H^1(M_{24}, \mathcal{C}_{11}^* \wedge \mathcal{C}_{11}^*)$ *has order* 2. □

The following assertion is a rather standard consequence of the above statement but a sketch of a proof might be helpful.

Corollary 1.2.4 *The following hold:*

(i) *there exists a unique indecomposable extension of $\mathcal{C}_{11}^*$ by $\mathcal{C}_{11}$;*
(ii) *the extension in (i) carries a non-singular invariant quadratic form of plus type.*

Proof. To prove (i) we need to enumerate (up to the obvious equivalence) the pairs (V, M) where $M \cong M_{24}$ and V is an M-module having $\mathcal{C}_{11}^*$ as a submodule and $\mathcal{C}_{11}$ as the corresponding factor module. We should keep in mind that exactly one such pair corresponds to the decomposable extension (where $V = \mathcal{C}_{11}^* \oplus \mathcal{C}_{11}$). We start with a 22-dimensional $GF(2)$-space V and enumerate the suitable subgroups M in the general linear group $G = GL(V) \cong GL_{22}(2)$. Let U_1 and U_2 be a pair of disjoint 11-dimensional subspaces in V. Then $N_G(U_1)$ is a semidirect product of

$$C = C_G(U_1) \cap C_G(V/U_1) \cong U_1 \otimes U_2^*$$

and

$$K = N_G(U_1) \cap N_G(U_2) \cong GL(U_1) \times GL(U_2).$$

Let M_0 be a subgroup of K isomorphic to M_{24} which acts on U_1 as on $\mathcal{C}_{11}^*$ and on U_2 as on $\mathcal{C}_{11}$. It is easy to see that up to conjugation in K such subgroup M_0 can be chosen uniquely. Then (V, M_0) is the pair corresponding to the decomposable extension and any other M which suits the requirements is a complement to C in CM_0. Since $C \cong \mathcal{C}_{11}^* \otimes \mathcal{C}_{11}^*$ (as an M_0-module), (1.2.3 (i)) shows that there is exactly one further complement M_1. The action of M_1 on V is indecomposable.

To establish (ii) we enumerate the triples (V, M, q) where V and M are as above and q is an M-invariant non-singular quadratic form on V. In this case we start with an orthogonal space (V, q) and enumerate the suitable subgroups M in the orthogonal group $H = O(V, q)$. Since $\mathcal{C}_{11}^*$ is irreducible and since it is not self-dual, the submodule in V isomorphic to $\mathcal{C}_{11}^*$ must be totally isotropic. Therefore, we choose U_1 and U_2 as in (i) and assume that each of them is totally isotropic with respect to q. Then $N_H(U_1)$ is a semidirect product of

$$D = C_H(U_1) \cong U_1 \wedge U_1$$

and

$$L = N_H(U_1) \cap N_H(U_2) \cong GL(U_1).$$

Take M_0 to be the subgroup in L isomorphic to M_{24} which acts on U_1 as on $\mathcal{C}_{11}^*$ (since the bilinear form associated with q establishes a duality between U_1 and U_2, M_0 acts on U_2 as on $\mathcal{C}_{11}$). By (1.2.3 (ii)) up to conjugation, $DL \cong (\mathcal{C}_{11}^* \wedge \mathcal{C}_{11}^*) : M_0$ contains, besides M_0, exactly one further complement which must be M_1. By (i) the pair (V, M_1) corresponds to the unique indecomposable extension. $\qquad \square$

In the next section the indecomposable extension of $\mathcal{C}_{11}^*$ by $\mathcal{C}_{11}$ will be constructed explicitly along with the invariant quadratic form on it.

Let U be a $GF(2)$-module for K and let $U : K$ denote the semidirect product of U and K with respect to the natural action. Then (since U is abelian) all the complements to U in $U : K$ are conjugate in the automorphism group of $U : K$. Therefore, $H^1(K, U)$ 'contributes' to the outer automorphism group of $U : K$. In fact this contribution takes place for any extension of U by K. In order to explain this phenomenon (which is probably well-known), we recall the notion of *partial semidirect product* (cf. p.27 in [G68]).

Let X, Y, Z be groups, let

$$\varphi_X : Z \to X,$$
$$\varphi_Y : Z \to Y$$

be monomorphisms whose images are normal in X and Y, respectively, and let

$$\alpha : X \to \text{Aut}(Y)$$

be a homomorphism such that for every $x \in X$ and $z \in Z$ the following equality holds

$$\varphi_X^{-1}(x^{-1}\varphi_X(z)x) = \varphi_Y^{-1}(\varphi_Y(z)^{\alpha(x)})$$

(notice that the equality implies that $\alpha(X)$ normalizes $\varphi_Y(Z)$). The usual semidirect product $S = Y : X$ of Y and X associated with the homomorphism α is the set

$$S = \{(x, y) \mid x \in X, y \in Y\}$$

together with the multiplication rule

$$(x_1, y_1) \cdot (x_2, y_2) = (x_1 x_2, y_1^{\alpha(x_2)} y_2).$$

The *partial semidirect product* of Y and X over Z is the quotient of S/Z_S where

$$S_Z = \{(\varphi_X(z), \varphi_Y(z^{-1})) \mid z \in Z\}$$

is a 'diagonal' subgroup of S isomorphic to Z.

Lemma 1.2.5 *Let U be a $GF(2)$-module for a group K and let E be an extension (split or non-split) of U by K. Then there is an injective homomorphism of $H^1(K, U)$ into $\mathrm{Out}(E)$.*

Proof. Let W be the largest indecomposable extension of U by a trivial module, so that $W/U \cong H^1(K, U)$ and let P be the partial semidirect product of W and E over U with respect to the homomorphism $\alpha : E \to GL(W)$ such that U acts trivially and $E/U \cong K$ acts naturally. Then E is (isomorphic to) a self-centralizing normal subgroup in P (so that $P/E \leq \mathrm{Out}(E)$) and $P/E \cong W/U \cong H^1(K, U)$, which gives the result. $\square$

The partial semidirect product of $C_{11}^* \cdot M_{24}$ and C_{12}^* over C_{11}^* is a non-split extension of C_{12}^* by M_{24} and we denote the extension by $C_{12}^* \cdot M_{24}$. The following result is not difficult to deduce from the data on M_{24} and its modules we have revealed already.

Lemma 1.2.6 *The following assertions hold:*

(i) $C_{12}^* \cdot M_{24} = \mathrm{Aut}(C_{11}^* \cdot M_{24})$;
(ii) $C_{12}^* \cdot M_{24}$ *is the only non-split extension of C_{12}^* by M_{24};*
(iii) *the Schur multiplier of $C_{11}^* \cdot M_{24}$ is trivial.* $\square$

We will make use of the following result of a combinatorial nature:

Lemma 1.2.7 *Let X be a subgroup of the general linear group $GL(C_{11})$ containing M_{24} and having on the element set of C_{11} the same orbits as M_{24} does (with lengths 1, 759, and 1288). Then $X = M_{24}$.*

Proof. The semidirect product $C_{11} : X$ acting on the cosets of X is an affine rank three permutation group. All such groups were classified in [Lie87] and the assertion can be deduced as a consequence of that classification (of course the assertion can be proved independently as a pleasant combinatorial exercise). □

1.3 Anti-heart module

The *intersection map* $C_{12} \times C_{12} \to C_{12}^*$, defined by

$$(u, v) \mapsto (u \cap v)^*$$

is bilinear. Since any two Golay sets intersect evenly the image of the intersection map is C_{11}^*. The intersection map (considered as a cocycle) determines an extension W_{24} of C_{12}^* by C_{12}

$$W_{24} = \{(u, v^*) \mid u \in C_{12}, v^* \in C_{12}^*\}$$

with

$$(u_1, v_1^*) + (u_2, v_2^*) = (u_1 + u_2, v_1^* + v_2^* + (u_1 \cap u_2)^*).$$

By (1.2.1), the permutation module V_{24} is an indecomposable extension of C_{12} by C_{12}^*. We call W_{24} the *anti-permutation* module. As the next lemma shows, the anti-permutation module is also indecomposable.

Lemma 1.3.1 *Let W_{24} be the above-defined anti-permutation module of M_{24}. Put*

$$W^{(1)} = \{(u, 0) \mid u \in V^{(1)}\},$$
$$W^{(23)} = \{(u, v^*) \mid u \in C_{12}, v^* \in C_{11}^*\},$$
$$S_{12}^* = \{(0, v^*) \mid v^* \in C_{12}^*\}, \quad S_{12} = \{(u, 0) \mid u \in C_{12}\}.$$

The following assertions hold:

(i) *$W^{(1)}$ and $W^{(23)}$ are the only submodules in W_{24} of dimension and co-dimension 1, respectively;*
(ii) *S_{12}^* is a submodule in W_{24} isomorphic to C_{12}^*;*
(iii) *$S_{11}^* := S_{12}^* \cap W^{(23)}$ is a submodule isomorphic to C_{11}^*;*
(iv) *S_{12} is M_{24}-invariant; it maps bijectively onto $W_{24}/S_{12}^* \cong C_{12}$, although S_{12} is not closed under the addition in W_{24};*
(v) *$W^{(23)}/W^{(1)}$ is an indecomposable extension of C_{11}^* by C_{11}.*

Proof. Since C_{12} and C_{12}^* are indecomposable containing one trivial composition factor each, (i) follows. Now (ii), (iii), and (iv) are easy to check using the definition of W_{24}. By (iv) and (1.1.1 (i)) the image of $\mathcal{S}_{12}$ in $W_{24}/W^{(1)}$ is the only M_{24}-invariant subset of $W^{(23)}/W^{(1)}$ which maps bijectively onto $W^{(23)}/\langle \mathcal{S}_{12}^*, W^{(1)} \rangle \cong \mathcal{C}_{11}$. By (iv), the image of $\mathcal{S}_{12}$ is not closed under the addition and the indecosability claimed in (v) follows. □

The quotient $V^{(23)}/V^{(1)}$ is the *heart* of the permutation module V_{24}. Therefore, we call $W^{(23)}/W^{(1)}$ the *anti-heart* module for M_{24} (and denote it by $\mathcal{A}(M_{24})$). Since the image of (u, v) under the intersection stays unchanged when u or v or both are substituted by their complements in $\mathcal{P}$ and since $u \cap v$ is even whenever $u, v \in \mathcal{C}_{12}$, the intersection map induces a bilinear map of $\mathcal{C}_{11} \times \mathcal{C}_{11}$ onto $\mathcal{C}_{11}^*$. For $u, v \in \mathcal{C}_{11}$, we denote the image of (u, v) under the induced intersection map by $(u \cap v)^*$. Then

$$\mathcal{A}(M_{24}) = \{(u, v^*) \mid u \in \mathcal{C}_{11}, v^* \in \mathcal{C}_{11}^*\}$$

and, subject to the notation we have adopted, the addition rule can be written exactly as that for W_{24}. Abusing the notation, we will denote by $\mathcal{S}_{11}^*$ and $\mathcal{S}_{11}$ the subsets $\{(0, v^*) \mid v^* \in \mathcal{C}_{11}^*\}$ and $\{(u, 0) \mid u \in \mathcal{C}_{11}\}$ in the anti-heart module. Then $\mathcal{S}_{11}^*$ is an irreducible submodule, while $\mathcal{S}_{11}$ is just an M_{24}-invariant subset.

Combining (1.2.4 (i)) and (1.3.1 (iv)), we obtain the following.

Lemma 1.3.2 *Every $GF(2)$-module for M_{24} having $\mathcal{C}_{11}^*$ as a submodule and $\mathcal{C}_{11}$ as the corresponding factor module is either the direct sum of $\mathcal{C}_{11}^*$ and $\mathcal{C}_{11}$ or the anti-heart module $\mathcal{A}(M_{24})$.* □

As shown in the next lemma the anti-permutation module $W^{(24)}$ carries a non-singular M_{24}-invariant quadratic form.

Lemma 1.3.3 *Let q_p and β_p be the $GF(2)$-valued forms on W_{24} defined by*

$$q_p(u, v^*) = \left(\frac{1}{4}|u| + |u \cap v|\right) \bmod 2;$$

$$\beta_p((u_1, v_1^*), (u_2, v_2^*)) = \left(\frac{1}{2}|u_1 \cap u_2| + |u_1 \cap v_2| + |u_2 \cap v_1|\right) \bmod 2.$$

Then q_p is a non-singular M_{24}-invariant quadratic form of plus type and β_p is the associated bilinear form.

Proof. Recall that for $v \subseteq \mathcal{P}$ by v^*, we denote the coset $v + \mathcal{C}_{12} \in \mathcal{C}_{12}^*$. Since the subspace $\mathcal{S}_{12}^*$ is totally singular with respect to q_p, the latter form is indeed of plus type. The reader can either check all the required properties of q_p and β_w directly, or consult Exercise 6.4 in [A94], or wait till Section 1.8 where q_p and β_p will reappear as the natural quadratic and bilinear forms on the Leech lattice modulo 2. □

The form q_p induces on the anti-heart module $\mathcal{A}(M_{24})$ a quadratic form q_h of plus type whose associated bilinear form β_h is induced by β_p. The values of q_h and β_h can be calculated using the right-hand sides of the expressions for q_p and β_p in (1.3.3), only the domain is different. By (1.2.4 (ii)), we obtain the following.

Lemma 1.3.4 *The form q_h is the only non-zero quadratic form on $\mathcal{A}(M_{24})$ which is M_{24}-invariant.* $\square$

For the purpose of the next chapter we deduce here a characterization of the anti-permutation module W_{24} and of the M_{24}-invariant quadratic and bilinear forms on W_{24}.

Lemma 1.3.5 *The anti-permutation module W_{24} is the only indecomposable $GF(2)$-module X for M_{24} of dimension 24 which contains C_{12}^* as a submodule, C_{12} as the corresponding factor-module, and $\mathcal{A}(M_{24})$ as a section.*

Proof. Since $\mathcal{A}(M_{24})$ is an extension of C_{11}^* by C_{11}, the order of $H^1(M_{24}, \mathcal{A}(M_{24}))$ is at most 2 by (1.2.2 (i), (iii)). Since $\mathcal{A}(M_{24})$ is self-dual, the assertions holds in view of discussions in the paragraph before (1.2.2). $\square$

Lemma 1.3.6 *The following assertions hold:*

(i) *M_{24} acts on $\mathcal{A}(M_{24})$ with trivial centralizer in $GL(\mathcal{A}(M_{24}))$;*

(ii) *the only non-identity element in $GL(W_{24})$ which commutes with the action of M_{24} is the transvection τ whose centre is the generator $(\mathcal{P}, 0)$ of $W^{(1)}$ and whose axis is $W^{(23)}$.*

Proof. Let X be the subset of $\mathcal{A}(M_{24})$ formed by the elements $(u, 0)$ where u is a partition of $\mathcal{P}$ into an octad and its complement. By (1.1.1 (i)), X is the only M_{24}-orbit in $\mathcal{A}(M_{24})$ on which it acts as on the set of octads. Since the action of M_{24} on the set of octads is primitive non-regular and $\mathcal{A}(M_{24})$ (being indecomposable) is spanned by X, assertion (i) follows. Let C be the centralizer of M_{24} in $GL(W_{24})$. By (i) C acts trivially on $W^{(23)}/W^{(1)} \cong \mathcal{A}(M_{24})$. Therefore, every element of C either centralizes $W^{(23)}$ or acts on $W^{(23)}$ by a transvection with centre $(\mathcal{P}, 0)$. Since M_{24} does not stabilize hyperplanes in $W^{(23)}$, we conclude that C centralizes $W^{(23)}$. Since $W^{(1)}$ is the only 1-dimensional subspace in $W^{(23)}$ centralized by M_{24}, and since the transvection τ indeed commutes with the action of M_{24}, (ii) follows. $\square$

Lemma 1.3.7 *Let τ be the trasvection on W_{24} with centre $(\mathcal{P}, 0)$ and axis $W^{(23)}$. Then:*

(i) *if q_p^τ is the image under τ of the quadratic form q_p, then $q_p^\tau = q_p + q_p^0$ where the support of q_p^0 is $W_{24} \setminus W^{(23)}$;*

(ii) β_p *is stable under* τ*;*

(iii) β_p *is the only non-zero* M_{24}*-invariant symplectic form on* W_{24}*;*

(iv) *a non-zero* M_{24}*-invariant quadratic form on* W_{24} *coincides with one of* q_p, q_p^0 *and* $q_p^\tau = q_p + q_p^0$.

Proof. If $(u, v^*) \in W_{24} \setminus W^{(23)}$, then $|v|$ is odd and $(u, v^*)^\tau = (\mathcal{P} \setminus u, v^*)$. In this case, $|u \cap v|$ and $|(\mathcal{P} \setminus u) \cap v|$ have different parity which implies the inequality

$$q_p^\tau(u, v^*) \neq q_p(u, v^*),$$

and proves (i). Since the bilinear form associated with q_p^0 is the zero form, the bilinear form β_p^τ associated with q_p^τ coincides with β_p. This proves the assertion (ii), which of course can also be checked directly. A symplectic form on W_{24} can be viewed as an involutory isomorphism of W_{24} onto its dual space W_{24}^*. From this viewpoint the composition of two such M_{24}-invariant forms is an automorphism of W_{24} (that is an element of $GL(W_{24})$) which commutes with the action of M_{24}. By (1.3.6), the τ-invariance of β_p established in (ii) implies its uniqueness claimed in (iii). Finally, since $W^{(23)}$ is the only M_{24}-invariant hyperplane in W_{24}, q_p^0 is the only M_{24}-invariant quadratic form whose associated bilinear form is constantly zero. Hence (iv) follows. $\qquad\square$

We conclude this section by the following lemma which can either be proved directly or deduced from the proof of (1.2.4 (ii)).

Lemma 1.3.8 *The direct sum* $\mathcal{C}_{11} \oplus \mathcal{C}_{11}^*$ *carries a unique non-zero* M_{24}*-invariant quadratic form* q_d *defined by*

$$q_d(u, v^*) = |u \cap v| \bmod 2. \qquad\square$$

1.4 Extraspecial extensions

A 2-group Q is said to be an *extraspecial 2-group* if its centre Z is of order 2 and Q/Z is elementary abelian. It is common to identify Z with the field of integers modulo 2 via

$$m \mapsto z^m,$$

where z is the generator of Z and also to consider Q/Z as a $GF(2)$-vector space. The following result is standard.

Lemma 1.4.1 *Let β_e and q_e be mappings defined via*

$$\beta_e : (aZ, bZ) \mapsto [a, b], \quad q_e : aZ \mapsto a^2$$

for $a, b \in Q$. Then β_e and q_e are well-defined, q_e is a non-singular quadratic form on Q/Z, and β_e the bilinear form associated with q_e. □

It follows from the classification of non-singular quadratic forms that $\dim(Q/Z) = 2n$ for some integer $n \geq 1$ and for a given n up to isomorphism there are exactly two non-singular quadratic forms on Q/Z, one of type plus and one of type minus. The type of the orthogonal space $(Q/Z, q_e)$ completely determines the isomorphism type of Q. Thus for every odd power of 2, there exist exactly two extraspecial groups of that order denoted by

$$2_+^{1+2n} \text{ and } 2_-^{1+2n}$$

respectively, where the subscript specifies the type of the form q_e.

Every automorphism of the orthogonal space $(Q/Z, q_e)$ is known to be induced by an automorphism of Q and therefore

$$\mathrm{Aut}(2_\varepsilon^{1+2n}) = Q/Z.O(Q/Z, q_e) \cong 2^{2n}.O_{2n}^\varepsilon(2).$$

It is also known (cf. [Gri73a]) that the extension of inner automorphism group by the outer one does not split if $n \geq 3$.

Until further notice, we will restrict ourselves to the extraspecial groups 2_+^{1+2n} of plus type.

Let U be a faithful $GF(2)$-module for a group K. Let q be a K-invariant non-singular quadratic form on U of plus type and let β be the associated bilinear form (this implies particularly that the dimension of U is even, say $2n$). Let $\mathcal{E}(K, U, q)$ denote the set of groups R taking up to isomorphism and possessing the following properties:

(E1) $Q \cong 2_+^{1+2n}$ is normal in R;

(E2) $R/Q \cong K$;

(E3) there is a surjective homomorphism $\chi : Q \to U$ with kernel $Z = Z(Q)$ such that the induced isomorphism of Q/Z onto U commutes with the action of K;

(E4) if $a \in Q$, then

$$q_e(a) = q(\chi(a)).$$

First we show that a *holomorph* of every extraspecial group 2_+^{1+2n} (i.e. an extension of Q by $\mathrm{Out}(Q)$) always exists.

Lemma 1.4.2 *Let* $Q \cong 2_+^{1+2n}$, *let* Z *be the centre of* Q, *let* q_e *be the above-defined quadratic form on* Q/Z, *and let* $O_{2n}^+(2)$ *be the outer automorphism group of* Q. *Then the family* $\mathcal{E}(O_{2n}^+(2), Q/Z, q_e)$ *is non-empty and a unique group in this family possesses a faithful* 2^n-*dimensional representation.*

Proof. It is known that Q possesses (up to equivalence) a unique faithful irreducible representation φ over real numbers with dimension 2^n. Furthermore, if V denotes the corresponding 2^n-dimensional real vector space, then $\mathrm{Im}(\varphi)$ is contained in $SL(V)$. The representation can be constructed as follows: (a) take a maximal elementary abelian subgroup E of Q (so that $|E| = 2^{n+1}$); (b) construct a linear representation ψ of E whose kernel does not contain Z (to achieve this consider a complement F to Z in E, make F to act trivially and make Z to act by $\{\pm 1\}$-scalars); and (c) induce ψ to the whole of Q. The uniqueness of φ implies that

$$N_{SL(V)}(\varphi(Q))/\varphi(Q) \cong \mathrm{Out}(Q) \cong O_{2n}^\varepsilon(2)$$

(compare Exercise 4 on p.64 in [A94]) so that $N_{SL(V)}(\varphi(Q))$ is an element of $\mathcal{E}(O_{2n}^+(2), Q/Z, q_e)$ which gives the existence assertion in (ii). The uniqueness assertion follows from the uniqueness of φ. $\qquad\square$

The group in (1.4.2), which possesses a faithful 2^n-dimensional representation, is known as the *standard holomorph* of Q; all other holomorphs are called *twisted*.

Lemma 1.4.3 *Let* U *be a faithful* $GF(2)$-*module for a group* K *of dimension* $2n$ *and suppose that* U *carries a non-singular quadratic form* q *of plus type. Then:*

(i) $\mathcal{E}(K, U, q)$ *is non-empty;*

(ii) $\mathcal{E}(K, U, q)$ *contains a unique group which possesses a faithful* 2^n-*dimensional representation over real numbers;*

(iii) *the isomorphism type of* R/Z *is independent of the choice of* $R \in \mathcal{E}(K, U, q)$, *where* $Z = Z(Q)$.

Proof. Let $Q \cong 2_+^{1+2n}$, $Z = Z(Q)$, and let $A \in \mathcal{E}(O_{2n}^+(2), Q/Z, q_e)$ (which exists by (1.4.2 (i))). Let

$$\alpha : A \to A/Q \cong \mathrm{Out}(Q) \cong O_{2n}^+(2),$$

$$\beta : A \to A/Z \cong \mathrm{Aut}(Q) \cong 2^{2n}.O_{2n}^+(2)$$

be the natural homomorphisms. Since the quadratic forms q and q_e are of the same dimension and type, there is an isomorphism

$$\eta : (U, q) \to (Q/Z, q_e)$$

of orthogonal spaces which induces an embedding

$$\zeta : K \to A/Q.$$

The preimage R of $\zeta(K)$ with respect to α is a member of $\mathcal{E}(K, U, q)$. If A possesses a 2^n-dimensional real representation (such A exists by (1.4.2)), then clearly so does the group R. In that case R is the subgroup of $N_{SL(V)}(\varphi(Q))$ such that $R/Q = \zeta(K)$. This proves (ii). Finally, R/Z is the preimage of $\zeta(K)$ with respect to β, so it is a subgroup of $\text{Aut}(Q)$ which is defined uniquely up to conjugation. Thus (iii) follows. $\qquad\square$

1.5 Parker loop $(\mathcal{L}, \circ)$

A *loop* is a pair $(L, \cdot)$ consisting of a set L and a binary multiplication law $\cdot$ such that L contains an identity element and every element of L has a right inverse and a left inverse. The groups are precisely the associative loops. The Parker loop is an extension (in the category of loops) of a group of order 2 by $\mathcal{C}_{12}$. In [C84] Conway used Parker's loop to construct an extension of a 2-group by M_{24}. Here we take an opposite route defining a loop structure on the cosets of a certain (non-normal) subgroup in a group from $\mathcal{E}(M_{24}, W_{24}, q_p)$.

Thus let $R \in \mathcal{E}(M_{24}, W_{24}, q_p)$. This means that (a) $Q = O_2(R) \cong 2_+^{1+24}$, (b) $R/Q \cong M_{24}$, and (c) there is a surjective homomorphism

$$\chi : Q \to W_{24}$$

with kernel $Z = Z(Q)$ such the induced isomorphism of Q/Z onto W_{24} commutes with the action of M_{24} and establishes an isomorphism of the orthogonal spaces $(Q/Z, q_e)$ and (W_{24}, q_p). Here q_e is the squaring map on Q and q_p is the quadratic form on W_{24} defined in (1.3.3). In Section 1.7 we will see that such R is unique up to isomorphism.

Since $\mathcal{S}_{12}^*$ is a maximal totally isotropic M_{24}-invariant subspaces in W_{24}, its preimage $\chi^{-1}(\mathcal{S}_{12}^*)$ is a maximal elementary abelian normal subgroup of order 2^{13} in Q which is normal in R. The corresponding quotient $R/\chi^{-1}(\mathcal{S}_{12}^*)$ is an extension of

$$Q/\chi^{-1}(\mathcal{S}_{12}^*) \cong W_{24}/\mathcal{S}_{12}^* \cong \mathcal{C}_{12}$$

by $R/Q \cong M_{24}$. Applying (1.2.2 (ii)), we conclude that

$$R/\chi^{-1}(\mathcal{S}_{12}^*) \cong \mathcal{C}_{12} : M_{24}$$

(the semidirect product with respect to the natural action) and by (1.2.2 (i)) all the M_{24}-complements in $R/\chi^{-1}(\mathcal{S}_{12}^*)$ are conjugate. Since $C_{\mathcal{C}_{12}}(M_{24})$ is of order 2, there are precisely 2^{11} such M_{24}-complements. Let M^α be the preimage in R of one of them, so that M^α is an extension of $\chi^{-1}(\mathcal{S}_{12}^*)$ by M_{24}. In view of (1.2.2 (iv)) this extension might or might not split, although by (1.2.2 (iii)) the subgroup $\chi^{-1}(\mathcal{S}_{12}^*)$ (considered as a module for $M^\alpha/\chi^{-1}(\mathcal{S}_{12}^*) \cong M_{24}$) splits into a direct sum

$$\chi^{-1}(\mathcal{S}_{12}^*) = Z \oplus K_{12}^*$$

so that the restriction of χ to K_{12}^* is an M^α-isomorphism onto $\mathcal{S}_{12}^* \cong \mathcal{C}_{12}^*$. If K_{11}^* is the codimension 1 submodule in K_{12}^* isomorphic to $\mathcal{C}_{11}^*$, then $\chi^{-1}(\mathcal{S}_{12}^*)/K_{11}^*$ is elementary abelian of order 4. This shows that there are precisely two choices for K_{12}^* (although K_{11}^* is determined by M^α uniquely).

By (1.1.3 (ii)), $M^\alpha/K_{11}^* \cong 2^2 \times M_{24}$ and therefore there exists a complement M^β to Z in M^α such that

$$M^\beta \cap Q = K_{12}^* \text{ and } M^\alpha = M^\beta \times Z.$$

If M^γ is the commutator subgroup of M^α, then

$$M^\alpha/M^\gamma \cong \chi^{-1}(\mathcal{S}_{12}^*)/K_{11}^* \cong 2^2.$$

It follows from the general properties of extraspecial groups that Q induces on $\chi^{-1}(\mathcal{S}_{12}^*)$ the group of all transvections with centre Z, in particular K_{12}^* is not normal in Q, and, since $Q \cap M^\beta = K_{12}^*$, the subgroup M^β is not normal in R either.

Consider the set $\mathcal{L} = \chi^{-1}(\mathcal{S}_{12})$ in more detail. For $u \in \mathcal{C}_{12}$ let $\widetilde{u}$ denote an element from $\mathcal{L}$ such that $\chi(\widetilde{u}) = (u, 0)$ (notice that $\widetilde{u}$ is determined up to multiplying by the generator z of Z).

For $v^* \in \mathcal{C}_{12}^*$ let $k(v^*)$, denote the unique element in K_{12}^* such that

$$\chi(k(v^*)) = (0, v^*).$$

Define on $\mathcal{L}$ a multiplication $\circ$ by the following rule

$$\widetilde{u} \circ \widetilde{w} = \chi^{-1}(\mathcal{S}_{12}) \cap \widetilde{u}\widetilde{w}M^\beta$$

for $\widetilde{u}, \widetilde{w} \in \mathcal{L}$.

It is quite clear that $\chi^{-1}(\mathcal{S}_{12})$ is a transversal of the set of left cosets of M^β in R and therefore $\circ$ is well defined.

Lemma 1.5.1 *Let* $\widetilde{u}, \widetilde{w} \in \mathcal{L}$. *Then*

$$\widetilde{u} \circ \widetilde{w} = \widetilde{u}\widetilde{w}\,k((u \cap w)^*).$$

Proof. By the definition, $\tilde{u} \circ \tilde{w}$ differs from the group product $\tilde{u}\tilde{w}$ by a right multiple $m \in M^\beta$. Since both products are in Q, m is also in Q and therefore it is in $M^\beta \cap Q = K^*_{12}$.

On the other hand, since $\chi(\tilde{u}) = (u, 0)$ and $\chi(\tilde{w}) = (w, 0)$, the multiplication rule in $W_{24} = \chi(Q)$ gives

$$\chi(\tilde{u}\tilde{w}) = (u + w, (u \cap w)^*) = (u + w, 0) + (0, (u \cap w)^*)).$$

Since $(u + w, 0) \in \mathcal{S}_{12}$ and $(0, (u \cap w)^*) \in \chi(K^*_{12})$, the result follows.   □

The intersection of two Golay sets is always even, therefore $k((u \cap w)^*)$ is in fact contained in K^*_{11}. The subgroup K^*_{11} being the intersection with Q of the commutator subgroup of M^α is uniquely determined by the latter. This gives the following:

Corollary 1.5.2 *The multiplication $\circ$ is uniquely determined by M^α and does not depend on the choice of the complement M^β to Z in M^α.*   □

Proposition 1.5.3 *The following assertions hold:*

(i) $(\mathcal{L}, \circ)$ *is a loop and if* $\tilde{u} \in \mathcal{L}$, *then* $\tilde{u}^{-1}$ *(the inverse of* $\tilde{u}$ *in* Q*) is both the left and the right inverse of* $\tilde{u}$ *in* $(\mathcal{L}, \circ)$;

(ii) *the mapping* $\varphi : \tilde{u} \mapsto u$ *is a homomorphism of* $(\mathcal{L}, \circ)$ *onto* $\mathcal{C}_{12}$ *with kernel* Z;

(iii) *if* $\tilde{u}, \tilde{v}, \tilde{w} \in \mathcal{L}$, *then the following equalities hold:*
   (a) $\tilde{u} \circ \tilde{u} = \frac{1}{4}|u| \bmod 2$;
   (b) $(\tilde{u} \circ \tilde{v}) \circ (\tilde{v} \circ \tilde{u})^{-1} = \frac{1}{2}|u \cap v| \bmod 2$;
   (c) $((\tilde{u} \circ \tilde{v}) \circ \tilde{w}) \circ (\tilde{u} \circ (\tilde{v} \circ \tilde{w}))^{-1} = |u \cap v \cap w| \bmod 2$;

(iv) *every 2-generated subloop in* $(\mathcal{L}, \circ)$ *is a group*;

(v) $(\mathcal{L}, \circ)$ *is the Parker loop.*

Proof. By (1.5.1), the identity element of R is left and right identity with respect to $\circ$ and $\tilde{u} \circ \tilde{u} = \tilde{u}\tilde{u}$ for every $\tilde{u} \in \mathcal{L}$. Since Q is extraspecial and by (E3), the squaring map in Q is

$$\tilde{u} \mapsto q_p(\chi(\tilde{u})) = \frac{1}{4}|u| \bmod 2,$$

(i) and (iii) (a) hold. The quotient of $(\mathcal{L}, \circ)$ over Z is isomorphic to the factor group

$$Q/\chi^{-1}(\mathcal{S}_{12}^*) = Q/\langle K_{12}^*, Z\rangle \cong C_{12}$$

which gives (ii).

Turning to (iii), start with following observation. Since $\mathcal{L}/Z$ (under the multiplication induced by ∘) is an elementary abelian 2-group, the images under φ of the left-hand sides in (iii) (a) to (c) are trivial. Therefore, the left-hand sides themselves are z-powers (recall that according to our convention z^m is written simply as m). Now (b) is immediate from (1.5.1), while (c) can be established in the following way

$$(\tilde{u} \circ \tilde{v}) \circ \tilde{w} = \widetilde{uv}\,k((u \cap v)^*) \circ \tilde{w} =$$
$$\widetilde{uv}\,k((u \cap v)^*)\tilde{w}k_1 = \widetilde{uvw}[k((u \cap v)^*), \tilde{w}]k_2,$$

while

$$\tilde{u} \circ (\tilde{v} \circ \tilde{w}) = \widetilde{uvw}k_3$$

for some $k_2, k_3 \in K_{11}^*$. Since K_{11}^* is disjoint from Z and since

$$[k((u \cap v)^*), \tilde{w}] = \beta((0, (u \cap v)^*), (w, 0)) = |u \cap v \cap w| \bmod 2$$

by (1.3.3), (c) follows. By (c) and since any two Golay sets intersect evenly, (iv) holds.

The left-hand sides in (a), (b), and (c) are the squares, the commutators, and the associators of $(\mathcal{L}, \circ)$. These three maps determine the Parker loop up to isomorphism. The claim was justified on p. 516 of [C84] and proved in more detail in Theorem 12.13 in [A94]. In any event, (v) follows.   □

1.6 Aut($\mathcal{L}$, ∘)

Some (but not all) automorphisms of Parker's loop $(\mathcal{L}, \circ)$ can be seen from the construction accomplished in the previous section. In fact, $\mathcal{L}$ can be viewed as the set of left cosets of M^β in R so that the loop product of two cosets is the coset containing the group product of their representatives from $\chi^{-1}(S_{12})$

$$\tilde{u}M^\beta \circ \tilde{v}M^\beta = \widetilde{uv}M^\beta.$$

This viewpoint enables us to define an action of $r \in R$ on $\mathcal{L}$ via left translations

$$r : \tilde{u}M^\beta \mapsto r\tilde{u}M^\beta.$$

Since $m\tilde{u}M^\beta = m\tilde{u}m^{-1}M^\beta$ for $m \in M^\beta$, the action of M^β by left translations is equivalent to its action on $\chi^{-1}(S_{12})$ by conjugation, and it is immediate that this action preserves the loop product.

Lemma 1.6.1 *The action of M^β on $\mathcal{L}$ by conjugation preserves the loop product $\circ$. Furthermore, $O_2(M^\beta) \cong C_{12}^*$ is the kernel of the action of M^β on $\mathcal{L}/Z$, while $M^\beta/O_2(M^\beta) \cong M_{24}$ is the stabilizer of the Golay code structure on $\mathcal{L}/Z$.*   □

Now let us turn to the full automorphism group of $(\mathcal{L}, \circ)$ which we denote by A. By (1.6.1), A contains M^β. Notice that the action of Z on $\mathcal{L}$ by conjugation is trivial. Therefore, the image of M^β in A is independent on the choice of M^β between the two complements to Z in M^α. At this stage we know that M^β is an extension of C_{12}^* by M_{24}, but don't know whether or not it splits. In (1.6.9) we will prove that the extension does not split, so that $M^\beta \cong C_{12}^* \cdot M_{24}$.

Lemma 1.6.2 *Let Y be the centre of $(\mathcal{L}, \circ)$. Then:*

(i) $Y = \{1, z, \widetilde{\mathcal{P}}, \widetilde{\mathcal{P}}z\}$ *is elementary abelian of order 4;*
(ii) *both Y and Z are characteristic in $(\mathcal{L}, \circ)$.*

Proof. It follows from the basic properties of the Golay code that $\emptyset$ and $\mathcal{P}$ are characterized among the Golay sets by the property that their intersection with any Golay set has size divisible by 4. In view of this observation, (i) is immediate from (1.5.3 (b)). Since Y is the centre, it is clearly characteristic. By (1.5.3 (a)), Z is generated by the squares of the elements from $\mathcal{L}$, therefore it is characteristic as well.   □

Put

$$\bar{\mathcal{L}} = \mathcal{L}/Z, \text{ and } \widehat{\mathcal{L}} = \mathcal{L}/Y.$$

By (1.5.3), $\bar{\mathcal{L}}$ and $\widehat{\mathcal{L}}$ are elementary abelian groups of order 2^{12} and 2^{11}, respectively. As above the elements of $\bar{\mathcal{L}}$ will be denoted by the Golay sets (although the full automorphism group of $(\mathcal{L}, \circ)$ is not going to preserve the Golay code structure on $\bar{\mathcal{L}}$).

The automorphisms from A, acting trivially on $\bar{\mathcal{L}}$ and $\widehat{\mathcal{L}}$, will be called *diagonal* and *central*, respectively. Let D and C denote the subgroups in A formed by the diagonal and central automorphisms.

Lemma 1.6.3 *The following assertions hold:*

(i) *both D and C are normal in A and $D \le C$;*
(ii) *both D and C/D are elementary abelian 2-groups.*

Proof. (i) is obvious since D and C are kernels of suitable homomorphisms, while (ii) follows from the observation that every element of $\bar{\mathcal{L}}$ has two preimages in $\mathcal{L}$ and every element of $\widehat{\mathcal{L}}$ has two preimages in $\bar{\mathcal{L}}$.   □

Lemma 1.6.4 *The subgroup D is the elementary abelian of order 2^{12} realized by the elements of $O_2(M^\beta)$ acting via conjugation.*

Proof. By (1.6.3 (ii)), D is the elementary abelian of exponent 2. In order to establish the assertion it is sufficient to show that for a non-identity element $d \in D$, the image in $\bar{\mathcal{L}}$ of $C_{\mathcal{L}}(d)$ is a hyperplane. Let $u, v, w \in \bar{\mathcal{L}}$ with $u + v = w$ (where the addition is the one induced by $\circ$). Let $\widetilde{u}, \widetilde{v}, \widetilde{w}$ be the corresponding preimages in $\mathcal{L}$ such that $\widetilde{u} \circ \widetilde{v} = \widetilde{w}$. Whenever d centralized $\widetilde{u}$ and $\widetilde{v}$ it centralized $\widetilde{w}$ which proves the claim. $\square$

Let us turn to $\bar{A} := A/D$. By (1.6.3), we know that $\bar{A}$ contains $M^\beta/O_2(M^\beta) \cong M_{24}$ as a subgroup. We know that up to isomorphism $(\mathcal{L}, \circ)$ is uniquely determined by the squaring, commutator, and associator maps. On the other hand, it is easy to check that the commutator and the associator maps are, respectively, the second and the third derived forms of the squaring map (compare discussion on p. 63 in [A94]). This gives the following:

Lemma 1.6.5 *$\bar{A}$ is the stabilizer in $\mathrm{Aut}(\bar{\mathcal{L}}) \cong GL_{12}(2)$ of the squaring map*

$$P : u \mapsto \frac{1}{4}|u| \bmod 2.$$

$\square$

Lemma 1.6.6 *The image $\bar{C} \cong C/D$ of C in $\bar{A}$ is elementary abelian of order 2^{11} canonically isomorphic to the dual space of $\widehat{\mathcal{L}} = \bar{\mathcal{L}}/\langle \mathcal{P} \rangle$.*

Proof. Since $\widehat{\mathcal{L}}$ is the quotient of $\bar{\mathcal{L}}$ over the 1-dimensional subspace generated by $\mathcal{P}$, every non-identity element from $\bar{C}$ acts on $\bar{\mathcal{L}}$ via a transvection with centre $\mathcal{P}$. On the other hand, since the values of the squaring map P on a Golay set and on its complement coincide, all these transvections preserve $\mathcal{P}$. Thus the result follows from (1.6.4). $\square$

Now it remains to identify the quotient $\widehat{A} = A/C$. By (1.6.4), the quotient is precisely the stabilizer in $GL(\widehat{\mathcal{L}}) \cong GL_{11}(2)$ of the map $\widehat{P}$ induced on $\widehat{\mathcal{L}}$ by the squaring map P.

Lemma 1.6.7 *$\widehat{A} \cong M_{24}$ coincides with the image of M^β.*

Proof. By (1.6.1) and the first paragraph in Section 1.2, M^β has two orbits on $\widehat{\mathcal{L}}$ with lengths 759 and 1288. The $\widehat{P}$-values on these orbits are different (0 and 1, respectively). Therefore, the assertion follows from (1.2.7). $\square$

Now with the chief factors of A being identified it remains to specify A up to isomorphism. The next lemma describes the relevant actions.

Lemma 1.6.8 *Let* Σ_1 *and* Σ_2 *be two elementary abelian groups of order* 2^{11} *each. Let* α *be a homomorphism of* M^β *into* $\mathrm{Aut}(\Sigma_1 \times \Sigma_2)$ *such that* (a) K_{11}^* *is the kernel of* α; (b) $M_{24} = O^2(M^\beta)/K_{11}^*$ *acts on* Σ_1 *and* Σ_2 *as it acts on* $\mathcal{C}_{11}$; (c) $O_2(M^\beta)/K_{11}^* \cong 2$ *permutes* Σ_1 *and* Σ_2 *(commuting with the action of* M_{24}). *Let* Σ_3 *be the unique (diagonal) subgroup of order* 2^{11} *in* $\Sigma_1 \times \Sigma_2$ *normalized by the image of* α. *Then* A *is the partial semidirect product of* $\Sigma_1 \times \Sigma_2$ *and* M^β *associated with* α *over* K_{11}^* *identified with* Σ_3.

Proof. For $y \in Y$ and $v^* \in \mathcal{C}_{12}^*$, define a mapping $\sigma(y, v^*)$ on $\mathcal{L}$ via

$$\sigma(y, v^*) : \widetilde{u} \mapsto \widetilde{u} y^{|u \cap v|}.$$

It is straightforward to check that $\sigma(y, v^*)$ is a (central) automorphism of $(\mathcal{L}, \circ)$ if either $y \in Z$ or $v^* \in \mathcal{C}_{11}^*$ or both. Then

$$\Sigma_i = \langle \sigma(y, v^*) \mid v^* \in \mathcal{C}_{11}^* \rangle,$$

where y is $\widetilde{\mathcal{P}}$, $\widetilde{\mathcal{P}}z$, or z for $i = 1, 2$, or 3, respectively, while

$$O_2(M^\beta) = K_{12}^* = \langle \sigma(z, v^*) \mid v^* \in \mathcal{C}_{12}^* \rangle.$$

Furthermore, if $v_1^*, v_2^* \in \mathcal{C}_{11}^*$, $y_1, y_2 \in Y$ and $v^* \in \mathcal{C}_{12}^* \setminus \mathcal{C}_{11}^*$, then

$$[\sigma(y_1, v_1^*), \sigma(y_2, v_2^*)] = 1$$

and

$$\sigma(z, v^*)\sigma(\widetilde{\mathcal{P}}, v_1^*) = \sigma(\widetilde{\mathcal{P}}z, v_1^*)\sigma(z, v^*).$$

These equalities imply the assertion. □

In order to accomplish the last bit in the identification of A, we apply a brilliant argument by Robert Griess from p.199 in [Gri87].

Lemma 1.6.9 $M^\beta \cong \mathcal{C}_{12}^* \cdot M_{24}$.

Proof. In view of (1.2.6 (ii)) all we have to show is that M^β does not split over K_{12}^*. Suppose to the contary that $M \cong M_{24}$ is a subgroup in M^β. Then, by (1.6.1), $\bar{\mathcal{L}}$ and $\mathcal{C}_{12}^*$ are isomorphic M-modules. Let d be a dodecad and let u, v be octads such that $u + v = d$. By (1.1.2 (ii)), there is an involution τ in the stabilizer $M(d)$ of d in M which permutes u and v. For a preimage $\widetilde{u}$ of u in $\mathcal{L}$, put $\widetilde{v} = \widetilde{u}^\tau$ and $\widetilde{d} = \widetilde{u}\widetilde{v}$. By (1.5.3 (a)), $\widetilde{u}$ and $\widetilde{v}$ are involutions, while $\widetilde{d}$

has order 4, so that $\widetilde{v} \circ \widetilde{u} = \widetilde{d}^{-1}$ and the subloop in $(\mathcal{L}, \circ)$ generated by $\widetilde{u}$ and $\widetilde{v}$ is isomorphic to the dihedral group D_8 of order 8. Since

$$\widetilde{d}^\tau = (\widetilde{u} \circ \widetilde{v})^\tau = \widetilde{u}^\tau \circ \widetilde{v}^\tau = \widetilde{v} \circ \widetilde{u} = \widetilde{d}^{-1},$$

we conclude that τ inverts $\widetilde{d}$. On the other hand, $M(d) \cong M_{12}$ does not contain subgroups of index 2 which is a contradiction.[1] $\square$

By (1.6.8), $A/D \cong \mathcal{C}_{11}^* : M_{24}$ and, by (1.2.2 (iii)) up to conjugacy, the latter contains two classes of M_{24}-complements. The subgroup M^β is the preimage of one of the complements. Denote the preimage of the other one by M^ε. Then the module $\bar{\mathcal{L}}$ under the action of M^ε splits into a direct sum of $\mathcal{C}_{11}$ and a 1-dimensional module. Since $\bar{\mathcal{L}}$ and D are dual to each other, D also splits under M^ε into a direct sum. We claim that M^ε does not split over D, so that $M^\varepsilon \cong \mathcal{C}_{11}^* \cdot M_{24} \times 2$. The claim comes from the following general principle whose elementary proof we leave out.

Lemma 1.6.10 *Let K be a group and U be a $GF(2)$-module for K which is a direct sum $U_1 \oplus U_2 \oplus \ldots \oplus U_m$ of irreducible submodules. Then the semidirect product of U and K with respect to the natural action does not contain subgroups isomorphic to a non-split extension of U_i by K for any $1 \le i \le m$.* $\square$

The situation similar to that in (1.6.10) with U being indecomposable is totally different. For instance, the semidirect of $L_3(2)$ and its indecomposable 6-dimensional $GF(2)$-module contains the non-split extension $2^3 \cdot L_3(2)$ (this fact has played an important role in [ISh05]).

In what follows we will mainly deal with the subgroup M^β of A. This can be characterized as the stabilizer of the Golay code structure on $\bar{\mathcal{L}}$. By this reason, we call M^β the *code automorphism group* of $(\mathcal{L}, \circ)$ and denote it by $A^{(g)}$ (where 'g' is for Golay).

1.7 Back to extraspecial extensions

In this section, we classify the extensions of the extraspecial group 2_+^{1+22} by M_{24} subject to the conditions stated in the introduction to the chapter.

Lemma 1.7.1 *The class $\mathcal{E}(M_{24}, W_{24}, q_p)$ contains a unique group R and R/Z does not split over Q/Z.* $\square$

---

[1] It was pointed to me by Dr. Ryuji Sasaki that there is an octad a, a 3-subset u, and an involution $\tau \in M_{24}$ such that $(a, u) +_{W_{24}} (a^\tau, u^\tau) = (d, 0)$. Therefore, the above proof applies directly to a group from $\mathcal{E}(M_{24}, \widetilde{W}_{24}, q_p)$.

Proof. In terms of Sections 1.5 and 1.6, R is generated by $\chi^{-1}(S_{12}) = \mathcal{L}$ and $M^\beta = A^{(g)}$. Therefore, R is a permutation group of the element set of Parker's loop generated by the code automorphisms together with the left translations

$$\lambda_{\widetilde{u}} : \widetilde{x} \mapsto \widetilde{u} \circ \widetilde{x}$$

taken for all $\widetilde{u} \in \mathcal{L}$. This proves the uniqueness assertion. The non-splitness also follows from the discussions in Sections 1.5 and 1.6. In fact, M^α is the preimage in R of a representative of the unique class of complements to $Q/\chi^{-1}(S_{12}^*)$ in $R/\chi^{-1}(S_{12}^*) \cong C_{12} : M_{24}$. On the other hand, $M^\alpha = M^\beta \times Z$ and, by (1.6.9), $M^\beta \cong C_{12}^* \cdot M_{24}$ (the non-split extension). $\qquad\square$

Next we analyse the groups in $\mathcal{E}(M_{24}, \mathcal{A}(M_{24}), q_h)$.

Lemma 1.7.2 *Let* $T \in \mathcal{E}(M_{24}, \mathcal{A}(M_{24}), q_h)$, $P = O_2(T)$, *and* $X = Z(P)$. *Then:*

(i) T/X *does not split over* P/X;
(ii) T *is unique up to isomorphism.*

Proof. By (1.4.3 (iii)), the isomorphism type of T/X does not depend on the choice of $T \in \mathcal{E}(M_{24}, \mathcal{A}(M_{24}), q_h)$ so in order to prove (i) it is sufficient to produce an example of T for which T/X does not split. Such an example can be constructed as a section of $R \in \mathcal{E}(M_{24}, W_{24}, q_p)$ in the following way. Let $\widetilde{\mathcal{P}}$ be a preimage of $(\mathcal{P}, 0) \in W_{24}$ with respect to the homomorphism $\chi :$ $Q \rightarrow W_{24}$, let $C = C_R(\widetilde{\mathcal{P}})$, and let $T = C/\langle\widetilde{\mathcal{P}}\rangle$. Then in terms of the previous section, we have (a) $M^\beta \cap C = M^\gamma \cong C_{11}^* \cdot M_{24}$, (b) the image P of $Q \cap C$ in T is the extraspecial group 2_+^{1+22}, and (c) P/Z is the anti-heart module for $M^\gamma/C_{11}^* \cong M_{24}$. Arguing as in the proof of (1.7.1), it is easy to show that the extension does not split. The uniqueness of T claimed in (ii) can be established in the following way. As in Section 1.5 we can define a loop structure on the set of cosets of (the image in T of) the subgroup M^γ. The resulting loop is precisely the quotient of the Parker loop $(\mathcal{L}, \circ)$ over the subloop of order 2 generated by $\widetilde{\mathcal{P}}$. Then T can be recovered as the permutation group of $\mathcal{L}/\langle\widetilde{\mathcal{P}}\rangle$ generated by the automorphism group of the factor loop (isomorphic to M^γ) together with the left translations. $\qquad\square$

Lemma 1.7.3 *The class* $\mathcal{E}(M_{24}, C_{11}^* \oplus C_{11}, q_d)$ *contains a single group which is a semidirect product of* 2_+^{1+22} *and* M_{24}.

Proof. A possible way of proving the assertion is to follow the strategy in Section 1.5: define a loop structure of the cosets in $S \in \mathcal{E}(M_{24}, C_{11}^* \oplus C_{11}, q_d)$ of a subgroup which is an extension of C_{11}^* by M_{24}. In the considered situation, the loop turns out to be just an elementary abelian group E (of order 2^{12}) and therefore S itself is a subgroup of the affine group $AGL(E)$. From this it is easy to conclude that S is bound to be the semidirect product as claimed. $\square$

Now we are ready to state the classification result.

Proposition 1.7.4 *Let T be a group such that:*

(a) $O_2(T) \cong 2_+^{1+22}$;
(b) $T/O_2(T) \cong M_{24}$;
(c) $O_2(T)/Z(T)$ *is an extension of* C_{11}^* *by* C_{11}.

Then either:

(i) *T is the only group in $\mathcal{E}(M_{24}, C_{11}^* \oplus C_{11}, q_d)$ and T splits over $O_2(T)$, or*
(ii) *T is the only group in $\mathcal{E}(M_{24}, \mathcal{A}(M_{24}), q_h)$ where $\mathcal{A}(M_{24})$ is the anti-heart module and T does not split over $O_2(T)$.*

Proof. The result follows from (1.3.2), (1.3.4), (1.3.8), (1.7.2), and (1.7.3). $\square$

1.8 Leech lattice and the monomial subgroup

The *Leech lattice* Λ can be produced in terms of the Golay code C_{12} as follows.

Consider the 24-element set $\mathcal{P}$ (equipped with a linear order) as a basis of a real vector space $\mathbb{R}^{24}$. Let Λ be the set of vectors $\lambda \in \mathbb{R}^{24}$ whose components λ_p, $p \in \mathcal{P}$ with respect to $\mathcal{P}$ satisfy the following three conditions either for $k = 0$ or for $k = 1$:

(Λ1) $\lambda_p = k \bmod 2$ for every $p \in \mathcal{P}$;
(Λ2) $\{p \mid \lambda_p = k \bmod 4\} \in C_{12}$;
(Λ3) $\sum_{p \in \mathcal{P}} \lambda_p = 4k \bmod 8$.

Considering $\mathcal{P}$ as an othonormal basis of the underlying vector space $\mathbb{R}^{24}$ (which means introducing the inner product)

$$(p, q)_\Lambda = \delta_{pq} \text{ for } p, q \in \mathcal{P}$$

we turn Λ into an even (integral) lattice. In order to make it unimodular we have to rescale the inner product by $\frac{1}{8}$. We are interested in the automorphism group of Λ, which by definition is the subgroup of $GL(\mathbb{R}^{24})$ which preserves both the set Λ and the form $(,)_\Lambda$ (as long as only the automorphism group is of concern, the particular rescaling is irrelevant).

The automorphism group of Λ is commonly denoted by Co_0; it is a non-split extension of the group of (± 1)-scalar transformations by the first Conway sporadic simple group Co_1.

The above construction of Λ exhibits considerable symmetry. In fact, we can modify the basis $\mathcal{P}$ preserving the conditions $(\Lambda 1)$ to $(\Lambda 3)$. These modifications include the reorderings of the elements of $\mathcal{P}$ according to permutations from $\mathrm{Aut}(\mathcal{C}_{12}) \cong M_{24}$ and the negatings of the elements of $\mathcal{P}$ contained in Golay sets. This can be formalized as follows to produce an important subgroup of Co_0 where the action in the basis $\mathcal{P}$ is monomial.

Lemma 1.8.1 *Let M_0 be the set of linear transformations of $\mathbb{R}^{24}$ presented in the basis $\mathcal{P}$ by the permutation matrices preserving $\mathcal{C}_{12}$. For $u \in \mathcal{C}_{12}$, let*

$$\varepsilon(u) = \mathrm{diag}((-1)^{|u \cap \{p\}|} \mid p \in \mathcal{P})$$

(also a linear transformation of $\mathbb{R}^{24}$ written in the basis $\mathcal{P}$). Let E_0 be the subgroup in $GL(\mathbb{R}^{24})$ generated by the elements $\varepsilon(u)$ taken for all $u \in \mathcal{C}_{12}$ and let F_0 be the subgroup in $GL(\mathbb{R}^{24})$ generated by M_0 and E_0. Then:

(i) $M_0 \cong M_{24}$;

(ii) $E_0 \cong \mathcal{C}_{12}$ *(elementary abelian group of order 2^{12})*;

(iii) M_0 *normalizes E_0 and $F_0 = E_0 M_0 \cong \mathcal{C}_{12} : M_{24}$ is the semidirect product with respect to the natural action;*

(iv) F_0 *preserves both the set Λ and the bilinear form $(\ ,\ )_\Lambda$, so that $F_0 \leq Co_0$;*

(v) *for every non-zero real α the subgroup F_0 is the stabilizer in Co_0 of the frame*

$$\mathcal{F}_\alpha := \{\pm\alpha\, p \mid p \in \mathcal{P}\};$$

(vi) $\varepsilon(\mathcal{P})$ *is the (-1)-scalar operator.*

Proof. We have obtained Λ following a standard construction of lattices from codes known as Construction C (cf. [CS88]). The assertions in the lemma are general features of that construction (see also the relevant chapters in [A94] and [Iv99]). $\qquad\square$

Put

$$\Lambda_m = \{\lambda \mid \lambda \in \Lambda, (\lambda, \lambda)_\Lambda = 16m\}.$$

Then Λ_0 consists of the zero vector, while Λ_m is empty for the odd ms (since Λ is even) as well as for $m = 2$ (there are no roots in Λ). The action of Co_0 is known to be transitive on Λ_2, Λ_3, Λ_4, Λ_5, and Λ_7 with stabilizers being Co_2, Co_3, and $2^{11} : M_{23}$, HS, and McL, respectively (cf. [Con71] and [A94]).

Let $\bar{\Lambda} = \Lambda/2\Lambda$ (the Leech lattice modulo 2). Then $\bar{\Lambda}$ is an elementary abelian group of order 2^{24} which is a faithful irreducible module for $Co_1 = Co_0/\{\pm 1\}$. For every element $\lambda \in \Lambda$, there is a unique $i \in \{0, 2, 3, 4\}$ such that $\lambda + 2\Lambda$ contains a vector from Λ_i. Furthermore, if $i < 4$, there are exactly two such vectors in the coset which are negatives of each other, while if $i = 4$, there are 48 such vectors which form the image of the frame $\mathcal{F}_8$ under an element of Co_0.

Adopting the bar convention for images in $\bar{\Lambda}$ of vectors and subsets of Λ, we have

$$\bar{\Lambda} = \bar{\Lambda}_0 \uplus \bar{\Lambda}_2 \uplus \bar{\Lambda}_3 \uplus \bar{\Lambda}_4$$

(where $\uplus$ stays for the disjoint union). In terms of (1.8.1), let M, E, and F be the images in Co_1 of M_0, E_0, and F_0, respectively. It is clear that F is the semidirect product with respect to the natural action of $E = E_0/\{\varepsilon(\emptyset)$, $\varepsilon(\mathcal{P})\} \cong \mathcal{C}_{11}$ and $M \cong M_{24}$.

Lemma 1.8.2 *The subgroup $F \cong \mathcal{C}_{11} : M_{24}$ (called the* monomial subgroup*) is the stabilizer in Co_1 of the coset*

$$\bar{\lambda}_4 := \sum_{p \in \mathcal{P}} 2p + 2\Lambda$$

contained in $\bar{\Lambda}_4$.

Proof. In view of (1.8.1 (iii), (iv)), it is sufficient to observe that the vector

$$8r - \sum_{p \in \mathcal{P}} 2p = 2\left(3r - \sum_{p \in \mathcal{P} \setminus \{r\}} p\right)$$

is contained in 2Λ for every $r \in \mathcal{P}$. □

Recall that Co_1 acts transitively on $\bar{\Lambda}_4$ and since $|\bar{\Lambda}_4|$ is odd, F contains a Sylow 2-subgroup of Co_1.

Lemma 1.8.3 *The Leech lattice modulo 2 $\bar{\Lambda}$ is isomorphic to W_{24} as a module for the complement $M \cong M_{24}$ of the monomial subgroup.*

Proof. Define a mapping

$$\omega : W_{24} \to \Lambda/2\Lambda$$

inductively using the following rules (where $u \in \mathcal{C}_{12}$, $p \in \mathcal{P}$, $v^*, v_1^*, v_2^* \in \mathcal{C}_{12}^*$)

$$\omega(u, 0) = \begin{cases} \sum_{p \in u} 2p + 2\Lambda & \text{if } |u| = 0 \bmod 8 \\ \sum_{p \notin u} 2p + 2\Lambda & \text{otherwise;} \end{cases}$$

$$\omega(0, \{p\}^*) = (3p - \sum_{r \in \mathcal{P} \setminus \{p\}} r) + 2\Lambda;$$

$$\omega(0, v_1^* + v_2^*) = \omega(0, v_1^*) + \omega(0, v_2^*);$$

$$\omega(u, v^*) = \omega(u, 0) + \omega(0, v^*).$$

A proof that ω is an isomorphism of M_{24}-modules can be achieved by straight-forward calculations. For example consider the case where u and w are octads intersecting in a pair $\{p, q\}$, so that in W_{24} we have

$$(u, 0) + (w, 0) = (d, \{p, q\}^*),$$

where

$$d = (u \cup w) \setminus \{p, q\}$$

is a dodecad. Then

$$\omega(u, 0) + \omega(w, 0) = \sum_{r \in u \setminus w} 2r + \sum_{s \in w \setminus u} 2s + 4p + 4q + 2\Lambda$$

$$= \sum_{r \in d} 2r + (4p - 4q) + 8q + 2\Lambda$$

(we are using (1.8.2) and its proof)

$$\sum_{r \in \mathcal{P} \setminus d} -2r + (4p - 4q) + 2\Lambda$$

(modulo 2Λ every Leech vector equals to its negative)

$$-\omega(d, 0) + \omega(0, \{p\}^*) - \omega(0, \{q\}^*) + 2\Lambda = \omega(d, \{p, q\}^*).$$

$\square$

The bilinear form $\frac{1}{8}(\ ,\ )_\Lambda$ on Λ taken modulo 2 induces on $\bar{\Lambda}$ a symplectic form $(\ ,\ )_{\bar{\Lambda}}$, while the mapping

$$\lambda \mapsto \frac{1}{16}(\lambda, \lambda)_\Lambda$$

taken modulo 2 induces on $\bar{\Lambda}$ a quadratic form $q_{\bar{\Lambda}}$ for which $(\ ,\ )_{\bar{\Lambda}}$ is the associated bilinear form. Thus

$$q_{\bar{\Lambda}}(\bar{\lambda}) = \begin{cases} 1 & \text{if } \bar{\lambda} \in \bar{\Lambda}_3; \\ 0 & \text{if } \bar{\lambda} \in \bar{\Lambda}_0 \uplus \bar{\Lambda}_2 \uplus \bar{\Lambda}_4. \end{cases}$$

Lemma 1.8.4 *The following assertions hold:*

(i) *the mapping $\omega : W_{24} \to \bar{\Lambda}$ in (1.8.3) establishes an isomorphism of the orthogonal spaces (W_{24}, q_p) and $(\bar{\Lambda}, q_{\bar{\Lambda}})$;*
(ii) *$q_{\bar{\Lambda}}$ is of plus type;*
(iii) *$q_{\bar{\Lambda}}$ is the unique Co_1-invariant quadratic form on $\bar{\Lambda}$.*

Proof. The assertion (i) can be checked directly by verifying it for every step of the inductive definition of ω. Since q_p is of plus type (cf. (1.6.3)), so is $q_{\bar{\Lambda}}$ and (ii) holds. As was mentioned, Co_1 acts transitively on $\bar{\Lambda}_i$ for $i = 0, 2, 3$, and 4. Therefore, a Co_1-invariant quadratic form on $\bar{\Lambda}$ must be constant on each of the $\bar{\Lambda}_i$s. Comparing the known sizes of the $\bar{\Lambda}_i$s from [Con71] with the possible numbers of isotropic vectors in a non-zero quadratic form on a 24-dimensional $GF(2)$-space, we can establish (iii). $\qquad\square$

The following lemma describes the action of $E \cong \mathcal{C}_{11}$ on $\bar{\Lambda}$:

Lemma 1.8.5 *Let $w, u \in \mathcal{C}_{12}$ and let $v^* \in \mathcal{C}_{12}^*$. Then*

$$\varepsilon(w) : \omega(u, 0) \mapsto \omega(u + (\frac{1}{2}|u \cap w| \bmod 2)\,\mathcal{P}, (u \cap w)^*),$$

$$\varepsilon(w) : \omega(0, v^*) \mapsto \omega((|v| \bmod 2)\,w + (|w \cap v| \bmod 2)\,\mathcal{P}, v^*)$$

and on the whole of $\bar{\Lambda}$ the action extends by linearity. $\qquad\square$

It is easily seen from the above lemma that the actions on $\bar{\Lambda}$ of $\varepsilon(w)$ and $\varepsilon(\mathcal{P} \setminus w)$ coincide, so that E_0 acts as $\mathcal{C}_{11} \cong \mathcal{C}_{12}/\langle \mathcal{P} \rangle$. The next lemma (which is a direct consequence of (1.8.5)) describes the composition series of $\bar{\Lambda}$ as a module for the monomial subgroup $F = EK \cong \mathcal{C}_{11} : M_{24}$.

Lemma 1.8.6 *Let $\bar{\Lambda}^{(1)} = \omega(W^{(1)}) = \langle \bar{\lambda} \rangle$, $\bar{\Lambda}^{(23)} = \omega(W^{(23)})$ and*

$$\bar{\Lambda}^{(12)} = \{\omega(u, v^*) \mid u \in \{\emptyset, \mathcal{P}\}, v^* \in \mathcal{C}_{11}^*\}.$$

Then:

(i) *$0 < \bar{\Lambda}^{(1)} < \bar{\Lambda}^{(12)} < \bar{\Lambda}^{(23)} < \bar{\Lambda}$ is the only composition series of $\bar{\Lambda}$ as a module for F;*
(ii) *the subspace $\bar{\Lambda}^{(12)}$ is maximal totally isotropic with respect to $q_{\bar{\Lambda}}$;*
(iii) *$\bar{\Lambda}^{(23)} = \bar{\lambda}_4^{\perp}$ (the orthogonal complement of $\bar{\Lambda}^{(1)}$ with respect to the symplectic form $(,)_{\bar{\Lambda}}$);*
(iv) *$\bar{\Lambda}^{(12)}$ is completely reducible under the action of $M \cong M_{24}$ and $E \cong \mathcal{C}_{11}$ acts on $\bar{\Lambda}^{(12)}$ by transvections with centre $\omega(\mathcal{P}, 0)$;*
(v) *ω induces an isomorphism*

$$\psi : \mathcal{A}(M_{24}) \to \bar{\Lambda}^{(23)}/\bar{\Lambda}^{(1)}$$

which commutes with the action of M and the action of E on $\bar{\Lambda}^{(23)}/\bar{\Lambda}^{(1)}$
is described by

$$\varepsilon(w) : \psi(u, v^*) \mapsto \psi(u, v^* + (u \cap w)^*)$$

for $u, w \in \mathcal{C}_{11}, v^* \in \mathcal{C}_{11}^*$. $\square$

For an element $p \in \mathcal{P}$, put

$$\bar{\lambda}_2 = (3p - \sum_{r \in \mathcal{P} \setminus \{p\}} r) + 2\Lambda$$

and

$$\bar{\lambda}_3 = (5p + \sum_{r \in \mathcal{P} \setminus \{p\}} r) + 2\Lambda.$$

Then $\bar{\lambda}_2 \in \bar{\Lambda}_2$ and $\bar{\lambda}_3 \in \bar{\Lambda}_3$, which explains the notations.

Lemma 1.8.7 *Let* $F(\bar{\lambda}_2)$ *and* $F(\bar{\lambda}_3)$ *be the stabilizers of* $\bar{\lambda}_2$ *and* $\bar{\lambda}_3$, *respectively in the monomial subgroup* $F \cong \mathcal{C}_{11} : M_{24}$. *Then:*

(i) $F(\bar{\lambda}_2) = F(\bar{\lambda}_3) \cong M_{23}$;
(ii) $\bar{\lambda}_2^{\perp} \cap \bar{\Lambda}^{(23)} = \bar{\lambda}_3^{\perp} \cap \bar{\Lambda}^{(23)}$ *represents the shortest F-orbit on the set of hyperplanes in* $\bar{\Lambda}^{(23)}$ *which do not contain* $\bar{\Lambda}^{(12)}$;
(iii) *the length of the orbit in* (ii) *is* $[F : F(\bar{\lambda}_2)] = 24 \cdot 2^{11}$;
(iv) $\bar{\lambda}_2^{\perp}, \bar{\lambda}_3^{\perp}$, *and* $\bar{\lambda}_4^{\perp} = \bar{\Lambda}^{(23)}$ *are the three hyperplanes in* $\bar{\Lambda}$ *containing* $\bar{\lambda}_2^{\perp} \cap \bar{\Lambda}^{(23)}$.

Proof. The result can be deduced from the description of the action of F on $\bar{\Lambda}$ given in (1.8.5). Alternatively, the assertions can be checked by comparing the known lengths of the F-orbits on $\bar{\Lambda}$ (cf. p. 155 in [Iv99]). Notice that $\bar{\lambda}_2 + \bar{\lambda}_3 = \bar{\lambda}_4$ in $\bar{\Lambda}$. $\square$

We conclude this section with some cohomological data.

Lemma 1.8.8 *The following assertion hold:*

(i) *the Schur multiplier of* Co_1 *is of order* 2;
(ii) $H^1(Co_1, \bar{\Lambda})$ *is trivial*;
(iii) $H^2(Co_1, \bar{\Lambda})$ *has order* 2.

Proof. Since the M_{24}-module $\mathcal{C}_{12}$ is indecomposable, $\{\varepsilon(\emptyset), \varepsilon(\mathcal{P})\}$ is not complemented in $F_0 = E_0 M_0 \cong \mathcal{C}_{12} : M_{24}$. Therefore Co_0 is a non-split extension of the group of order 2 by Co_1. It is well known that Co_0 is the largest perfect central extension of Co_1 (cf. [CCNPW]) which gives (i). The triviality of $H^1(Co_1, \bar{\Lambda})$ was established in [Gri82a], Lemma 2.11. The assertion (iii) has been believed to be true for quite a while. By (1.8.4 (i)) and (1.7.1), a group $X \in \mathcal{E}(Co_1, \bar{\Lambda}, q_{\bar{\lambda}})$ contains as a subgroup the unique group $Y \in \mathcal{E}(M_{24}, W_{24}, q_p)$. Since $O_2(X) = O_2(Y)$ and since $Y/Z(Y)$ is a non-split extension of W_{24} by M_{24}, we conclude that $X/Z(X)$ does not split over $O_2(X)/Z(X)$. Therefore, $H^2(Co_1, \bar{\Lambda})$ is non-trivial (this was known since [Gri73a]). Nowadays it is possible to check the order of $H^2(Co_1, \bar{\Lambda})$ computationally. This was accomplished by Derek Holt. In fact the whole of Co_1 is still too large, but Derek has computed $H^2(F, \bar{\Lambda})$ to be of order 2. Since the monomial subgroup contains a Sylow 2-subgroup of Co_1 and in view of the lower bound this gives (iii). □

1.9 Hexacode

In the next chapter we will study another important subgroup of Co_1. For this study to go smoothly we need a better setting for the structure of the sextet stabilizer in the Mathieu group M_{24}.

Let $(V_{24}, \mathcal{P}, \mathcal{C}_{12})$ be the full name of the Golay code, let $\mathcal{T}$ be a 4-subset of $\mathcal{P}$, and let

$$\mathcal{S} = \mathcal{S}(\mathcal{T}) = \{w \mid w \subseteq \mathcal{P}, |w| = 4, \mathcal{T}^* = w^*\}$$

be the sextet containing $\mathcal{T}$. Then $\mathcal{S}$ is a partition of $\mathcal{P}$ into six 4-subsets called *tetrads* such that the union of any two distinct tetrads from $\mathcal{S}$ is an octad in $\mathcal{C}_{12}$. The following result is rather standard in the theory of the Golay code (cf. Lemma 2.10.2 in [Iv99]).

Lemma 1.9.1 *Let $M(\mathcal{S})$ be the stabilizer of $\mathcal{S}$ in $M \cong M_{24}$. Then*

$$M(\mathcal{S}) \cong 2^6 : 3 \cdot S_6$$

and the structure of $M(\mathcal{S})$ can be described as follows. Let $Q_{\mathcal{S}} = O_2(M(\mathcal{S}))$, let $K_{\mathcal{S}} = O_{2,3}(M(\mathcal{S}))$, let $X_{\mathcal{S}}$ be a Sylow 3-subgroup in $K_{\mathcal{S}}$, and let $Y_{\mathcal{S}} = N_{M(\mathcal{S})}(X_{\mathcal{S}})$. Then:

(i) *$Q_{\mathcal{S}}$ is elementary abelian of order 2^6;*
(ii) *$X_{\mathcal{S}}$ is of order 3, the action of $X_{\mathcal{S}}$ on $Q_{\mathcal{S}}$ is fixed-point free and $K_{\mathcal{S}} = Q_{\mathcal{S}} X_{\mathcal{S}}$;*
(iii) *$Y_{\mathcal{S}}$ is a complement to $Q_{\mathcal{S}}$ in $M(\mathcal{S})$;*
(iv) *$M(\mathcal{S})$ acting on the set of tetrads in $\mathcal{S}$ induces the full symmetric group S_6 of this set and $K_{\mathcal{S}}$ is the kernel of the action;*

(v) Y_S *does not split over* X_S *and* $C_{Y_S}(X_S) \cong 3 \cdot A_6$ *is a perfect central extension of* A_6;

(vi) *if* Ω *is the set of elements of* $\mathcal{P}$ *fixed by* X_S, *then* Ω *is of size* 6, Ω *contains an element from every tetrad in* S, *and* Ω *is not contained in an octad of* $\mathcal{C}_{12}$;

(vii) Y_3 *is the setwise stabilizer of* Ω *in* M;

(viii) *if* $\mathcal{O}$ *is an octad which is the union of a pair of tetrads from* S *and* $M(\mathcal{O}) = Q_{\mathcal{O}} K_{\mathcal{O}} \cong 2^4 : L_4(2)$ *is the stabilizer of* $\mathcal{O}$ *in* M, *then* $Q_{\mathcal{O}} \cap Q_S$ *is elementary abelian of order* 2^2 *on which* X_S *acts fixed-point freely.* $\square$

Let $Y \cong 3 \cdot S_6$ be the group denoted by Y_S in (1.9.1), let $X = O_3(Y)$, let $S = Y/X \cong S_6$, and let $A = C_Y(X)/X \cong A_6$. Then Y is the unique extension of a group of order 3 by S_6 whose commutator subgroup is a perfect central extension of a group of order 3 by A_6. The group Y will be called the *hexacode group*. The outer automorphism group of S_6 is well known to be of order 2 and because of the uniqueness of Y, every automorphism of S extends to an automorphism of Y. Because of the non-splitness, the identity is the only automorphism of Y which centralizes Y/X, so that the outer automorphism group of Y is also of order 2. Because of the identification of Y and Y_S, the subgroup Q_S can be considered as a 6-dimensional $GF(2)$-module for Y. This module is called the *hexacode module* of Y and we denote it by $\mathcal{H}$. The fixed-point free action of X induces on $\mathcal{H}$ a $GF(4)$-vector space structure preserved by $C_Y(X)$ (an element from $Y \setminus C_Y(X)$ acts on $\mathcal{H}$ as a Frobenius automorphism).

Let $\Omega^{\diamond}$ and $\mathcal{H}^{\diamond}$ denote the images of Ω and $\mathcal{H}$ under an outer automorphism of Y. Then the stabilizer in S of a 2-subset in Ω is the stabilizer of a partition of $\Omega^{\diamond}$ into three pairs.

Let $V_{1\backslash 4\backslash 1}$ and $V^{\diamond}_{1\backslash 4\backslash 1}$ be the $GF(2)$-permutation modules of Y (or rather of S) acting on Ω and $\Omega^{\diamond}$, respectively. Then

$$0 < V_1 < V_{1\backslash 4} < V_{1\backslash 4\backslash 1}$$

is the only composition series of $V_{1\backslash 4\backslash 1}$ where V_1 and $V_{1\backslash 4}$ are formed by the improper and even subsets, respectively and S is the full stabilizer of the symplectic space structure on the heart $U_4 := V_{1\backslash 4}/V_1$ of the permutation module, exhibiting the famous isomorphism

$$S_6 \cong Sp_4(2).$$

The structure of the permutation module $V^{\diamond}_{1\backslash 4\backslash 1}$ is similar and its heart will be denoted by $U^{\diamond}_4$.

Lemma 1.9.2 *The following assertions hold:*

(i) *both U and $U^\diamond$ are self-dual modules;*
(ii) *each of the following groups*

$$H^1(A_6, U_4), \ H^1(S_6, U_4), \ H^1(A_6, U_4^\diamond), \ H^1(S_6, U_4^\diamond)$$

has order 2;
(iii) *$H^2(A_6, U_4)$ is trivial, while $H^2(S_6, U_4)$ has order 2;*
(iv) *both $H^1(Y, \mathcal{H})$ and $H^2(Y, \mathcal{H})$ are trivial.*

Proof. The assertion (i) follows from existence of the invariant symplectic forms on U_4 and on $U_4^\diamond$, while (ii) is well known (the standard reference is [JP76]). The indecomposable extension of U_4 by S_6 is the 'third' subgroup of index 2 in $V_{4\backslash 1} : S_6$ other than $U_4 : S_6$ and $V_{4\backslash 1} : A_6$. In any extension K of $\mathcal{H}$ by Y, the normalizer $N_K(X)$ is a representative of the unique class of complements to $O_2(K)$. This proves (iv). $\qquad\square$

Lemma 1.9.3 *The following assertions hold:*

(i) *$\mathcal{H}$ and $\mathcal{H}^\diamond$ are dual $GF(2)$-modules for Y;*
(ii) *Y acts on the set of non-zero elements of $\mathcal{H}$ with two orbits $\mathcal{H}_{18}$ and $\mathcal{H}_{45}$ of lengths 18 and 45, respectively;*
(iii) *there are surjections*

$$\mathcal{H}_{18} \to \Omega^\diamond \text{ and } \mathcal{H}_{45} \to \binom{\Omega}{2},$$

which commute with the action of Y and whose fibers the X-orbits.

Proof. The proof can be deduced from the following model for the hexacode adopted in [Iv99]. Consider $\mathcal{H}$ as a 3-dimensional $GF(4)$-module, and let $\Sigma = \Sigma L(\mathcal{H}) \cong \Sigma L_3(4)$ be the extension of $SL(\mathcal{H})$ by the Frobenius automorphism. Then Y is the stabilizer in Σ of a hyperoval in the projective plane (P, L) of order 4 associated with Σ. In these terms

$$P = \Omega^\diamond \bigcup \binom{\Omega}{2} \text{ and } L = \Omega \bigcup \binom{\Omega^\diamond}{2}.$$

Furthermore, a suitable diagram automorphism of Σ normalizes Y inducing on it an outer automorphism. $\qquad\square$

1.10 Centralizer–commutator decompositions

It is a general feature of coprime actions that a $GF(2)$-module V for the hexacode group $Y \cong 3 \cdot S_6$ possesses the direct sum decomposition

$$V = C_V(X) \oplus [V, X].$$

In this section we calculate such decompositions for V being V_{24}, C_{12}, C_{12}^*, and W_{24}.

As in Section 1.9 let S be the sextet stabilized by $M(S) \cong 2^6 : 3 \cdot S_6$ such that Y is a complement to $O_2(M(S))$ in $M(S)$. By (1.9.1 (vii)), Y is the stabilizer in M of the 6-subset Ω of $\mathcal{P}$ consisting of the elements fixed by X. By (1.9.1 (vi)), Ω is not contained in any octad and it forms a transversal of the set of tetrads in S. For $\omega \in \Omega$, let S_ω denote the tetrad in S containing ω and put $T_\omega = S_\omega \setminus \{\omega\}$. Then

$$\{T_\omega \mid \omega \in \Omega\}$$

is the set of X-orbits on $\mathcal{P}$ of length 3.

Lemma 1.10.1 *Let $v \in V_{24}$. Then:*

(i) $v \in C_{V_{24}}(X)$ *if and only if there are subsets Δ and Ξ of Ω such that v is the union of Δ with $\cup_{\omega \in \Xi} T_\omega$;*

(ii) $v \in [V_{24}, X]$ *if and only if $v \cap \Omega = \emptyset$ and $|v \cap T_\omega|$ is even (which means 0 or 2) for every $\omega \in \Omega$.*

Proof. The assertion (i) is obvious. To establish (ii) notice that $v \in [V, X]$ if and only if $v + x(v) + x^2(v)$ is the zero vector in V, where x is a generator of X. □

The following lemma is a direct consequence of (1.10.1).

Lemma 1.10.2 *The following assertions hold:*

(i) $C_{V_{24}}(X)$ *is the direct sum of two copies of the $GF(2)$-permutation module $V_{1\backslash 4\backslash 1}$ of Y on Ω;*

(ii) $[V_{24}, X] = P/C_P(X)$ *where P is the 18-dimensional $GF(2)$-permutation module P of Y on $\mathcal{P} \setminus \Omega$ and $C_P(X)$ is generated by the subsets T_ω taken for all $\omega \in \Omega$.* □

Since V_{24} is an extension of the Golay code module C_{12} by its dual C_{12}^*, it follows from (1.10.2) that each of

$$C_{C_{12}}(X), \;\; C_{C_{12}^*}(X), \;\; [C_{12}, X], \;\; [C_{12}^*, X]$$

is 6-dimensional (with the obvious duality relation).

Lemma 1.10.3 *For a subset* Δ *of* Ω *define* $\Delta^{(g)}$ *to be the union of* Δ *with* $\cup_{\omega \in \Delta} T_\omega$ *or with* $\cup_{\omega \in \Omega \setminus \Delta} T_\omega$ *depending on whether* $|\Delta|$ *is even or odd. Then:*

(i) $\{\Delta^{(g)} \mid \Delta \subseteq \Omega\} = C_{\mathcal{C}_{12}}(X)$;

(ii) $C_{\mathcal{C}_{12}}(X)$ *is isomorphic to the* $GF(2)$*-permutation module of* Y *on* Ω.

Proof. Let Δ be a 5-subset in Ω. Since $(\mathcal{P}, \mathcal{B})$ is a Steiner system of type $S(5, 8, 24)$, Δ is contained in a unique octad u, and u is clearly stabilized by X. Therefore, $u = \Delta \cup T_\omega$ for some $\omega \in \Omega$. Since any two octads have even intersection and since the union of any two tetrads from $\mathcal{S}$ is an octad, we conclude that ω must be the unique element of Ω outside Δ, that is $u = \Delta^{(g)}$. Since the symmetric difference of any two Golay sets is again a Golay set, (i) follows. Now (ii) comes from (i) in view of the paragraph before the lemma. $\square$

Since the permutation modules are self-dual, we have the following isomorphisms of Y-modules

$$C_{\mathcal{C}_{12}}(X) \cong C_{\mathcal{C}_{12}^*}(X) \cong V_{1 \setminus 4 \setminus 1}.$$

Lemma 1.10.4 $[\mathcal{C}_{12}, X]$ *and* $\mathcal{H}$ *are isomorphic* Y*-modules.*

Proof. Let $\mathcal{O}$ be an octad contained in $[\mathcal{C}_{12}, X]$. By (1.10.2) there is a pair $\alpha, \beta \in \Omega$ such that $\mathcal{O} \cap (S_\alpha \cup S_\beta) = \emptyset$ and $|\mathcal{O} \cap T_\gamma| = 2$ for every $\gamma \in \Omega \setminus \{\alpha, \beta\}$. Since (a) u has three images under X, (b) there are 15 pairs in Ω transitively permuted by Y, and (c) dim $[\mathcal{C}_{12}, X] = 6$, we conclude that the Y-orbit of $\mathcal{O}$ has length 45 (we denote this orbit by $\mathcal{K}_{45}$). Let $M(\mathcal{O})$ be the stabilizer of $\mathcal{O}$ in M. By (1.9.1 (viii)), $|Q_\mathcal{O} \cap Q_\mathcal{S}| = 4$, and, if $\iota(q)$ denotes the unique octad in the X-orbit of $\mathcal{O}$ which is stabilized by $q \in (Q_\mathcal{O} \cap Q_\mathcal{S})^{\#}$, then

$$\iota : Q_\mathcal{O} \cap Q_\mathcal{S} \to \langle x(u) \mid x \in X \rangle$$

is an $(M(\mathcal{S}) \cap M(\mathcal{O}))$-isomorphism. It is easy to see that ι extends to the isomorphism between $[\mathcal{C}_{12}, X]$ and $\mathcal{H}$ we are after. $\square$

Lemma 1.10.5 *The following are isomorphisms of* Y*-modules*

$$\mathcal{C}_{12} \cong V_{1 \setminus 4 \setminus 1} \oplus \mathcal{H}, \quad \mathcal{C}_{12}^* \cong V_{1 \setminus 4 \setminus 1} \oplus \mathcal{H}^{\diamond}.$$

Proof. The assertion is a direct consequence of (1.10.3), (1.10.4), and the duality between $\mathcal{C}_{12}$ and $\mathcal{C}_{12}^*$. $\square$

The following result can be deduced either from the above considerations or from Section 3.8 in [Iv99].

Lemma 1.10.6 *The module* $\mathcal{C}_{12}$ *possesses a unique* $M(\mathcal{S})$*-composition series*

$$0 < C_{\mathcal{C}_{12}}(Q_\mathcal{S}) < [\mathcal{C}_{12}, Q_\mathcal{S}] < \mathcal{C}_{12},$$

where $C_{\mathcal{C}_{12}}(Q_S) \cong V_{1\backslash 4}$ *is a hyperplane in* $C_{\mathcal{C}_{12}}(X)$

$$[\mathcal{C}_{12}, Q_S]/C_{\mathcal{C}_{12}}(Q_S) \cong [\mathcal{C}_{12}, X] \cong \mathcal{H},$$

and $\mathcal{C}_{12}/[\mathcal{C}_{12}, Q_S]$ *is* 1-*dimensional.* $\qquad\square$

Let us turn to the module W_{24}. Recall that $W_{24} = \{(u, v^*) \mid u \in \mathcal{C}_{12}, v^* \in \mathcal{C}_{12}^*)\}$ as an M_{24}-set and

$$(u_1, v_1^*) + (u_2, v_2^*) = (u_1 + u_2, v_1^* + v_2^* + (u_1 \cap u_2)^*).$$

It is clear that

$$C_{W_{24}}(X) = \{(u, v^*) \mid u \in C_{\mathcal{C}_{12}}(X), v^* \in C_{\mathcal{C}_{12}^*}(X)\}.$$

Lemma 1.10.7 $C_{W_{24}}(X)$ *is the direct sum of two copies of the permutation module of* Y *on* Ω.

Proof. By (1.10.3 (iii)), $C_{W_{24}}(X)$ is an extension of $C_{\mathcal{C}_{12}^*}(X) \cong V_{1\backslash 4\backslash 1}$ by $C_{\mathcal{C}_{12}}(X) \cong V_{1\backslash 4\backslash 1}$. Thus is suffices to produce a Y-invariant complement to $C_{\mathcal{C}_{12}^*}(X)$. We easily check that

$$\{(u, (1 + |u| \bmod 2)(\Omega \backslash u)^*) \mid u \in C_{\mathcal{C}_{12}}(X)\}$$

does the job. $\qquad\square$

A direct calculation in W_{24} shows that for $u \in [\mathcal{C}_{12}, X]$ and a generator x of X, the equality

$$(u, 0) + (x(u), 0) + (x^2(u), 0) = (0, \eta(u)^*)$$

holds, where

$$\eta(u) = \{\omega \mid \omega \in \Omega, |u \cap T_\omega| = 2\}.$$

Lemma 1.10.8 *The following assertions hold:*

(i) $[W_{24}, X] = \{(u, \eta(u)^* + v^*) \mid u \in [\mathcal{C}_{12}, X], v^* \in [\mathcal{C}_{12}^*, X]\}$;
(ii) $[W_{24}, X]$ *is an indecomposable extension of* $\mathcal{H}^\diamond$ *by* $\mathcal{H}$.

Proof. (i) follows from the paragraph before the lemma. To justify the incomposability, notice that $\{(u, \eta(u)^*) \mid u \in [\mathcal{C}_{12}, X]\}$ is the only Y-invariant subset in $[W_{24}, X]$ on which Y acts as it does on $\mathcal{H}$. Since this orbit (together with the zero vector) is not closed under the addition in W_{24}, (ii) follows. $\qquad\square$

The extension $[W_{24}, X]$ in (1.10.8 (ii)) can be obtained by adjoining the field automorphism to the $GF(4)$-module induced from a non-trivial 1-dimensional module of a suitable index 6 subgroup in $Y' \cong 3 \cdot A_6$ (the subgroup is isomorphic to $3 \times A_5$).

The mapping

$$\eta : \mathcal{H} \to V_{1\backslash 4}$$

defined before (1.10.8) commutes with the action of Y. The mapping η is not a homomorphism of modules but it is a coboundary associated with an extension of $V_{1\backslash 4}$ by $\mathcal{H}$ and it will play an important role in the subsequent exposition.

Lemma 1.10.9 *Let q_p and β_p be M_{24}-invariant quadratic and associated bilinear forms on W_{24} as in (1.3.3). Then:*

(i) $W_{24} = C_{W_{24}}(X) \oplus [W_{24}, X]$ *is an orthogonal decomposition with respect to β_p;*

(ii) *both summands in (i) are non-singular 12-dimensional orthogonal spaces of plus type with respect to the restricted forms;*

(iii) *the $\mathcal{H}^\diamond$-submodule in $[W_{24}, X]$ and a direct summand of $C_{W_{24}}(X)$ isomorphic to $V_{1\backslash 4\backslash 1}$ are totally singular.*

Proof. First it is easy to check that the restriction of q_p to $[W_{24}, X]$ is non-singular. Thus the orthogonal complement of $[W_{24}, X]$ is 12-dimensional disjoint from $[W_{24}, X]$. Therefore the orthogonal complement must be precisely $C_{W_{24}}(X)$ and the latter must also be non-singular, which proves (i). Since $\mathcal{S}_{12}^*$ is a maximal totally singular subspace in W_{24}, (1.10.5) gives (ii) and (iii). $\qquad\square$

1.11 Three bases subgroup

In Section 1.8 the Leech lattice Λ has been constructed in terms of the stabilizer $F_0 \cong \mathcal{C}_{12} : M_{24}$ of the frame $\mathcal{F}_8$ defined in (1.8.1 (v)) and formed by the preimages in Λ_4 of the element

$$\bar{\lambda}_4 = \sum_{p \in \mathcal{P}} 2p + 2\Lambda$$

from $\bar{\Lambda}_4$. In this section we analyse the stabilizer in Co_1 of a triple of such frames.

Let

$$\bar{\mu}_4 = \sum_{q \in \mathcal{T}} 4q + 2\Lambda,$$

where $\mathcal{T}$ is a 4-subset of $\mathcal{P}$ and let $\bar{\nu}_4 = \bar{\lambda}_4 + \bar{\mu}_4$. By (1.8.2), $F = E : M \cong \mathcal{C}_{11} : M_{24}$ is the stabilizer of $\bar{\lambda}_4$ in Co_1.

Lemma 1.11.1 *The following assertions hold:*

(i) $\{\bar{\mu}_4, \bar{\nu}_4\}$ *is an E-orbit;*
(ii) *the stabilizer of $\bar{\mu}_4$ in M coincides with the stabilizer $M(\mathcal{S}) \cong 2^6 : 3 \cdot S_6$ of the sextet $\mathcal{S}(v)$;*
(iii) *every permutation of $\{\bar{\lambda}_4, \bar{\mu}_4, \bar{\nu}_4\}$ is realized by a suitable element of Co_1.*

Proof. The assertions are standard properties of the Leech lattice (modulo 2) and of the Conway group Co_1. The proofs can be found for instance in Sections 4.6 and 4.7 in [Iv99]. Notice that for a 4-subset w of $\mathcal{P}$ the vector

$$\sum_{r \in w} 4r + \sum_{q \in v} 4q$$

belongs to 2Λ if and only if $w \in \mathcal{S}(v)$.   □

Our next goal is to specify the isomorphism type of the *three bases subgroup* T in Co_1 which is the set-wise stabilizer of

$$\bar{\Theta} := \{\bar{\lambda}_4, \bar{\mu}_4, \bar{\nu}_4\}$$

and to describe the action of T on $\bar{\Lambda}$. For $\bar{\theta} \in \bar{\Theta}$, let $F^{(\bar{\theta})}$ be the stabilizer of $\bar{\theta}$ in Co_1, so that $F^{(\bar{\lambda}_4)} = F$ is the monomial subgroup and let $E^{(\bar{\theta})} = O_2(F^{(\bar{\theta})})$. Let T^+ be the centralizer of $\bar{\Theta}$ in Co_1, so that

$$T/T^+ \cong S_3$$

by (1.11.1 (iii)). The next lemma specifies the isomorphism type of T^+.

Lemma 1.11.2 *Let Y be a complement to $O_2(M(\mathcal{S}))$ in the stabilizer $M(\mathcal{S})$ of $\bar{\Theta}$ in M, and let $X = O_3(Y)$. For $\bar{\theta} \in \bar{\Theta}$, let $W^{(\bar{\theta})} = E^{(\bar{\theta})} \cap T^+$. Then:*

(i) $Y \cong 3 \cdot S_6$ *is the hexacode group;*
(ii) T^+ *is a semidirect product of Y and the subgroup $O_2(T^+)$ of order 2^{16};*
(iii) $Z(O_2(T^+))$ *coincides with $C_{O_2(T^+)}(X)$ and has order 2^4;*
(iv) *if A is a maximal normal elementary abelian subgroup in $O_2(T^+)$, then $|A| = 2^{10}$ and $A = W^{(\bar{\theta})}$ for some $\bar{\theta} \in \bar{\Theta}$;*
(v) $W^{(\bar{\lambda}_4)} W^{(\bar{\mu}_4)} = O_2(T^+)$ *and $W^{(\bar{\lambda}_4)} \cap W^{(\bar{\mu}_4)} = C_{O_2(T^+)}(X)$;*
(vi) *there is a unique Y-isomorphism*

$$\alpha : U_4 \to C_{O_2(T^+)}(X);$$

(vii) *for every $\bar{\theta} \in \bar{\Theta}$ there is a unique Y-isomorphism*

$$\beta_{\bar{\theta}} : \mathcal{H} \to [W^{(\bar{\theta})}, X];$$

(viii) *for* $h, k \in \mathcal{H}$ *the equality*

$$[\beta_{\bar{\lambda}_4}(h), \beta_{\bar{\mu}_4}(k)] = \alpha(\vartheta(h+k))$$

    holds, where $\vartheta : \mathcal{H} \to U_4$ *is the composition of the mapping* η *as in*
    (1.10.8) *and the natural homomorphism* $V_{1\backslash 4} \to U_4$;
(ix) *the conditions* (i) *to* (viii) *uniquely determine the isomorphism type*
    of T^+.

Proof. Since F is the semidirect product of $E \cong C_{11}$ and $M \cong M_{24}$ with respect to the natural action, the assertions from (i) to (iii) follow from (1.11.1 (ii)) and (1.9.1). The structure of C_{11} as a module for Y and the triality symmetry of T^+ implied by (1.11.1 (iii)) give (iv) to (vii). The uniqueness of the isomorphisms follow from the absolute irreducibility of the Y-modules U_4 and $\mathcal{H}$. To prove (vii) notice first that Y acting on the direct sum

$$U_4 \oplus \mathcal{H}$$

has exactly one orbit of length 18 (contained in $\mathcal{H}$) and two orbits of length 45: one of them, the orbit $\mathcal{H}_{45}$, is contained in $\mathcal{H}$, while the other one is

$$\{h\vartheta(h) \mid h \in \mathcal{H}_{45}\}.$$

The set $\{\beta_{\bar{\lambda}_4}(h)\beta_{\bar{\mu}_4}(h) \mid h \in \mathcal{H}_{45}\}$ is a Y-orbit of length 45 inside $W(\bar{\nu}_4)$. If this orbit were $\beta_{\bar{\nu}_4}(\mathcal{H}_{45})$, then $\beta_{\bar{\lambda}_4}(h)$ and $\beta_{\bar{\mu}_4}(k)$ would commute for all $h, k \in \mathcal{H}$, which is impossible since $O_2(T^+)$ is clearly non-abelian. Therefore, for every $h \in \mathcal{H}$ the equality

$$\beta_{\bar{\lambda}_4}(h)\beta_{\bar{\mu}_4}(h) = \beta_{\bar{\nu}_4}(h)\vartheta(h)$$

holds. In view of the addition rule in $U_4 \oplus \mathcal{H}$, this gives the commutator relation in (viii). Now (ix) is rather clear. $\square$

Lemma 1.11.3 *The outer automorphism group of* T^+ *is isomorphic to* $S_3 \times 2$.

Proof. By (1.9.2) and (1.11.2), T^+ contains exactly two classes of complements to $O_2(T^+)$, both intersecting $Z(O_2(T^+))Y$. Since T^+ is a semidirect product of $O_2(T^+)$ and Y, the automorphism of $Z(O_2(T^+))Y$ which centralizes both $Z(O_2(T^+))$ and $Z(O_2(T^+))Y/Z(O_2(T^+))$ and permutes the classes of $3 \cdot S_6$-complements extends to an automorphism of T^+.

If τ is a permutation of $\bar{\Theta}$, then by (1.11.2 (vi), (vii), (viii)) the mapping which centralizes both Y and $C_{O_2(T^+)}(X)$ and maps $\beta_{\bar{\theta}}(h)$ onto $\beta_{\tau(\bar{\theta})}(h)$ for every $h \in \mathcal{H}$ extends to an automorphism of T^+. On the other hand, both U_4 and $\mathcal{H}$ are absolutely irreducible Y-modules and therefore the action of Y on $W^{(\bar{\theta})}$ has a trivial centralizer in the general linear group of $W^{(\bar{\theta})}$ (isomorphic to

$GL_{10}(2)$). Hence 12 is an upper bound on the order of the outer automorphism group of T^+ and the result follows.   □

Since the centre of T^+ is trivial, T is a subgroup of $\mathrm{Aut}(T^+)$ that contains $\mathrm{Inn}(T^+)$ and where the image in $\mathrm{Out}(T^+)$ is isomorphic to S_3. By (1.11.3) there are two S_3-subgroups in $\mathrm{Out}(T^+)$; one of them stabilizes the classes of complements to $O_2(T^+)$, while the other one does not. By (1.10.5), $C_E(X) \cong V_{4\backslash 1}$ is an indecomposable extension of $U_4 \cong C_{O_2(T^+)}(X)$. This gives the following characterization of the isomorphism type of T.

Lemma 1.11.4 T *is the subgroup of* $\mathrm{Aut}(T^+)$ *that contains the inner automorphisms and that maps onto the* S_3*-subgroup in* $\mathrm{Out}(T^+)$ *permuting the two classes of complements to* $O_2(T^+)$ *in* T^+.   □

Next we turn to the action of T on $\bar{\Lambda}$.

Lemma 1.11.5 *Under the action of* $T \cong 2^{4+12} \cdot (3 \cdot S_6 \times S_3)$, *the module* $\bar{\Lambda}$ *possesses a unique composition series*

$$0 < \bar{\Lambda}^{(2)} < \bar{\Lambda}^{(8)} < \bar{\Lambda}^{(16)} < \bar{\Lambda}^{(22)} < \bar{\Lambda},$$

where:

(i) $\bar{\Lambda}^{(2)}$ *is generated by* $\bar{\Theta}$ *and is isomorphic to the natural module of* $T/T^+ \cong S_3 \cong SL_2(2)$;

(ii) $\bar{\Lambda}^{(8)} = C_{\bar{\Lambda}}(Z(O_2(T)))$ *and* $\bar{\Lambda}^{(8)}/\bar{\Lambda}^{(2)}$ *is isomorphic to* $\mathcal{H}^\circ$ *(as a module for the hexacode group* $Y \cong 3 \cdot S_6$);

(iii) $\bar{\Lambda}^{(16)}/\bar{\Lambda}^{(8)}$ *is isomorphic to the tensor product of* U_4 *and* $\bar{\Lambda}^{(2)}$;

(iv) $\bar{\Lambda}^{(22)}/\bar{\Lambda}^{(16)}$ *is isomorphic to* $\mathcal{H}$;

(v) $\bar{\Lambda}/\bar{\Lambda}^{(22)}$ *is isomorphic to (the dual of)* $\bar{\Lambda}^{(2)}$.

Proof. Since $F \cap T = E : M(\mathcal{S})$, (1.10.6) and (1.8.6) enable us to deduce a composition series of $\bar{\Lambda}$ under the actions of $F \cap T$ and T^+. Since the composition series of $F^{(\bar{\mu}_4)}$ is different from that of F, the two composition factors of T^+ isomorphic to U_4 must be permuted by T. The rest is now clear since $\bar{\Lambda}$ is a self-dual module.   □

2

The Monster amalgam $\mathcal{M}$

In this chapter we construct the Monster amalgam $\mathcal{M}$. We give an abstract definition of $\mathcal{M}$ and then construct $\mathcal{M}$ explicitly. The uniqueness comes as an immediate byproduct of the construction.

2.1 Defining the amalgam

In this section we give an abstract definition of the Monster amalgam and prove that it exists and is unique up to isomorphism.

The *Monster amalgam* $\mathcal{M}$ is formed by a triple of groups G_1, G_2, and G_3 subject to the following conditions $(\mathcal{M}1)$ to $(\mathcal{M}3)$, where $G_{ij} = G_i \cap G_j$, $G_{123} = G_1 \cap G_2 \cap G_3$, $Q_i = O_2(G_i)$, and $Z_i = Z(Q_i)$.

$(\mathcal{M}1)$ $G_1 \in \mathcal{E}(Co_1, \bar{\Lambda}, q_{\bar{\Lambda}})$ in terms of Section 1.4, which means that

$$Q_1 \cong 2_+^{1+24}, \quad G_1/Q_1 \cong Co_1$$

and there is a homomorphism

$$\chi : Q_1 \to \bar{\Lambda}$$

with kernel Z_1, such that the induced isomorphism between Q_1/Z_1 and $\bar{\Lambda}$ commutes with the action of Co_1 and establishes an isomorphism of the orthogonal spaces $(Q_1/Z_1, q_e)$ and $(\bar{\Lambda}, q_{\bar{\Lambda}})$, where q_e is induced by the squaring map on Q_1.

$(\mathcal{M}2)$ $G_{12} = N_{G_1}(Z_2)$, where

$$Z_2 = \chi^{-1}(\langle \sum_{p \in \mathcal{P}} 2p + 2\Lambda \rangle);$$

41

(by (1.8.2), G_{12} is the preimage in G_1 of the monomial subgroup $F \cong \mathcal{C}_{11} : M_{24}$ of $G_1/Q_1 \cong Co_1$)

$$[G_2 : G_{12}] = 3 \text{ and } Z(G_2) = 1.$$

($\mathcal{M}3$) $G_3 = \langle G_{13}, G_{23} \rangle$; $G_{13} = N_{G_1}(Z_3)$, $G_{23} = N_{G_2}(Z_3)$, where

$$Z_3 = \chi^{-1}(\langle \sum_{p \in \mathcal{P}} 2p + 2\Lambda, \sum_{q \in \mathcal{T}} 4q + 2\Lambda \rangle)$$

for a 4-subset $\mathcal{T}$ in $\mathcal{P}$ (so that G_{13} is the preimage in G_1 of the three base subgroup $T \cong 2^{4+12}.(3 \cdot S_6 \times S_3)$ of $G_1/Q_1 \cong Co_1$).

($\mathcal{M}4$) If G_3^s is the largest subgroup in G_{123} which is normal in G_3, then $C_{G_3}(G_3^s) = Z_3$ and $G_3/G_3^s \cong L_3(2)$.

Definition 2.1.1 *The* Monster group G *is a faithful generating completion of a Monster amalgam. This means that G is a group which contains subgroups G_1, G_2, and G_3 satisfying the conditions ($\mathcal{M}1$) to ($\mathcal{M}4$) and G is generated by these three subgroups.*

It might be appropriate to go into some detail of the formal definition of an amalgam and its completions. To define $\mathcal{M}$ we take three finite groups $(G_1, \odot_1)$, $(G_2, \odot_2)$, $(G_3, \odot_3)$, where the G_is are the element sets and the $\odot_i$s are the product operators. The element sets have pairwise non-empty intersections and whenever $x, y \in G_{ij} = G_i \cap G_j$ for $1 \leq i, j \leq 3$, the equality

$$x \odot_i y = x \odot_j y$$

holds. So to say, the group operations coincide on the intersections. In particular, G_{ij} is a subgroup in both $(G_i, \odot_i)$ and $(G_j, \odot_j)$. A *completion* of $\mathcal{M}$ is a pair $((G, \odot), \psi)$ where $(G, \odot)$ is a group and

$$\psi : G_1 \cup G_2 \cup G_3 \to G$$

is a map, such that whenever $x, y \in G_i$ for $1 \leq i \leq 3$ the equality

$$\psi(x \odot_i y) = \psi(x) \odot \psi(y)$$

holds (i.e. the restriction of ψ to G_i is a homomorphism.) The map ψ is called the *completion map* and $(G, \odot)$ is the *completion group*. A completion is said to be:

(a) *faithful* if ψ is injective;
(b) *generating* if the image of ψ generates $(G, \odot)$;
(c) *universal* if for any other completion $((H, \diamond), \chi)$ there is a surjective homomorphism $\eta : G \to H$ such that χ is the composition of ψ and η.

The universal completion is necessarily generating; faithful completion exists if and only if the universal completion is faithful. The isomorphism between amalgams and completions is given the obvious meaning. In particular, all the universal completions of a given amalgam are isomorphic. When defining $\mathcal{M}$ at the beginning of the section we suppress the explicit reference to the group operators and we will continue doing so in what follows.

A priori it is not clear that at least one Monster amalgam exists. We will show that up to isomorphism there is a unique Monster amalgam $\mathcal{M}$. The next obvious question is whether $\mathcal{M}$ possesses a faithful completion. This is not an easy question at all and it will be answered affirmatively by producing a faithful 196 883-dimensional representation of $\mathcal{M}$. Eventually we will show that the only faithful completion of $\mathcal{M}$ is the universal one and the completion group is the Monster as we know it, that is a non-abelian finite simple group in which G_1 is the centralizer of an involution.

2.2 The options for G_1

Let G_1 be a group satisfying the conditions in $(\mathcal{M}1)$ (i.e. $G_1 \in \mathcal{E}(Co_1, \bar{\Lambda}, q_{\bar{\Lambda}})$). By (1.4.3) such a group G_1 exists and the isomorphism type of G_1/Z_1 is uniquely determined. Let $G_1^{(s)}$ be the unique group in $\mathcal{E}(Co_1, \bar{\Lambda}, q_{\bar{\Lambda}})$ which is a subgroup of the standard holomorph of $Q_1 \cong 2_+^{1+24}$. Then $G_1^{(s)}$ is the only group satisfying $(\mathcal{M}1)$ which possesses a faithful 2^{12}-dimensional representation over real numbers. By (1.8.8 (iii)) and its proof, G_1/Z_1 is the only non-split extension of $\bar{\Lambda} \cong 2^{24}$ by Co_1 which we denote by $2^{24} \cdot Co_1$.

Since G_1 is a perfect group and since G_1/Z_1 is uniquely determined, all the possible G_1s are quotients of the universal perfect central extension $\widehat{G}_1$ of G_1/Z_1. Let $\widehat{Y}_1$ be the intersection of the kernels of all the homomorphisms of $\widehat{G}_1$ onto groups from $\mathcal{E}(Co_1, \bar{\Lambda}, q_{\bar{\Lambda}})$. Then the G_1s are quotients of $\widetilde{G}_1 := \widehat{G}_1/\widehat{Y}_1$. Our nearest goal is to identify $\widetilde{G}_1$. Let $\widetilde{Q}_1$ be the preimage of Q_1/Z_1 with respect to the natural homomorphism of $\widetilde{G}_1$ onto G_1/Z_1 and let $\widetilde{Z}_1$ be the kernel of that homomorphism, so that $\widetilde{Z}_1 = Z(\widetilde{G}_1)$.

Lemma 2.2.1 $\widetilde{Z}_1$ *is an elementary abelian 2-group.*

Proof. The image of $\widetilde{Z}_1$ under every homomorphism of $\widetilde{G}_1$ onto a group from $\mathcal{E}(Co_1, \bar{\Lambda}, q_{\bar{\Lambda}})$ is of order 2. Since the intersection of the kernels of these homomorphisms is trivial, the result follows. $\square$

Lemma 2.2.2 *The commutator subgroup of* $\widetilde{Q}_1$ *is of order 2.*

Proof. Since $\widetilde{Z}_1$ is the centre of $\widetilde{G}_1$, the squaring map on $\widetilde{Q}_1$ induces a well-defined map σ of $\widetilde{Q}_1/\widetilde{Z}_1 \cong \bar{\Lambda}$ into $\widetilde{Z}_1$. We claim that the image of σ is of order 2. To prove this observe that Co_1 acts primitively on the set $\bar{\Lambda}_3$ of non-isotropic vectors in $\bar{\Lambda}$ with respect to the form $q_{\bar{\Lambda}}$ and that by the universality principle every automorphism of $Q_1/Z_1 \cong 2^{24} \cdot Co_1$ lifts to an automorphism of $\widetilde{G}_1$. The quotient $\widetilde{Q}_1/\mathrm{Im}(\sigma)$ has exponent 2 and therefore it is abelian. Hence $\mathrm{Im}(\sigma)$ coincides with the commutator subgroup of $\widetilde{Q}_1$ and the result follows. $\square$

Let $\widetilde{Y}_1 = \mathrm{Im}(\sigma)$ be the commutator subgroup of $\widetilde{Q}_1$. Then $\widetilde{Q}_1/\widetilde{Y}_1$ is a $GF(2)$-module for $\widetilde{G}_1/\widetilde{Q}_1 \cong Co_1$ and this module is an extension of the trivial module $\widetilde{Z}_1/\widetilde{Y}_1$ by $Q_1/Z_1 \cong \bar{\Lambda}$. Since $\bar{\Lambda}$ is a self-dual Co_1-module, $\widetilde{Q}_1/\widetilde{Y}_1$ is semi-simple by (1.8.8 (ii)). Let $\widetilde{P}_1$ be the preimage in $\widetilde{Q}_1$ of a complement to $\widetilde{Z}_1/\widetilde{Y}_1$ in $\widetilde{Q}_1/\widetilde{Y}_1$. Then $\widetilde{G}_1/\widetilde{P}_1$ is a perfect central extension of $\widetilde{G}_1/\widetilde{Q}_1 \cong Co_1$. By (1.8.8 (i)), the order of $\widetilde{Q}_1/\widetilde{P}_1$ is at most 2.

Proposition 2.2.3 *The following assertions hold:*

(i) *$\widetilde{Z}_1$ has order four;*
(ii) *$\mathcal{E}(Co_1, \bar{\Lambda}, q_{\bar{\Lambda}}) = \{G_1^{(s)}, G_1^{(t)}\}$;*
(iii) *the minimal dimensions of faithful representations over real numbers of the groups $G_1^{(s)}$ and $G_1^{(t)}$ are 2^{12} and $24 \cdot 2^{12}$, respectively;*
(iv) *$\widetilde{G}_1$ is the universal perfect central extension of $G_1/Z_1 \cong 2^{24} \cdot Co_1$.* $\square$

Proof. The upper bound four for the order of $\widetilde{Z}_1$ was established in the paragraph before the proposition. Consider the subdirect product X of $G_1^{(s)}$ and Co_0 over their homomorphisms onto Co_1

$$X = \{(g_1, g_2) \mid g_1 \in G_1^{(s)}, g_2 \in Co_0, \varphi_1(g_1) = \varphi_2(g_2)\},$$

where $\varphi_1 : G_1^{(s)} \to G_1^{(s)}/Q_1 \cong Co_1$ and $\varphi_2 : Co_0 \to Co_0/Z(Co_0) \cong Co_1$ are the natural homomorphisms. Then X is a perfect group and has centre of order four. Moreover, if z_1 and z_2 are the generators of the centres of $G_1^{(s)}$ and Co_0, respectively (considered in the obvious way as elements of X), then

$$G_1^{(t)} := X/\langle z_1 z_2 \rangle$$

is a group from $\mathcal{E}(Co_1, \bar{\Lambda}, q_{\bar{\Lambda}})$. Thus $\widetilde{G}_1 = X$ and $\mathcal{E}(Co_1, \bar{\Lambda}, q_{\bar{\Lambda}}) = \{G_1^{(s)}, G_1^{(t)}\}$. To establish (ii) it only remains to show that $G_1^{(s)}$ and $G_1^{(t)}$ are not isomorphic. This can be achieved by showing that the outer automorphism group of X is trivial using (1.8.8 (ii)) and the triviality of the outer automorphism group of Co_0 [CCNPW]. In order to prove (iii) consider the 2^{12}-dimensional representation

$$\alpha : G_1^{(s)} \to GL(T)$$

of $G_1^{(s)}$ and the natural 24-dimensional representation

$$\beta : Co_0 \to GL(\Lambda_{\mathbf{R}}),$$

where $\Lambda_{\mathbf{R}} = \Lambda \otimes \mathbf{R}$ is the 24-dimensional real vector space containing the Leech lattice Λ. Let

$$\gamma : G_1^{(s)} \times Co_0 \to GL(T \otimes \Lambda_{\mathbf{R}})$$

be the representation defined in terms α and β in the usual manner. Then both $\gamma(z_1)$ and $\gamma(z_2)$ (-1)-scalar operators. Therefore, $\gamma(z_1 z_2)$ is the identity operator and $\gamma(X) \cong G_1^{(t)}$. The fact that $24 \cdot 2^{12}$ is the minimal degree of a faithful real (or even complex) representation of $G_1^{(t)}$ and the uniqueness of the representation can be established using the Clifford theory [CR62] and the result that the only minimal faithful representation of Co_0 is the natural action on $\Lambda \otimes \mathbf{R}$ [CCNPW]. Therefore (iii) holds. The assertion (iv) is easy to prove using the uniqueness of the Co_1-invariant quadratic form $q_{\bar{\Lambda}}$ on $\bar{\Lambda}$. $\square$

2.3 Analysing G_{12}

By (2.2.3 (ii)), we have precisely two candidates $G_1^{(s)}$ and $G_1^{(t)}$ for G_1. Both $G_1^{(s)}$ and $G_1^{(t)}$ are quotients of the universal perfect central extension $\tilde{G}_1$ of $G_1/Z_1 \cong 2^{24} \cdot Co_1$. In what follows G_1 denotes either of these two groups.

Let z be the generator of Z_1, let x and y be the elements in Q_1 such that

$$\chi(x) = \chi(y) = \sum_{p \in \mathcal{P}} 2p + 2\Lambda,$$

and let $Z_2 = \{1, x, y, z\}$. Then $Z_2 = \chi^{-1}(\bar{\Lambda}^{(1)})$, in particular Z_2 is elementary abelian of order 4. By ($\mathcal{M}2$), we have

$$G_{12} = N_{G_1}(Z_2).$$

By (1.8.2), this shows that G_{12} is the preimage in G_1 of the monomial subgroup

$$F = E : M \cong \mathcal{C}_{11} : M_{24}$$

of $Co_1 = G_1/Q_1$ and therefore $O_2(G_{12})$ is the preimage of E. By (1.8.5 (iii)), an element from Q_1 centralizes Z_2 if its image under χ is contained in $\bar{\Lambda}^{(23)}$ and it permutes x and y (centralizing z) otherwise. Put

$$G_2^s = C_{G_1}(Z_2), \quad R_2 = \chi^{-1}(\bar{\Lambda}^{(12)}), \quad \text{and} \quad Q_2 = O_2(G_2^s).$$

Lemma 2.3.1 *The following assertions hold:*

(i) G_2^s *is the only index 2 subgroup in* G_{12}, $G_2^s = [G_{12}, G_{12}]$, *and* G_2^s *is perfect;*

(ii) Z_2 *is the preimage in* G_{12} *of* $Z(G_{12}/Z_1) = Z(G_2^s/Z_1)$, *in particular* Z_2 *is characteristic in both* G_{12} *and* G_2^s;

(iii) $Q_2 \cap Q_1 = \chi^{-1}(\bar{\Lambda}^{(23)})$ *and* $Q_1 Q_2 = O_2(G_{12})$;

(iv) $G_2^s/Q_2 \cong G_{12}/O_2(G_{12}) \cong F/E \cong M_{24}$;

(v) $Z_2 = Z(Q_2) = Z(G_2^s)$;

(vi) R_2 *is an elementary abelian self-centralizing normal subgroup in* G_{12} *of order* 2^{13};

(vii) Q_2 *centralizes* R_2/Z_2, *while* $G_2^s/Q_2 \cong M_{24}$ *acts on* R_2/Z_2 *as on the irreducible Todd module* $\mathcal{C}_{11}^*$;

(viii) Q_2 *induces on* R_2 *the subgroup of* $GL(R_2)$ *generated by all the transvections where the centres are in* Z_2 *and the axes contain* Z_2;

(ix) G_2^s/R_2 *splits over* Q_2/R_2.

Proof. Since $F \cong \mathcal{C}_{11} : M_{24}$ is perfect centreless and since it acts faithfully on $\bar{\Lambda}^{(23)}$, (i), (ii), and (iii) follow from (1.8.6). The subgroup Z_2 is contained in $Z(Q_2)$ and by (1.8.6 (iii)), $Z_2 = Z(Q_1 \cap Q_2)$. Since the action of F on $\bar{\Lambda}^{(23)}$ is faithful, (v) follows. Now (vi) and (vii) are direct consequences of (1.8.5 (ii), (iv)). Let X be the centralizer of Z_2 in the general linear group $GL(R_2)$, and let Y be the image of G_2^s in $GL(R_2)$. By (v) and (vi), $Y \leq X$ and $X \cong G_2^s/R_2$. It follows from the standard properties of $GL(R_2)$ that:

(A) $|O_2(X)| = 2^{22}$;

(B) $O_2(X) = C_X(R_2/Z_2)$;

(C) there is a (Levi) complement L to $O_2(X)$ in X;

(D) $L \cong L_{11}(2)$ acts faithfully on R_2/Z_2.

By (iii), (vi), (vii), and the order consideration, the properties (A) to (D) imply that $O_2(Y) = O_2(X)$. Let K be an M_{24}-subgroup in L, where the action on R_2/Z_2 coincides with that of G_2^s. We already know that Y contains the kernel of the action of X on R_2/Z_2, therefore K is also contained in Y and (ix) holds. $\square$

Lemma 2.3.2 *The following assertions hold:*

(i) G_2^s *is normal in* G_2 *and* $G_2/G_2^s \cong S_3$ *acts faithfully on* $Z_2^\#$;

(ii) Q_2 *(as defined before (2.3.1)) is precisely* $O_2(G_2)$;

(iv) R_2 *is the preimage in* Q_2 *of* $Z(Q_2/Z_2)$.

Proof. Let N be the largest subgroup in G_{12} which is normal in G_2. Since $[G_2 : G_{12}] = 3$ and since by (2.3.1) G_2^s is the only index 2 subgroup in G_{12}, either $N = G_{12}$ or $N = G_2^s$. In any event, G_2^s is normal in G_2. Since G_2^s contains Q_2, it follows from (2.3.1 (v)) that $Z_2 = Z(G_2^s)$, which means that G_2 acts on $Z_2^{\#}$. Since G_{12} induces on $Z_2^{\#}$ an action of order 2 and since G_2 is centreless, (i) follows. By (i), $O_2(G_2)$ is contained in G_2^s, which gives (ii). Finally, (iii) follows from (2.3.1 (vii)) and the fact that F acts faithfully on $\bar{\Lambda}^{(23)}/\bar{\Lambda}^{(1)}$. $\qquad\qquad\square$

By (2.3.1 (i)) and (2.3.2), G_2^s is the smallest normal subgroup in G_2 such that the corresponding factor group is solvable.

Throughout the current and next sections we are maintaining a symmetry between the elements of $Z_2^{\#}$. We denote by a, b, and c three pairwise distinct elements of $Z_2^{\#}$, so that

$$\{a, b, c\} = \{x, y, z\}.$$

Let $X^{(a)}$ be the subgroup of Q_2/R_2 whose action on R_2 is generated by the transvections with centre a and axes containing Z_2. Then by (2.3.1 (viii), (ix))

$$G_2^s/R_2 = (X^{(a)} \times X^{(b)}) : K = (X^{(a)} \times X^{(c)}) : K,$$

where $K \cong M_{24}$ and $X^{(a)} \cong X^{(b)} \cong X^{(c)} \cong \mathcal{C}_{11}$ (as K-modules). Let M^{δ} be the preimage of K in G_2^s. Then $Q_1 M^{\delta}/Q_1$ is a complement to $Q_1 Q_2/Q_1 \cong \mathcal{C}_{11}$ in $G_{12}/Q_1 \cong \mathcal{C}_{11} : M_{24}$. Thus, choosing K to be a suitable complement to $X^{(x)}$ in $X^{(x)} K$, we can and do assume that the action of M^{δ} on $Q_1/Z_1 \cong \bar{\Lambda}$ coincides with that of the subgroup $M \cong M_{24}$ of coordinate permutations in the basis $\mathcal{P}$.

Lemma 2.3.3 *The subgroup M^{δ} is a direct product*

$$M^{\delta} = Z_2 \times [M^{\delta}, M^{\delta}],$$

where $[M^{\delta}, M^{\delta}] \cong \mathcal{C}_{11}^* \cdot M_{24}$ is the unique non-split extension of $\mathcal{C}_{11}^*$ by M_{24}.

Proof. Let M^0 be the preimage in G_1 of the subgroup $M \cong M_{24}$ in $Co_1 \cong G_1/Q_1$. By (1.8.3), $Q_1/Z_1 \cong \bar{\Lambda}$ (as a module for $M^0/Q_1 \cong M_{24}$) is isomorphic to the anti-permutation module W_{24}. Therefore, M^0 is the unique group from $\mathcal{E}(M_{24}, W_{24}, q_p)$ and the result follows from (1.7.1). $\qquad\square$

In what follows, M_{24}-modules will be considered also as M^{δ}-modules in the obvious manner.

Lemma 2.3.4 *Let $Q_2^{(a)}$ be the preimage of $X^{(a)}$ in Q_2. Then:*

(i) $Q_2^{(a)} = C_{Q_2}(R_2/\langle a \rangle)$ *and $Q_2^{(a)}$ is normal in G_2^s;*

(ii) $Q_2^{(a)} Q_2^{(b)} = Q_2$ *and $Q_2^{(a)} \cap Q_2^{(b)} = R_2$;*

(iii) $Q_2^{(z)} = Q_1 \cap Q_2 = \chi^{-1}(\bar{\Lambda}^{(23)})$;

(iv) $[Q_2^{(a)}, Q_2^{(a)}] = \langle a \rangle$;

(v) $Q_2^{(a)}/\langle b \rangle \cong 2_+^{1+22}$;

(vi) $Q_2^{(a)}/R_2$ and $\mathcal{C}_{11}$ are isomorphic M^δ-modules.

Proof. Recall that $X^{(a)}$ was defined as the subgroup in Q_2/R_2, which acts on R_2 by transvections with centre a and axis Z_2. Therefore, the assertions (i) to (iii) follow from (2.3.1 (viii)) and the standard properties of $GL(R_2)$. Notice that:

(A) $\langle a \rangle$ is the commutator subgroup of $Q_2^{(a)}$;

(B) $Z_2 = \langle a, b \rangle$ is the centre of $Q_2^{(a)}$;

(C) $Q_2^{(a)}/Z_2$ is elementary abelian.

Hence (v) follows. Finally, (vi) holds by (2.3.1 (vii)) since $Q_2^{(a)}/R_2$ is dual to R_2/Z_2. $\qquad\square$

Lemma 2.3.5 *Put* $N^{(a)} = Q_2^{(a)} M^\delta$. *Then* $N^{(a)}/\langle b \rangle$ *is isomorphic to the unique group in* $\mathcal{E}(M_{24}, \mathcal{A}(M_{24}), q_h)$.

Proof. By (2.3.4 (i), (v)), $Q_2^{(a)}/\langle b \rangle$ is a normal subgroup in $N^{(a)}/\langle b \rangle$ isomorphic to the extraspecial group 2_+^{1+22}, while by (2.3.3) we have

$$N^{(a)}/Q_2^{(a)} \cong M^\delta/R_2 \cong M_{24}.$$

By (2.3.1 (vii)) and (2.3.4 (vi)), $Q_2^{(a)}/Z_2$ (as an M^δ-module) is an extension of $\mathcal{C}_{11}^*$ by $\mathcal{C}_{11}$. By (1.7.4), this implies that $N^{(a)}/\langle b \rangle$ is either the unique group in $\mathcal{E}(M_{24}, \mathcal{C}_{11}^* \oplus \mathcal{C}_{11}, q_d)$ or the unique group in $\mathcal{E}(M_{24}, \mathcal{A}(M_{24}), q_h)$. By (2.3.3), M^δ/Z_2 does not split over R_2/Z_2, and therefore the latter possibility holds. $\qquad\square$

Let $p \in \mathcal{P}$, and let $g = g(p)$ be an element of Q_1 such that

$$\chi(g) = \left(3p - \sum_{r \in \mathcal{P} \setminus \{p\}} r \right) + 2\Lambda.$$

for some $p \in \mathcal{P}$. Then $\chi(g) \notin \bar{\Lambda}^{(23)}$ and hence by (1.8.6 (iii)) g conjugates x onto y. On the other hand, $\chi(g)$ is isotropic with respect to $q_{\bar{\Lambda}}$, therefore

$$G_{12} = \langle G_2^s, g \rangle \text{ and } g^2 = 1.$$

Since $Q_1/Z_1 \cong \bar{\Lambda}$ is the anti-permutation M-module, there are exactly 24 images of $\chi(g)$ under M and if $\bar{\Theta}$ is the subspace in $\bar{\Lambda}$ spanned by these images, then

$$\bar{\Theta} = \omega(\mathcal{S}_{12}^*), \ \bar{\Theta} \cap \bar{\Lambda}^{(12)} = \omega(\mathcal{S}_{11}^*),$$

and $\bar{\Theta}$ is isomorphic to C_{12}^* as a module for M (recall that $\omega : W_{24} \to \bar{\Lambda}$ is the isomorphism of M-modules defined in (1.8.3)). Since M and M^δ act on $\bar{\Lambda}$ in the same way, and since M^δ contains $R_2 = \chi^{-1}(\bar{\Lambda}^{(12)})$, we conclude that g normalizes M^δ and centralizes M^δ/R_2. This is recorded in the next lemma where $\bar{G}_{12} = G_{12}/Z_2$ and the bar convention is adopted for the images in $\bar{G}_{12}$ of elements and subgroups of G_{12}.

Lemma 2.3.6 *Let $g \in Q_1$ be as in the paragraph preceding the lemma. Then:*

(i) *$\bar{G}_{12} = \langle \bar{G}_2^s, \bar{g} \rangle$;*
(ii) *$\bar{g}$ is of order 2;*
(iii) *g centralizes $\bar{Q}_2^{(z)}$ and acting by conjugation it establishes an M^δ-module isomorphism between $\bar{Q}_2^{(x)}$ and $\bar{Q}_2^{(y)}$;*
(iv) *$\bar{g}$ normalizes $\bar{M}^\delta$ and centralizes $\bar{M}^\delta/\bar{R}_2 \cong M_{24}$;*
(v) *$\langle \bar{M}^\delta, \bar{g} \rangle \cong C_{12}^* \cdot M_{24}$.* $\square$

2.4 G_2^s/Z_2 and its automorphisms

In this section we continue to follow the bar convention for elements and subgroups of $\bar{G}_{12} = G_{12}/Z_2$ introduced before (2.3.6). It is worth emphasizing that $\bar{G}_{12}$ is a section of G_1/Z_1 and therefore its isomorphism type is independent of the choice of G_1 between $G_1^{(s)}$ and $G_1^{(t)}$.

By (2.3.5), for every $a \in Z_2^{\#}$ there is an $\bar{M}^\delta$-module isomorphism

$$\psi^{(a)} : \mathcal{A}(M_{24}) \to \bar{Q}_2^{(a)}.$$

Lemma 2.4.1 *The following assertions hold for all $u, u_1, u_2 \in C_{11}$, and $v^*, v_1^*, v_2^* \in C_{11}^*$:*

(i) *the restriction ψ of $\psi^{(a)}$ to $\mathcal{S}_{11}^* = \{(0, w^*) \mid w^* \in C_{11}^*\}$ is independent of the choice of $a \in Z_2^{\#}$ and $\mathrm{Im}(\psi) = \bar{R}_2$ (instead of $\psi((0, w^*))$ we simply write $\psi(w^*)$);*
(ii) *$\psi^{(a)}(u, v^*)\psi^{(b)}(u, v^*) = \psi^{(c)}(u, 0)$;*
(iii) *$[\psi^{(a)}(u_1, v_1^*), \psi^{(b)}(u_2, v_2^*)] = \psi((u_1 \cap u_2)^*)$.*

Proof. Since $Q_2^{(a)} \cap Q_2^{(b)} = R_2$ by (2.3.4 (ii)) and $\bar{R}_2$ is a C_{11}^*-submodule in $\bar{Q}_2^{(a)}$ by (2.3.1 (vii)), (i) follows. It is easy to see that $\pi = \psi^{(a)}(u, 0)\psi^{(b)}(u, 0)$ acts on R_2 as a transvection with centre c, therefore $\pi \in \bar{Q}_2^{(c)}$. On the other hand, π is centralized by the stabilizer of u in $\bar{M}^\delta$. By (1.1.1) this means that π is bound to be $\psi^{(c)}(u, 0)$. If $a = z$, then the commutator relation in (iii) follows from the way $E \cong C_{11}$ acts on $\bar{\Lambda}^{(23)}/\bar{\Lambda}^{(1)} \cong \bar{Q}_2^{(z)}$ (cf. (1.8.6 (v))).

Applying (ii) it is easy to expand the validity of the relation to the generic a. Alternatively we can employ (1.1.1 (ii)) to show that the only non-trivial $\bar{M}^\delta$-invariant commutator map

$$\bar{Q}_2^{(a)} \times \bar{Q}_2^{(b)} \to \bar{R}_2$$

is the one in (iii). $\square$

Lemma 2.4.2 *The group $\bar{G}_2^s$ possesses the following properties which specify $\bar{G}_2^s$ up to isomorphism:*

  (i) $\bar{G}_2^s = \bar{Q}_2^{(a)} \bar{Q}_2^{(b)} \bar{M}^\delta$;
 (ii) $\bar{Q}_2^{(a)}$ *and* $\bar{Q}_2^{(b)}$ *are normal in* $\bar{G}_2^s$, *and* $\bar{M}^\delta \cong \mathcal{C}_{11}^* \cdot M_{24}$;
(iii) $\bar{Q}_2^{(a)} \cap \bar{Q}_2^{(b)} = O_2(\bar{M}^\delta) = \bar{R}_2 = Z(\bar{Q}_2)$;
 (iv) $\bar{Q}_2^{(a)} \bar{Q}_2^{(b)} = \bar{Q}_2 = O_2(\bar{G}_2^s)$;
  (v) *there are $\bar{M}^\delta$-module isomorphisms $\psi^{(a)}$ and $\psi^{(b)}$ of $\mathcal{A}(M_{24})$ onto $\bar{Q}_2^{(a)}$ and $\bar{Q}_2^{(b)}$ where the restrictions to $\mathcal{S}_{11}$ are equal (and denoted by ψ);*
 (vi) *for all $u_1, u_2 \in \mathcal{C}_{11}$ and $v_1^*, v_2^* \in \mathcal{C}_{11}^*$, the equality*

$$[\psi^{(a)}(u_1, v_1^*), \psi^{(b)}(u_2, v_2^*)] = \psi((u_1 \cap u_2)^*)$$

   holds.

Proof. The assertion (i) is by (2.3.1 (iv)) and (2.3.4 (ii)), while (ii) is by (2.3.4 (i)) and (2.3.3). The assertions (iii) and (iv) hold by (2.3.4 (ii)) and by the definition of M^δ. The assertion (v) holds by (2.3.5) and (2.4.1 (i)). Finally, (vi) is (2.4.1 (ii), (iii)). These properties determine $\bar{G}_2^s$ up to isomorphism. In fact, $\bar{Q}_2$ can be obtained as a partial semidirect product of $\bar{Q}_2^{(a)}$ and $\bar{Q}_2^{(b)}$ (both being copies of $\mathcal{A}(M_{24})$) over the common subgroup

$$\psi^{(a)}(\mathcal{S}_{11}) = \psi^{(b)}(\mathcal{S}_{11}) = \mathrm{Im}(\psi),$$

with respect to the action described by the commutator relation in (vi). The whole of $\bar{G}_2^s$ is the partial semidirect product of $\bar{Q}_2$ and $\bar{M}^\delta$ over $O_2(\bar{M}^\delta)$ identified with $\bar{Q}_2^{(a)} \cap \bar{Q}_2^{(b)}$ (both isomorphic to $\mathcal{C}_{11}^*$ as M^δ-modules). The action of $\bar{M}^\delta$ on $\bar{Q}_2$ is defined by the natural action on the anti-heart module $\mathcal{A}(M_{24})$ through the isomorphisms $\psi^{(a)}$ and $\psi^{(b)}$. $\square$

It appears instructive to show explicitly how the conditions in (2.4.4 (v)) and (2.4.4 (vi)) imply the fact that the third 'diagonal' normal subgroup of order 2^{22} in $\bar{G}_2^s$ (which is the subgroup $\bar{G}_2^{(c)}$) is also isomorphic to the anti-heart module for $\bar{G}_2^s / \bar{Q}_2$. The diagonal subgroup is generated by the products $\psi^{(a)}(u, 0) \psi^{(b)}(u, 0)$ taken for all $u \in \mathcal{C}_{11}$. If $u_1, u_2 \in \mathcal{C}_{11}$, then the commutator relation in (2.4.4 (vi)) gives the equality

$$\psi^{(a)}(u_1, 0)\psi^{(b)}(u_1, 0)\psi^{(a)}(u_2, 0)\psi^{(b)}(u_2, 0) =$$

$$\psi^{(a)}(u_1, 0)\psi^{(a)}(u_2, 0)\psi^{(b)}(u_1, 0)\psi^{(b)}(u_2, 0)\psi((u_1 \cap u_2)^*).$$

Applying twice the addition rule in $\mathcal{A}(M_{24})$ transfered into $\bar{Q}_2^{(a)}$ and $\bar{Q}_2^{(b)}$ we see that the latter expression is equal to

$$\psi^{(a)}(u_1 + u_2, 0)\psi^{(b)}(u_1 + u_2, 0)\psi((u_1 \cap u_2)^*)^3.$$

Since the powers of elements from $\mathrm{Im}(\psi)$ should be read modulo 2, we conclude that

$$\psi^{(c)} : (u, v^*) \mapsto \psi^{(a)}(u, 0)\psi^{(b)}(u, 0)\psi(v^*)$$

is indeed an $\bar{M}^\delta$-module isomorphism of $\mathcal{A}(M_{24})$ onto the diagonal subgroup in $\bar{Q}_2$.

The characterization of $\bar{G}_2^s$ in (2.4.2) immediately suggests some automorphisms of that group.

Lemma 2.4.3 *Let π be a permutation of $Z_2^{\#}$ and let μ be an outer automorphism of $\bar{M}^\delta$ which centralizes both $O_2(\bar{M}^\delta)$ and $\bar{M}^\delta/O_2(\bar{M}^\delta)$. Then:*

(i) *the mapping which centralizes $\bar{M}^\delta$ and sends $\psi^{(a)}(u, v^*)$ onto $\psi^{(\pi(a))}(u, v^*)$ for all $a \in Z_2^{\#}$, $u \in \mathcal{C}_{11}$, and $v^* \in \mathcal{C}_{11}^*$ extends to a unique automorphism of $\bar{G}_2^s$;*

(ii) *the mapping which centralizes $\bar{Q}_2$ and permutes the elements of $\bar{M}^\delta$ according to μ extends to a unique automorphism of $\bar{G}_2^s$.*

Proof. The assertion (i) follows from the total symmetry between a, b, and c in (2.4.1). The group $\bar{G}_2^s$ can be represented as a partial semidirect product of $\bar{Q}_2$ and $\bar{M}^\delta$ over $\bar{R}_2$. On the other hand, by (2.4.2) we have $O_2(\bar{M}^\delta) = Z(\bar{Q}_2) = \bar{R}_2$. Therefore, $O_2(\bar{M}^\delta)$ is the kernel of the homomorphism of $\bar{M}^\delta$ into $\mathrm{Aut}(\bar{Q}_2)$ involved in the definition of the partial semidirect product. Hence μ can indeed be extended to a unique automorphism of $\bar{G}_2^s$. $\qquad\square$

The automorphisms of $\bar{G}_2^s$ constructed in (2.4.2 (i)) and (2.4.2 (ii)) will be denoted by π and μ, respectively.

Lemma 2.4.4 *The automorphisms constructed in (2.4.2) generate in $\mathrm{Aut}(\bar{G}_2^s)$ a subgroup*

$$A \cong Sym(Z_2^{\#}) \times \mathrm{Out}(\mathcal{C}_{11}^* \cdot M_{24}) \cong S_3 \times 2$$

which maps isomorphically onto $\mathrm{Out}(\bar{G}_2^s)$.

Proof. Every automorphism in (2.4.3 (i)) commutes with the automorphism μ in (2.4.3 (ii)) which gives the isomorphism type of A. In order to justify the isomorphism $A \cong \text{Out}(\bar{G}_2^s)$ it is sufficient to show that an arbitrary automorphism α of $\bar{G}_2^s$ can be multiplied by inner automorphisms and by automorphisms constructed in (2.4.3) to be reduced to the identity automorphism. First, multiplying α by a suitable automorphism π from (2.4.3 (i)) we can make α to normalize $\bar{Q}_2^{(a)}$ for every $a \in Z_2^\#$. Second, since $\bar{G}_2^s/\bar{R}_2 \cong (\mathcal{C}_{11} \times \mathcal{C}_{11}) : M_{24}$ contains a single class of M_{24}-complements (cf. 1.2.2 (iii)) and $\bar{M}^\delta/\bar{R}_2$ is one of the complements, we can make α to normalize $\bar{M}^\delta$. By (1.2.6 (i)) the outer automorphism group of $\bar{M}^\delta$ is of order 2 generated by μ. Therefore, adjusted by a power of μ and by the inner automorphism induced by a suitable element of $\bar{M}^\delta$, α can be made to centralize $\bar{M}^\delta$. Finally, by (1.3.6 (i))

$$C_{GL(\mathcal{A}(M_{24}))}(\bar{M}^\delta) = 1,$$

and this forces α to be the identity automorphism. □

By (2.3.2 (i)) and in view of the definition of the $Q_2^{(a)}$'s in (2.3.4), the image of G_2 in $\text{Out}(\bar{G}_2^s)$ is isomorphic to S_3 and acts faithfully on $\{\bar{Q}_2^{(a)} \mid a \in Z_2^\#\}$. By (2.4.4 (ii)), $\text{Out}(\bar{G}_2^s)$ contains two S_3-subgroups which we have to distinguish. An outer automorphism of $\bar{G}_2^s$ will be called *standard* if its image in $\text{Out}(\bar{G}_2^s)$ coincides with that of an automorphism constructed in (2.4.3 (i)). The remaining outer automorphisms will be called *twisted*. In these terms $\text{Out}(\bar{G}_2^s)$ contains a *standard* S_3-subgroup and a *twisted* S_3-subgroup.

Lemma 2.4.5 *The image of G_2 in $\text{Out}(\bar{G}_2^s)$ is the twisted S_3-subgroup.*

Proof. The assertion follows from (2.3.6 (v)) and the paragraph preceding the lemma. □

2.5 Assembling G_2 from pieces

By (2.4.2) and (2.4.5) the quotient G_2/Z_2 is identified uniquely up to isomorphism as the subgroup in $\text{Aut}(\bar{G}_2^s)$ generated by the inner and twisted automorphisms (notice that $\bar{G}_2^s$ being centreless maps bijectively into $\text{Aut}(\bar{G}_2^s)$). On the other hand, G_2^s is $C_{G_1}(Z_2)$ where G_1 is either $G_1^{(s)}$ or $G_1^{(t)}$. In this section we bring these two pieces together to build the whole of G_2.

Our strategy can be outlined as follows. We consider a perfect central extension $\widehat{G}_2^s$ of $\bar{G}_2^s$ (with centre $\widehat{Z}_2$) such that:

(A) every automorphism of $\bar{G}_2^s$ can be lifted to an automorphism of $\widehat{G}_2^s$,

(B) $C_{G_1}(Z_2)$ is a quotient of $\widehat{G}_2^s$ both for $G_1 = G_1^{(s)}$ and for $G_1 = G_1^{(t)}$.

By (2.3.1 (i), (v)), G_2^s is perfect with centre Z_2; therefore, $\widehat{G}_2^s$ satisfying (A) and (B) exists (for instance we can take the universal perfect central extension of $\bar{G}_2^s$).

The group G_2 must be the extension of $\widehat{G}_2^s$ by (a lifting of) the twisted S_3-subgroup of automorphisms, which is factorized over the kernel of the homomorphism of $\widehat{G}_2^s$ onto $C_{G_1}(Z_2)$.

Although $\widehat{G}_2^s$ can be taken to be the universal perfect central extension of $\bar{G}_2^s$, we are not doing this in order to avoid unnecessary eliminations afterward. Instead, we impose the following conditions (adopting the hat convention for the preimages in $\widehat{G}_2^s$ of subgroups from G_2^s or $\bar{G}_2^s$):

(C) $\widehat{R}_2$ is an elementary abelian 2-group;
(D) for every $a \in Z_2^{\#}$ the following holds: if

$$\zeta^{(a)} : \bar{Q}_2^{(a)} \to \widehat{Z}_2$$

is the mapping induced by the squaring map, then $\mathrm{Im}(\zeta^{(a)})$ has order 2 and $\zeta^{(a)}$ is the image under $\psi^{(a)}$ of the quadratic form q_h on the anti-heart module $\mathcal{A}(M_{24})$.

Lemma 2.5.1 *Let $\widehat{G}_2^s$ be the largest perfect central extension of $\bar{G}_2^s$ satisfying (C) and (D). Then $\widehat{G}_2^s$ also satisfies (A) and (B).*

Proof. For (B) the claim is clear since $C_{G_1}(Z_2)$ satisfies (C) and (D). Indeed, (C) holds by (2.3.6), while (D) holds by (2.3.4 (iv), (v)) and (2.3.5). In order to show that (A) holds, it is sufficient to show that the kernel K of the homomorphism onto $\widehat{G}_2^s$ of the universal perfect central extension of $\bar{G}_2^s$ is $\mathrm{Aut}(\bar{G}_2^s)$-stable. For (C) to hold, K must contain the squares of all the preimages of elements from $\bar{R}_2$. By (2.4.2), $\bar{R}_2 = Z(O_2(\bar{G}_2^s))$, therefore the set of squares are $\mathrm{Aut}(\bar{G}_2^s)$-stable. Let $m_1, m_1, n \in \mathcal{A}(M_{24})$ be such that $q_h(m_1) = q_h(m_2) = 1$, $q_h(n) = 0$. Then for (D) to hold, K must contain the square of every preimage of $\psi^{(a)}(n)$ and the product of the squares of the preimages of $\psi^{(a)}(m_1)$ and $\psi^{(a)}(m_2)$. By (2.4.2), the set of these squares and products is also $\mathrm{Aut}(\bar{G}_2^s)$-stable. Therefore, (A) holds since $\widehat{G}_2^s$ is defined to be the largest perfect central extension subject to the conditions (C) and (D). $\square$

Our next goal is to show that the order of $\widehat{Z}_2$ is 2^4 and to describe the way $\mathrm{Out}(\bar{G}_2^s)$ acts on $\widehat{Z}_2$.

Lemma 2.5.2 *For $a \in Z_2^{\#}$, let $\widehat{a}$ denote the generator of $\mathrm{Im}(\zeta^{(a)})$ as in (D). Then the following assertions hold:*

(i) $\widehat{a}, \widehat{b}$, and $\widehat{c}$ are pairwise distinct and $\widehat{a}\widehat{b}\widehat{c} = 1$;

(ii) $\widehat{Y}_2 := [\widehat{R}_2, \widehat{Q}_2]$ *is elementary abelian of order* 2^2 *generated by* $\widehat{a}, \widehat{b}$
   and $\widehat{c}$;

(iii) $|\widehat{Z}_2| = 2^4$.

Proof. By (C) and (D), $\widehat{R}_2$ is the kernel of the action of $\widehat{Q}_2^{(a)}$ on $\widehat{R}_2$ by conjugation and the action is via transvections with centre $\widehat{a}$. It is easy to deduce from the equality in (2.4.1 (ii)) that $\widehat{a}\widehat{b} = \widehat{c}$. Furthermore, $\widehat{a}, \widehat{b},$ and $\widehat{c}$ are pairwise distinct since their images in G_2^s are such. Therefore, (i) holds. Since $\widehat{Q}_2 = \widehat{Q}_2^{(a)}\widehat{Q}_2^{(b)}$, (ii) holds. By the definition, $\widehat{R}_2/\widehat{Y}_2$ is the centre of $\widehat{Q}_2/\widehat{Y}_2$ and (as a module for $\widehat{G}_2^s/\widehat{Q}_2 \cong M_{24}$) the quotient $\widehat{R}_2/\widehat{Y}_2$ is an extension of the trivial module $\widehat{Z}_2/\widehat{Y}_2$ by $\widehat{R}_2/\widehat{Z}_2 \cong \mathcal{C}_{11}^*$. Since $H^1(M_{24}, \mathcal{C}_{11})$ is trivial by (1.2.2 (i)) and since $\mathcal{C}_{11}^*$ is dual to $\mathcal{C}_{11}$, there is an $\widehat{M}^\delta$-submodule $\widehat{X}_2$ in $\widehat{R}_2$ containing $\widehat{Y}_2$ such that

$$\widehat{R}_2/\widehat{Y}_2 = \widehat{Z}_2/\widehat{Y}_2 \oplus \widehat{X}_2/\widehat{Y}_2.$$

The quotient $\widehat{G}_2^s/\widehat{X}_2$ is a perfect central extension of

$$\bar{G}_2^s/\bar{R}_2 \cong (\mathcal{C}_{11} \oplus \mathcal{C}_{11}) : M_{24}.$$

By (1.1.3 (ii)) and (1.2.2), $\widehat{G}_2^s/\widehat{X}_2$ is a quotient of $(\mathcal{C}_{12} \oplus \mathcal{C}_{12}) : M_{24}$ which gives the upper bound 2^2 for $|\widehat{Z}_2/\widehat{Y}_2|$. To show that this upper bound is attained consider the subdirect product of G_2^s and $(\mathcal{C}_{12} \oplus \mathcal{C}_{12}) : M_{24}$ with respect to the natural homomorphisms onto $(\mathcal{C}_{11} \oplus \mathcal{C}_{11}) : M_{24} \cong G_2^s/R_2$ (2.3.1 (viii), (ix)). The constructed subdirect product is clearly a perfect central extension of $\widehat{G}_2^s$ which satisfies (C) and (D) and where the centre is of order 2^4. This completes the proof of (iii). $\qquad\square$

When outlining our strategy of constructing G_2 it was implicit that every automorphism of $\bar{G}_2^s$ possesses a unique lifting to an automorphism of $\widehat{G}_2^s$. This is indeed the case since every automorphism of $\widehat{G}_2^s$ centralizing $\widehat{G}_2^s/\widehat{Z}_2$ centralizes every subgroup of odd order in $\widehat{G}_2^s$ and hence the whole of $\widehat{G}_2^s$, because the latter is a perfect group.

By (2.4.4 (iv)), (2.5.1) and the previous paragraph, if we put $A = \mathrm{Out}(\bar{G}_2^s)$, then $\mathrm{Aut}(\widehat{G}_2^s)$ is the semidirect product of $\mathrm{Inn}(\widehat{G}_2^s) \cong \bar{G}_2^s$ and $A \cong S_3 \times 2$ with respect to the natural action. Therefore, $\widehat{Z}_2$ carries a well-defined structure of a $GF(2)$-module for A. We are going to show that $\widehat{Z}_2$ is a faithful A-module. By (2.5.2), it is immediate that the standard S_3-subgroup acts faithfully. The fact that the automorphism μ from (2.4.3 (ii)) acts non-trivially is less obvious. To establish this fact we have to introduce some further notation.

For B being a subgroup of A and $\widehat{X}$ being a B-submodule in $\widehat{Z}_2$, let $\widehat{G}_2^s(B, \widehat{X})$ denote the semidirect product of $\widehat{G}_2^s/\widehat{X}$ and B with respect to the natural action. Alternatively, this group is the quotient of $\widehat{G}_2^s(B, 0)$ over $\widehat{X}$.

In what follows $\widehat{G}_2^s$ will be identified with the corresponding subgroup of $\widehat{G}_2^s(B, 0)$, where B will be clear from the context.

Lemma 2.5.3 *Let* $\widetilde{M}^\delta = \langle [\widehat{M}^\delta, \widehat{M}^\delta], \mu \rangle$, $\widetilde{Q}_2^{(a)} = [\widetilde{Q}_2^{(a)}, \widehat{M}^\delta]$, *and* $\widetilde{N}^{(a)} = \langle \widetilde{M}^\delta, \widetilde{Q}_2^{(a)} \rangle$ *be subgroups of* $\widehat{G}_2^s(A, 0)$. *Then the following assertions hold:*

(i) $\widetilde{M}^\delta \cong \mathcal{C}_{12}^* \cdot M_{24}$;

(ii) $\widetilde{Q}_2^{(a)} \cap \widehat{Z}_2 \cong 2^2$;

(iii) $\widetilde{N}^{(a)} \in \mathcal{E}(M_{24}, W_{24}, q_p)$;

(iv) μ *acts non-trivially on* $\widetilde{Q}_2^{(a)} \cap \widehat{Z}_2$.

Proof. Since the Schur multiplier of $\bar{M}^\delta \cong \mathcal{C}_{11}^* \cdot M_{24}$ is trivial by (1.2.6 (iii)), we have

$$\widehat{M}^\delta = \widehat{Z}_2 \times [\widehat{M}^\delta, \widehat{M}^\delta] \cong \widehat{Z}_2 \times \bar{M}^\delta$$

and (i) follows from the definition of μ. The assertion (ii) has been (implicitly) established within the proof of (2.5.2). Since μ centralizes $\widehat{M}^\delta / O_2(\widehat{M}^\delta) \cong M_{24}$, $\widetilde{P} := \langle \widetilde{Q}_2^{(a)}, \mu \rangle$ is precisely $O_2(\widetilde{N}^{(a)})$. Since μ is of order 2, the image of the squaring map $\widetilde{P} \to \widehat{Z}_2$ is generated by $\widehat{a}$. We claim that $\widetilde{P}/\langle \widehat{a} \rangle$ and W_{24} are isomorphic $\widehat{M}^\delta$-modules. In fact, the image of $\widetilde{Q}_2^{(a)} \cap \widehat{Z}_2$ is a 1-dimensional submodule, the image of $\widetilde{Q}_2^{(a)}$ is a submodule of codimension 1, and the corresponding quotient is $\bar{Q}_2^{(a)} \cong \mathcal{A}(M_{24})$. Furthermore, by (i) the module $\widetilde{P}/\langle \widehat{a} \rangle$ possesses $\mathcal{C}_{12}^*$ as a submodule and by the proof of (2.5.2) it possesses $\mathcal{C}_{12}$ as a factor module. Thus the claim follows from (1.3.5). Now to complete the proof of (iii) it only remains to show that the quadratic form on $\widetilde{P}/\langle \widehat{a} \rangle$ induced by the squaring map on $\widetilde{P}$ is precisely q_p. To achieve this it is sufficient to apply (1.3.7) and the fact that μ has order 2 (this implies that the submodule isomorphic to $\mathcal{C}_{12}^*$ is totally isotropic). Now (iv) is immediate from the observation that the image of μ is not contained in $\widetilde{Q}_2^{(a)}/\langle \widehat{a} \rangle$ which is the orthogonal complement of $(\widetilde{Q}_2^{(a)} \cap \widehat{Z}_2)/\langle \widehat{a} \rangle$ with respect to the bilinear form β_p associated with q_p. $\qquad \square$

Thus $\widehat{Z}_2$ is indeed a faithful $GF(2)$-module for $A \cong S_3 \times 2$. Since A contains a normal subgroup S of order 3 acting on $\widehat{Z}_2$ fixed-point freely, the image of A in $GL(\widehat{Z}_2)$ is determined uniquely up to conjugation. Identifying A with its image we have

$$N_{GL(\widehat{Z}_2)}(S) \cong \Gamma L_2(4) \cong 3 : S_5$$

and A is the semidirect product of S and an elementary abelian subgroup of order 2^2 from S_5 containing one even involution μ (centralizing S) and two odd involutions (inverting S). In particular, the actions on $\widehat{Z}_2$ of the standard and twisted S_3-subgroups are isomorphic. Let T be the twisted S_3-subgroup. Then $\widehat{Z}_2$ contains exactly three proper T-submodules $\widehat{Y}_i, i = 0, 1, 2$ of dimension 3 each, where $\widehat{Y}_2 = [\widehat{R}_2, \widehat{Q}_2]$ was defined in (2.5.2 (ii)). The element μ centralizes $\widehat{Y}_2$ permuting $\widehat{Y}_0$ and $\widehat{Y}_1$. If D is a subgroup of order 2 in T, then D fixes in $\widehat{Z}_2$ exactly three 1-dimensional subspaces which are

$$\widehat{U}_i = C_{\widehat{Y}_i}(D)$$

for $i = 0, 1, 2$.

Lemma 2.5.4 $G_2 \cong \widehat{G}_2^s(T, \widehat{Y}_0) \cong \widehat{G}_2^s(T, \widehat{Y}_1)$.

Proof. By (2.4.5), the image of G_2 in $\mathrm{Out}(\bar{G}_2^s)$ is the twisted S_3-subgroup T. By (2.5.2 (iii)), G_2^s is a quotient of $\widehat{G}_2^s$ over a subgroup $\widehat{X}$ of order 2^2 in $\widehat{Z}_2$. Since $\widehat{X}$ must be T-invariant, $\widehat{X} = \widehat{Y}_i$ for $i = 0, 1$, or 2. Since $Z_2 = [R_2, Q_2]$ by (2.3.1 (viii)), $i \neq 2$. Finally, since the image of S in G_2 acts on Z_2 fixed-point freely, by a Frattini argument G_2 splits over G_2^s. Since μ permutes Y_0 and Y_1, both the remaining candidates for G_2 are isomorphic. $\qquad\square$

Now we only have to specify the choice between $i = 0$ and $i = 1$ in (2.5.4). In order to achieve this we turn back to the group G_1. Let $\widetilde{G}_1$ be the universal perfect central extension of G_1, let $\widetilde{Z}_2$, $\widetilde{G}_{12}$, and $\widetilde{G}_2^s$ be the preimages in $\widetilde{G}_1$ of the subgroups Z_2, G_{12}, and G_2^s in G_1 and let $\widetilde{g}$ be a preimage of the element $g \in Q_1$ as in (2.3.6).

Lemma 2.5.5 *The following assertions hold:*

  (i) $\widetilde{G}_2^s$ *is a perfect central extension of* $\bar{G}_2^s$ *by the centre* $\widetilde{Z}_2$ *of order* 2^3 *satisfying conditions* (C) *and* (D);
 (ii) $\widetilde{g}$ *can be taken to be of order 2 and its image in* $A = \mathrm{Out}(\bar{G}_2^s)$ *is an involution from the twisted* S_3-*subgroup;*
(iii) $\widetilde{G}_{12}$ *is a semidirect product of* $\widetilde{G}_2^s$ *and* $\langle\widetilde{g}\rangle$ *(assuming that* $\widetilde{g}$ *is of order 2).*

Proof. It follows from the realization of $\widetilde{G}_1$ as a subdirect product of G_1 and Co_0 that $\widetilde{G}_{12}$ is a subdirect product of $G_{12} \cong 2_+^{1+24}.(\mathcal{C}_{11} : M_{24})$ and $\mathcal{C}_{12} : M_{24}$ over their homomorphisms onto $\mathcal{C}_{11} : M_{24}$. From this viewpoint the assertions are rather clear. $\qquad\square$

By (2.5.5), we conclude that $\widetilde{G}_{12}$ is $\widehat{G}_2^s(D, \widehat{U})$ where D is a twisted subgroup of order 2 (generated by the image of $\widetilde{g}$) and $\widehat{U}$ is a 1-dimensional subspace in $\widehat{Z}_2$ centralized by D. Since $\widehat{Z}_2$ maps surjectively onto Z_2 and so far we have maintained the complete symmetry between $\widehat{Y}_0$ and $\widehat{Y}_1$, we can assume without loss of generality that $\widehat{U} = \widehat{U}_0$. On the other hand, G_{12} is the quotient of $\widetilde{G}_{12}$ over a subgroup of order 2 in $\widetilde{Z}_2$ centralized by D. The 1-dimensional D-submodules in $\widetilde{Z}_2 = \widehat{Z}/\widehat{U}_0$ are the images of

$$\widehat{Y}_0, \ \langle \widehat{U}_0, \widehat{U}_1 \rangle \text{ and } \langle \widehat{U}_0, \widehat{U}_2 \rangle.$$

The latter subgroup is the commutator subgroup of $\widetilde{Q}_1$ and it can't possibly be the kernel of the homomorphism of $\widetilde{G}_1$ onto G_1. Therefore, we have

$$\{N_{G_1^{(s)}}(Z_2), N_{G_1^{(t)}}(Z_2)\} = \{\widehat{G}_2^s(D, \widehat{Y}_0), \widehat{G}_2^s(D, \langle \widehat{U}_0, \widehat{U}_1 \rangle)\}.$$

Thus $\widehat{G}_2^s(D, \widehat{Y}_0)$ (isomorphic by (2.5.4) to an index 3 subgroup in G_2) is also isomorphic to exactly one of the groups $N_{G_1^{(s)}}(Z_2)$ and $N_{G_1^{(t)}}(Z_2)$.

Proposition 2.5.6 *The amalgam* $\{G_1, G_2\}$ *satisfying* $(\mathcal{M}1)$ *and* $(\mathcal{M}2)$ *exists and is unique up to isomorphism.* $\square$

Proof. The paragraph before the proposition implies the existence of the amalgam and it also specifies the choice of G_1 up to isomorphism. The isomorphism type of G_2 is specified by (2.5.4) and the paragraph preceding that lemma. The subgroup G_{12} is uniquely specified in G_1 and in G_2 up to conjugation in these groups. To prove the uniqueness of the whole amalgam using Goldschmidt's theorem (cf. [Gol80] and (8.3.2) in [ISh02]) we need to calculate the automorphism group of G_{12}. By (2.4.4), we have

$$\mathrm{Aut}(\bar{G}_{12}) \cong \bar{G}_2^s : \langle T, \mu \rangle$$

although μ maps $\widehat{Y}_0$ (which is the kernel of the homomorphism of $\widehat{Z}_2$ onto Z_2) onto $\widehat{Y}_1$, so there is no way μ can be extended to an automorphism of G_{12}. The only non-trivial automorphism of G_{12} which is identical on $\bar{G}_2^s$ maps the generator t of T onto the product tz. Although this automorphism is the inner one performed by y acting via conjugation. Thus all the automorphisms of G_{12} are inner and the amalgam is indeed unique by Goldschmidt's theorem. $\square$

2.6 Identifying $\{G_1, G_2\}$

In this section we accomplish the task left unfinished in the previous section by showing that in the unique amalgam $\{G_1, G_2\}$ satisfying $(\mathcal{M}1)$ and $(\mathcal{M}2)$ the former of the group is in fact $G_1^{(t)}$. We achieve this by showing that the dimension of a faithful representation of the group G_1 from the amalgam is at least $24 \cdot 2^{12}$ and applying (2.2.3).

Let U be a real vector space and let

$$\rho : G_1 \to GL(U)$$

be a faithful irreducible representation. Then the restriction of ρ to $Q_1 \cong 2_+^{1+24}$ is a direct sum of some m copies of the unique faithful irreducible representation

$$\omega : Q_1 \to GL(X)$$

of dimension 2^{12}, so that $\dim(U) = m \cdot 2^{12}$.

The representation ω can be obtained by inducing to Q_1 of a linear (1-dimensional) representation of $R_2 \cong 2^{13}$ where the kernel is a hyperplane in R_2 disjoint from Z_1. If $\mathcal{K}$ denotes the set of all the 2^{12} hyperplanes in R_2 disjoint from Z_1, then (since the action of Q_1 on $\mathcal{K}$ via conjugation is transitive) we have

$$X = \bigoplus_{H \in \mathcal{K}} C_X(H).$$

An element $r \in R_2$ acts on $C_X(H)$ as the identity operator if $r \in H$ and as the (-1)-scalar operator in $r \notin H$. A half of the hyperplanes in $\mathcal{K}$ contain y and the other half contain z. Therefore, $C_X(y)$ is 2^{11}-dimensional and hence

$$\dim(C_U(y)) = \frac{1}{2}\dim(U) = m \cdot 2^{11}.$$

Since G_2^s centralizes y, it acts on $C_U(y)$ (and clearly y is in the kernel of the action). By (2.3.4) this means that $Q_2^{(y)}$ acts on $C_U(y)$ as an elementary abelian group $Q_2^{(y)}/\langle y \rangle$ of order 2^{23}. Consider the decomposition of U into eigenspaces of $Q_2^{(y)}$

$$U = C_U(Q_2^{(y)}) \bigoplus_{D \in \mathcal{D}} C_U(D),$$

where $\mathcal{D}$ is the set of preimages in $Q_2^{(y)}$ of the hyperplanes in $Q_2^{(y)}/\langle y \rangle$. Since the action of R_2 on $C_X(y)$ is faithful, so is the action of R_2 on $C_U(y)$. Hence there exists $D \in \mathcal{D}$ such that $R_2 \not\leq D$ and $C_U(D) \neq 0$. This means that the dimension of $C_U(y)$ is at least the length of the G_2^s-orbit containing D. Thus all we need is the following:

Lemma 2.6.1 *A shortest orbit of G_2^s on the set of index 2 subgroups in $Q_2^{(y)}$ containing y and not containing R_2 is unique and its length is $24 \cdot 2^{11}$.*

Proof. Since G_2 contains G_2^s as a normal subgroup and permutes transitively the non-identity elements of Z_2, the actions of G_2^s on $Q_2^{(y)}/\langle y \rangle$ and on $Q_2^{(z)}/\langle z \rangle$ are isomorphic. In turn, the latter action is isomorphic to the action of the monomial subgroup $F \cong C_{11} : M_{24}$ on $\bar{\Lambda}^{(23)}$ (cf. (2.3.1 (iii), (iv)) and (2.3.4 (iii))). Therefore, in order to complete the proof it suffices to apply (1.8.7).   □

Lemma 2.6.2 *Let* $\{G_1, G_2\}$ *be the unique amalgam satisfying* $(\mathcal{M}1)$ *and* $(\mathcal{M}2)$. *Then* $G_1 \cong G_1^{(t)}$.

Proof. The result is by (2.2.3 (iii)), (2.6.1), and the discussions prior the latter lemma.   □

2.7 Conway's realization of G_2

In [C84], the group $\widehat{G}_2^s(T, 0)$ has been realized as a group of permutations of three copies of (the element set of) the Parker loop $(\mathcal{L}, \circ)$. In this section we reproduce this realization and extend it to the whole of $\widehat{G}_2^s(A, 0)$. This will bridge Conway's work with the strategy we follow here.

Let $(\mathcal{L}, \circ)$ be the Parker loop as in Section 1.5. Let $Z = \{1, z\}$ be the subloop generated by the squares, so that the projection

$$\pi : \mathcal{L} \to \bar{\mathcal{L}} = \mathcal{L}/Z$$

is a loop homomorphism onto $\bar{\mathcal{L}} \cong \mathcal{C}_{12}$. The elements of $\mathcal{L}$ are denoted by $\widetilde{u}$ where $u \in \mathcal{C}_{12}$ so that

$$\pi : \widetilde{u} \mapsto u$$

and $\pi^{-1}(u) = \{\widetilde{u}, \widetilde{u} \circ z\}$ so that $\widetilde{u}$ is determined up to multiplication by z. Up to isomorphism the loop $(\mathcal{L}, \circ)$ is determined by the projection π together with the squaring, commuting, and associating rules in (1.5.3 (iii)) which are reproduced below

$$\widetilde{u}^{-1} = \widetilde{u} \circ z^{\frac{1}{4}|u|};$$

$$\widetilde{u} \circ \widetilde{v} = \widetilde{v} \circ \widetilde{u} \circ z^{\frac{1}{2}|u \cap v|};$$

$$(\widetilde{u} \circ \widetilde{v}) \circ \widetilde{w} = \widetilde{u} \circ (\widetilde{v} \circ \widetilde{w}) \circ z^{|u \cap v \cap w|}$$

for $u, v, w \in \mathcal{C}_{12}$ (since z commutes with every element of the loop and associates with every pair of elements, we have saved a few brackets). By the above equalities (compare (1.5.2)), the centre of $(\mathcal{L}, \circ)$ is $Y = \{1, z, \widetilde{\mathcal{P}}, \widetilde{\mathcal{P}} \circ z\}$, which is an elementary abelian group of order 4.

The group D of diagonal automorphisms of $(\mathcal{L}, \circ)$ is the kernel of the action on $\bar{\mathcal{L}}$ of the full automorphism group of $(\mathcal{L}, \circ)$ and it consists of the automorphisms $d(v^*)$ taken for all $v^* \in \mathcal{C}_{12}^*$ and defined via

$$d(v^*) : \widetilde{u} \mapsto \widetilde{u} \circ z^{|u \cap v|}.$$

In terms of Sections 1.5 and 1.6, the automorphism $d(v^*)$ is performed by the conjugation via the element $k(v^*)$ (compare the proof of (1.6.8)). The code automorphism group $A^{(g)}$ is a subgroup in the automorphism group of $(\mathcal{L}, \circ)$ characterized by the properties that it contains D and $\bar{\mathcal{L}}$, as a module for $A^{(g)}/D \cong M_{24}$ is $\mathcal{C}_{12}$ (rather than the direct sum of $\mathcal{C}_{11}$ and a 1-dimensional module). Then by (1.6.9) we have $A^{(g)} \cong \mathcal{C}_{12}^* \cdot M_{24}$. An automorphism from $A^{(g)}$ is said to be *even* if it centralizes Y. If

$$A^{(e)} = C_{A^{(g)}}(Y)$$

is the group of all even code automorphisms, then $A^{(e)}$ is the commutator subgroup of $A^{(g)}$ isomorphic to $\mathcal{C}_{11}^* \cdot M_{24}$. The subgroup

$$D^{(e)} := D \cap A^{(e)}$$

of even diagonal automorphisms consists of the elements $d(v^*)$ taken for all even $v \subseteq \mathcal{P}$.

For $\widetilde{u} \in \mathcal{L}$, the left and right translation (maps) induced by $\widetilde{u}$ are the permutations of $\mathcal{L}$ defined, respectively, by

$$\lambda_{\widetilde{u}}(\widetilde{x}) = \widetilde{u} \circ \widetilde{x} \text{ and } \rho_{\widetilde{u}}(\widetilde{x}) = \widetilde{x} \circ \widetilde{u}^{-1},$$

where $\widetilde{x} \in \mathcal{L}$. In the next lemma we summarize some basic properties of these translations which are immediate consequence of the squaring, commuting, and associating rules. Notice that the left and right translations induced by $\widetilde{u}$ coincide if and only if $\widetilde{u} \in Y$. If $\widetilde{u} \in Y$, then the corresponding translation (left or right) will be denoted simply by $\widetilde{u}$.

Lemma 2.7.1 *The following equalities hold for all $u, w \in \mathcal{C}_{12}$:*

(i) $\rho_{\widetilde{u}}^2 = \lambda_{\widetilde{u}}^2 = z^{\frac{1}{4}|u|}$;

(ii) $[\rho_{\widetilde{u}}, \rho_{\widetilde{w}}] = [\lambda_{\widetilde{u}}, \lambda_{\widetilde{w}}] = z^{\frac{1}{2}|u \cap w|}$;

(iii) $\rho_{\widetilde{u}}\rho_{\widetilde{w}} = \rho_{(\widetilde{u} \circ \widetilde{w})}d((u \cap w)^*)$;

(iv) $\lambda_{\widetilde{u}}\lambda_{\widetilde{w}} = \lambda_{(\widetilde{u} \circ \widetilde{w})}d((u \cap w)^*)$;

(v) $[\rho_{\widetilde{u}}, \lambda_{\widetilde{w}}] = d((u \cap w)^*)$. $\square$

The following lemma deals with certain subgroups of the symmetric group of $\mathcal{L}$ generated by right and/or left translations together with the (even) code automorphisms. We follow notation from Sections 2.4 and 2.5.

Lemma 2.7.2 *Put*

$$Q^{(l)} = \langle \lambda_{\widetilde{u}} \mid \widetilde{u} \in \mathcal{L} \rangle, \quad Q^{(r)} = \langle \rho_{\widetilde{u}} \mid \widetilde{u} \in \mathcal{L} \rangle \text{ and } Q^{(t)} = \langle \lambda_{\widetilde{u}}\rho_{\widetilde{u}} \mid \widetilde{u} \in \mathcal{L} \rangle.$$

Then:

(i) $Q^{(l)} \cap A^{(g)} = Q^{(r)} \cap A^{(g)} = Q^{(t)} \cap A^{(g)} = D^{(e)}$;
(ii) $\langle Q^{(l)}, A^{(g)} \rangle$ *and* $\langle Q^{(r)}, A^{(g)} \rangle$ *are isomorphic to* $\mathcal{E}(M_{24}, W_{24}, q_p)$;
(iii) $[Q^{(l)}, Q^{(l)}] = [Q^{(r)}, Q^{(r)}] = Z$;
(iv) $Q^{(t)}$ *is an elementary abelian of order* 2^{22} *isomorphic to* $\mathcal{A}(M_{24})$, *as a module for* $A^{(g)}/D \cong M_{24}$ *and* $Q^{(t)}$ *is disjoint from* Y;
(v) $\langle Q^{(l)}, Q^{(r)}, A^{(e)} \rangle / Y \cong \bar{G}_2^s$;
(vi) $\langle Q^{(l)}, Q^{(r)}, A^{(g)} \rangle \cong \widehat{G}_2^s(\langle \mu \rangle, \widehat{U})$ *for an order 4 subgroup* $\widehat{U}$ *in* $\widehat{Z}_2$.

Proof. Assertions (i), (ii), and (iii) are direct consequences of the proof of (1.7.1) and the obvious symmetry between right and left translations. Alternatively, the assertions along with (vi) can be deduced directly from the relations in (2.7.1). To establish (v) we apply the characterization in (2.4.2) with $\bar{Q}_2^{(a)} = Q^{(l)}/Y$, $\bar{Q}_2^{(b)} = Q^{(r)}/Y$, $\bar{M}^\delta = A^{(e)}$, $\bar{R}_2 = D^{(e)}$, and with $\psi^{(a)}$ and $\psi^{(b)}$ being the mappings

$$(u, v^*) \mapsto \rho_{\widetilde{u}} \, d(v^*) Y \text{ and } (u, v^*) \mapsto \lambda_{\widetilde{u}} \, d(v^*) Y,$$

respectively. Then the key commutator relation in (2.4.2 (vi)) follows from (2.7.1 (v)). Furthermore, $\bar{Q}^{(c)}$ corresponds to $Q^{(t)}Y/Y$. The subgroup in (vi) is easily seem to be a perfect central extension of the group in (v) which satisfies the condition (C) and (D) in Section 2.5 extended by the automorphism μ. Therefore, the assertion (vi) follows from the structure of $\widehat{G}_2^s$ and its automorphism group. $\qquad\square$

In terms of the notation introduced within the proof of the above lemma, we can specify the assertion (vi) by saying that

$$\widehat{U} = [\widehat{Q}_2^{(c)}, \widehat{M}^\delta].$$

The above construction shows that if $\widetilde{Z}_2^{(c)} = [\widetilde{Q}_2^{(c)}, \widehat{M}^\delta]$, then the quotient $\widehat{G}_2^s / \widehat{Z}_2^{(c)}$ can be realized as a group of permutations of $\mathcal{L}$ generated by the left and right translations together with the even code automorphisms. In this realization, the preimage of the 'diagonal' group $\bar{G}_2^{(c)}$ is generated by the 'two-sided' translations

$$\tau_{\widetilde{u}}(\widetilde{x}) = \widetilde{u} \circ \widetilde{x} \circ \widetilde{u}^{-1}$$

(the associating rule in $(\mathcal{L}, \circ)$ enables us to drop the brackets). Since $\widehat{Z}_2 = \widehat{Z}_2^{(c)} \widehat{Z}_2^{(a)}$, the whole of $\widehat{G}_2^s$ can be realized as a permutation group of two copies of $\mathcal{L}$. When we take three copies (as Conway did), the triality symmetry of $\widehat{G}_2^s$ appears most vividly.

Thus consider three copies of Parker's loop $(\mathcal{L}^\alpha, \circ)$ where

$$\alpha \in \{a, b, c\} = Z_2^\#$$

and fix a cyclic ordering $\sigma = (a, b, c)$ of $Z_2^\#$. Then for every $\tilde{u} \in \mathcal{L}$ and $\alpha \in Z_2^\#$ define a permutation $q_{\tilde{u}}^{(\alpha)}$ which acts on the set-theoretical union

$$\mathcal{T} = \mathcal{L}^a \cup \mathcal{L}^b \cup \mathcal{L}^c$$

by the following rule

$$q_{\tilde{u}}^{(\alpha)}(\tilde{x}) = \begin{cases} \tilde{u} \circ \tilde{x} \circ \tilde{u}^{-1} & \text{if } \tilde{x} \in \mathcal{L}^\alpha; \\ \tilde{u} \circ \tilde{x} & \text{if } \tilde{x} \in \mathcal{L}^{\sigma(\alpha)}; \\ \tilde{x} \circ \tilde{u}^{-1} & \text{if } \tilde{x} \in \mathcal{L}^{\sigma^{-1}(\alpha)}. \end{cases}$$

The permutation group of $\mathcal{T}$ generated by all the $q_{\tilde{u}}^{(\alpha)}$s together with the even code automorphisms of $\mathcal{L}$ acting conformably on the three copies is precisely $\widehat{G}_2^s$. The automorphisms of $\widehat{G}_2^s$ can be obtained by adjoining all the code automorphisms (still acting conformably) and the symmetric group of $Z_2^\#$ permuting the three copies of $\mathcal{L}$.

One of the advantages of our treatment is that it does not require the explicit formulas.

2.8 Introducing G_3

According to $(\mathcal{M}3)$, Z_3 is the preimage under

$$\chi : Q_1 \to \bar{\Lambda}$$

of the subgroup in $\bar{\Lambda}$ generated by

$$\bar{\lambda} = \sum_{p \in \mathcal{P}} 2p + 2\Lambda \text{ and } \bar{\mu} = \sum_{q \in \mathcal{T}} 4q + 2\Lambda,$$

where $\mathcal{T}$ is a 4-subset in $\mathcal{P}$. In terms of $(1.11.5)$, $Z_3 = \chi^{-1}(\bar{\Lambda}^{(2)})$, in particular Z_3 is elementary abelian of order 2^3. As suggested in $(\mathcal{M}3)$, we put

$$G_{13} = N_{G_1}(Z_3), \ G_{23} = N_{G_2}(Z_3), \ G_{123} = N_{G_{12}}(Z_3).$$

Let

$$G_3^s = C_{G_{12}}(Z_3) \text{ and } Q_3 = O_2(G_3^s).$$

Since Z_3 contains both Z_1 and Z_2, G_3^s coincides with the centralizers of Z_3 in G_1 and G_2. We start with the following:

Lemma 2.8.1

(i) $|Q_3| = 2^{39}$ *and* $G_3^s/Q_3 \cong 3 \cdot S_6$ *is the hexacode group;*

(ii) $Z_3 = Z(Q_3) = Z(G_3^s);$

(iii) $G_{13}/G_3^s \cong G_{23}/G_3^s \cong S_4$, $G_{123}/G_3^s \cong D_8;$

(iv) *the images of* G_{13}, G_{23}, *and* G_{123} *in* $GL(Z_3) \cong L_3(2)$ *are the stabilizers of the* 1-*subspace* Z_1, *the* 2-*subspace* Z_2, *and the flag* $\{Z_1, Z_2\}$, *respectively.*

Proof. It follows from the definition of Z_3 that G_{13} is the preimage in G_1 of the three bases subgroup $T \cong 2^{4+12} \cdot (3 \cdot S_6 \times S_3)$ of $Co_1 \cong G_1/Q_1$ (cf. Section 1.11). Furthermore, by (1.11.5), $G_3^s \cap Q_1 = \chi^{-1}(\bar{\Lambda}^{(22)})$ and $G_3^s Q_1/Q_1 = T^+$. This gives (i) and the isomorphism $G_{13}/G_3^s \cong S_4$. Since T^+ acts faithfully on $\bar{\Lambda}^{(22)}$ and $Z_3 = Z(\chi^{-1}(\bar{\Lambda}^{(22)}))$, the assertion (ii) follows. In terms of (2.4.2), Z_3 is the preimage in R_2 of $\psi(v^*)$, in particular $G_{23}G_2^s = G_2$ and G_{23} induces S_3 on Z_2. This completes the proof of (iii) and (iv). $\qquad \square$

Let us adopt the bar convention for the images in the automorphism group of G_3^s of normal extensions of G_3^s. By (2.8.1 (ii)), we have $\bar{G}_3^s \cong G_3^s/Z_3$. Let $\bar{G}_3$ denote the subgroup in $\text{Aut}(G_3^s)$ generated by $\bar{G}_{13}$ and $\bar{G}_{23}$. Let

$$\bar{C}_Z = C_{\bar{G}_3}(Z_3).$$

As an immediate consequence of (2.8.1 (iii), (iv)) we have the following:

Lemma 2.8.2 $\bar{G}_3/\bar{C}_Z \cong L_3(2)$. $\qquad \square$

Let $R_3 = \chi^{-1}(\bar{\Lambda}^{(8)})$, so that R_3 is the preimage in R_2 of the subgroup in $\bar{R}_2$ generated by the elements $\psi(w^*)$ taken for the $\mathcal{P}$-subsets w intersecting evenly every tetrad in the sextet $\mathcal{S}(v)$. Let

$$\bar{C}_R = C_{\bar{G}_3}(R_3/Z_3).$$

Lemma 2.8.3 $\bar{G}_3/\bar{C}_R \cong G_3^s/Q_3 \cong 3 \cdot S_6$ *(the hexacode group).*

Proof. By (1.11.5 (ii)), R_3/Z_3 is centralized by Q_3 and it is isomorphic to $\mathcal{H}^\diamond$ as a module for the hexacode group G_3^s/Q_3. On the one hand, this action is clearly normalized by that of $\bar{G}_3$. The module $\mathcal{H}^\diamond$ is not stable under outer automorphisms of G_3^s/Q_3 and $\mathcal{H}^\diamond$ is an absolutely irreducible $(3 \cdot S_6)$-module. Therefore, the action of G_3^s on R_3/Z_3 is self-normalizing in $GL(R_3/Z_3) \cong GL_6(2)$ and the result follows. $\qquad \square$

Let

$$\bar{C}_3 = \bar{C}_Z \cap \bar{C}_R \text{ and } \widehat{G}_3 = \bar{G}_3/\bar{C}_3.$$

By (2.8.2), (2.8.3), and the homomorphism theorem, we have the following.

Lemma 2.8.4 $\bar{C}_3$ *contains* $\bar{Q}_3$ *and* $\widehat{G}_3 \cong L_3(2) \times 3 \cdot S_6$. $\qquad\qquad\square$

Lemma 2.8.5 *Let* $T_3 = C_{Q_3}(R_3)$. *Then* Q_3/T_3 *is centralized by* $\bar{C}_3$ *and, as a module for* $\widehat{G}_3$, *it is isomorphic to the tensor product*

$$Z_3 \otimes \mathcal{H}.$$

Proof. Let S be the stabilizer of Z_3 in $GL(R_3) \cong L_9(2)$. Then on the one hand G_3/T_3 is a subgroup of S, on the other hand

$$S \cong (Z_3 \otimes R_3/Z_3) : (GL(Z_3) \times GL(R_3/Z_3))$$

and $O_2(S)$ (which is the tensor product of Z_3 and R_3/Z_3) is generated by the transvections where the centres are contained in Z_3 and axes contain Z_3. Furthermore, $O_2(S) = C_S(Z_3) \cap C_S(R_3/Z_3)$. The action of $\widehat{G}_3$ on R_3 is clearly faithful and this action is easily seen to project (isomorphically) onto a subgroup in $S/O_2(S)$ acting irreducibly on $O_2(S)$. By (1.11.5), $\chi^{-1}(\bar{\Lambda}^{(22)})$ acts on R_3 by the tranvections with centre Z_1 where the axes contain Z_3. Since $\widehat{G}_3$ acts transitively on $Z_3^{\#}$, we conclude that the image of Q_3/T_3 coincides with the whole of $O_2(S)$ and the result follows. $\qquad\qquad\square$

Notice that G_3/T_3 splits over Q_3/T_3. This can be deduced either from the proof of (2.8.5) or from the fact that $O_3(\widehat{G}_3)$ acts on Q_3/T_3 fixed-point freely. It is also implicit in the proof of (2.8.5) that $\bar{C}_3/\bar{T}_3$ splits over $\bar{Q}_3/\bar{T}_3$ (since the kernel of the action on R_3 is a complement).

By (2.8.5) and its proof, Q_3/T_3 (considered as a module for $G_3^s/Q_3 \cong 3 \cdot S_6$) contains exactly seven irreducible submodules (each isomorphic to $\mathcal{H}$) indexed by the non-identity elements of Z_3. For $a \in Z_3^{\#}$, let $Q_3^{(a)}$ denote the preimage in Q_3 of the irreducible G_3^s/Q_3-submodule in Q_3/T_3. Then $Q_3^{(a)}$ acts on R_3 by transvections with centre a (and axes containing Z_3) and $Q_3^{(z)} = Q_1 \cap Q_3$.

Lemma 2.8.6 *In the notation introduced in the paragraph before the lemma both* $Q_3^{(a)}/R_3$ *and* T_3/Z_3 *are elementary abelian 2-groups.*

Proof. It appears easiest to prove the assertion in terms of $\bar{G}_2^s = G_2^s/Z_2$. Since $\widehat{G}_3$ acts transitively on Z_3, without loss of generality we assume that $a \in Z_2$. In terms of Section 2.4, Z_3/Z_2 is generated by $\psi(\mathcal{T}^*)$ (where $\mathcal{T}$ is the 4-subset of $\mathcal{P}$ as in ($\mathcal{M}3$)); R_3/Z_2 is generated by the elements $\psi(w^*)$ taken for all $\mathcal{P}$-subsets w intersecting evenly every tetrad in the sextet $\mathcal{S}(\mathcal{T})$. Furthermore, it is easy to deduce from (1.10.5) that if $\mathcal{B}(\mathcal{T})$ denote the set of (15) octads which are unions of pairs of tetrads in $\mathcal{S}(\mathcal{T})$, and $\mathcal{E}(\mathcal{T})$ denote the set of octads intersecting evenly every tetrad in $\mathcal{S}(\mathcal{T})$ (compare Lemma 2.14.1 in [Iv99]), then

$$T_3/Z_2 = \langle \psi^{(a)}(u_1, w_1^*), \psi^{(b)}(u_2, w_2^*) \mid u_1, u_2 \in \mathcal{B}(\mathcal{T}), w_1, w_2 \subseteq \mathcal{P} \rangle;$$

$$Q_2^{(a)}/Z_2 = \langle \psi^{(a)}(u_1, w_1^*), \psi^{(b)}(u_2, w_2^*) \mid$$
$$u_1 \in \mathcal{E}(\mathcal{T}), u_2 \in \mathcal{B}(\mathcal{T}), w_1, w_2 \subseteq \mathcal{P} \rangle.$$

Now the result follows from the commutator relation in (2.4.1 (iii)) and the following two easy observations: (a) if $u_1, u_2 \in \mathcal{B}(\mathcal{T})$, then $u_1 \cap u_2$ is either an element of $\mathcal{C}_{12}$ or a tetrad from $\mathcal{S}(\mathcal{T})$ (so that $\psi((u_1 \cap u_2)^*) \in Z_3/Z_2$ in any event); (b) if $u_1 \in \mathcal{E}(\mathcal{T})$ and $u_2 \in \mathcal{B}(\mathcal{T})$, then the set $u_1 \cap u_2$ intersects evenly every tetrad in $\mathcal{S}(\mathcal{T})$ (so that $\psi((u_1 \cap u_2)^*) \in R_3/Z_2$).   □

Lemma 2.8.7 T_3/R_3 *is centralized by* $\bar{C}_3$ *and, as a module for* $\widehat{G}_3 \cong L_3(2) \times 3 \cdot S_6$, *it is isomorphic to the tensor product*

$$Z_3^* \otimes U_4.$$

Proof. It is immediate from (1.10.6) and the proof of (2.8.6) that every composition factor of T_3/R_3 (considered as G_3^s/Q_3-module) is isomorphic to the natural 4-dimensional symplectic module U_4. Furthermore, R_2/R_3 is the only irreducible submodule in T_3/R_3 stabilized by $\bar{G}_{23}$. Since $\widehat{G}_3$ induces the full linear group on the dual module Z_3^* of Z_3 (whose non-zero vectors are the 2-subspaces in Z_3), we get the structure of T_3/R_3 as claimed.   □

We summarize the essence of the results established in this section in the following lemma:

Lemma 2.8.8 *The action of* $\bar{G}_3$ *on* Q_3 *preserves the chief series*

$$Z_3 < R_3 < T_3 < Q_3,$$

where each chief factor is centralized by $\bar{C}_3$ *and the following are isomorphisms of* $\widehat{G}_3$-*modules*

$$R_3 \cong Z_3 \oplus \mathcal{H}^{\circ}, \quad T_3/R_3 \cong Z_3^* \otimes U_4, \quad Q_3/T_3 \cong Z_3 \otimes \mathcal{H}.$$

□

2.9 Complementing in G_3^s

Aiming to calculate the automorphism group $\bar{A}_3$ of G_3^s and to specify $\bar{G}_3$ as a subgroup in $\bar{A}_3$ we analyse in this section which extensions describing the structure of G_3^s split and enumerate the relevant classes of complements. We start with the following:

Lemma 2.9.1 G_3^s *splits over* Q_3.

Proof. Let D be the normalizer of a Sylow 7-subgroup in $\bar{C}_Z$. Then D projects onto a Frobenius subgroup of order 21 in $\widehat{G}_3$ and by (2.8.8) and a Frattini argument, $C_{G_3^s}(D)$ is an extension of $R_3/Z_3 \cong \mathcal{H}^\diamond$ by $G_3^s/Q_3 \cong 3 \cdot S_6$. Since $O_3(G_3^s/Q_3)$ acts on $\mathcal{H}^\diamond$ fixed-point freely, the splitness follows. $\square$

Let $Y \cong 3 \cdot S_6$ be a complement to Q_3 in G_3^s, and let $Y_Z = YZ_3$. Then Y_Z is said to be a *quasi-complement* in G_3^s, since

$$\bar{Y}_Z = \bar{Y}$$

is a complement to $\bar{Q}_3$ in $\bar{G}_3^s$ and Y_Z is the full preimage of $\bar{Y}$ in G_3^s. Let $Y_Z' \cong 3 \cdot A_6$ be the commutator subgroup of Y_Z, and let $y \in Y \setminus Y'$ be an involution. Then for every $a \in Z_3$, the subgroup

$$Y_a = \langle Y_Z', ya \rangle$$

is a complement to Z_3 in Y_Z. Since $\bar{Y}_a = \bar{Y}$, it is easy to see that the mapping which centralizes $Q_3 Y'$ and permutes y with ya extends to a unique (outer) automorphism $\xi^{(a)}$ of G_3^s called a *central automorphism*. The following result is now rather clear.

Lemma 2.9.2 *The group*

$$\Xi = \{\xi^{(a)} \mid a \in Z_3\}$$

of central isomorphisms is canonically isomorphic to Z_3 and acts regularly on the set of complements to Q_3 in G_3^s contained in a given quasi-complement. $\square$

Next we enumerate the classes of quasi-complements in G_3^s.

Lemma 2.9.3 *The group $\bar{G}_3^s$ contains precisely eight conjugacy classes of complements to $\bar{Q}_3$ and $\bar{G}_3$ permutes these classes transitively.*

Proof. As above by Y we denote a complement to Q_3 in G_3^s and let $X = O_3(Y)$. Then Y project isomorphically into $\widehat{G}_3$ and by (2.8.8) each irreducible chief factor of Y inside $\bar{Q}_3$ is isomorphic either to $\mathcal{H}$, or to $\mathcal{H}^\diamond$, or to U_4. By (1.9.2), $\mathcal{H}$ and $\mathcal{H}^\diamond$ are dual and both $H^1(Y, \mathcal{H})$ and $H^1(Y, \mathcal{H}^\diamond)$ are trivial. Therefore, only the chief factors isomorphic to U_4 might contribute to the number of classes of quasi-complements. Since U_4 is a self-dual module with first cohomology of order 2 (cf. (1.9.2 (ii))) and since there are precisely three chief factors of Y isomorphic to U_4 (all of them are inside T_3/R_3), there are at most eight classes of quasi-complements. In order to prove the lemma it remains to show that the upper bound on the number of classes is attained and that $\bar{G}_3$ permutes the eight classes transitively. We are going to justify both the assertions simultaneously.

By (2.8.8), T_3/R_3 contains seven irreducible Y-factor modules isomorphic to U_4 and indexed by the non-identity elements of Z_3. This means that precisely one of these factor modules is stabilized by G_{13} and in fact this factor is $T_3/(T_3 \cap Q_1)$ (compare (1.11.2) and (1.11.5)). Since Y centralizes Z_3, it is contained in $G_1 \cap G_3^s$ and hence Y projects isomorphically onto a complement

$$\text{to } O_2(G_1 \cap G_3^s)/Q_1 \cong 2^{4+12} \text{ in } (G_1 \cap G_3^s)/Q_1 \cong 2^{4+12} : 3 \cdot S_6.$$

By (1.11.4), $T \cong G_{13}/Q_1$ permutes the two classes of the subgroup in $T^+ \cong (G_1 \cap G_3^s)/Q_1$ isomorphic to the hexacode group. Therefore, if $g \in G_{13} \setminus G_3^s$ and if g has even order, then YQ_1/Q_1 and $Y^g Q_1/Q_1$ are not conjugate in $(G_1 \cap G_3^s)/Q_1$. Since $\bar{G}_3$ permutes transitively the non-identity elements in Z_3, every irreducible Y-factor module in T_3/R_3 produces a new class of quasi-complements and all the classes are permuted transitively by $\bar{G}_3$. So indeed both the assertions hold.   □

As a direct consequence of the above proof and a basis property of $L_3(2)$, we obtain the following.

Lemma 2.9.4 *The normalizer in* $\bar{G}_3/\bar{C}_Z \cong L_3(2)$ *of a class of complements to* $\bar{Q}_3$ *in* $\bar{G}_3^s$ *is a Frobenius group of order* 21.   □

We are going we show that the action on the classes of complements to Q_3 in G_3^s of the full automorphism group $\bar{A}_3$ of G_3^s coincides with that of $\bar{G}_3$.

Lemma 2.9.5 *The stabilizer in* $\bar{A}_3 = \mathrm{Aut}(G_3^s)$ *of the class on complements to* Q_3 *in* G_3^s *containing* Y *stabilizes a polarity of the projective plane of order* 2 *associated with* Z_3 *which sends every point onto a line that is non-incident to the point.*

Proof. Let W be the preimage in T_3 of a Y-irreducible submodule in T_3/R_3 and suppose that W corresponds to Z_2 (considered as an element of Z_3^*) in the sense of (2.8.8). Then W is the only such preimage stabilized by G_{23} and it follows from (2.3.1 (vii)) and (the dual version of) (1.10.6) that $W = R_2$. This means that W is elementary abelian. Since R_2/Z_2 and C_{11}^* are isomorphic M_{24}-modules, by the dual version of (1.10.6) $C_{W/Z_2}(X)$ is the indecomposable extension of the trivial 1-dimensional module by U_4 (that is $C_{W/Z_2}(X) \cong V_{1\backslash 4}$). Since U_4 is a self-dual module and $H^1(Y, U_4)$ is of order 2 by (1.9.2 (ii)), we conclude that

$$[Y, C_W(X)] \cap Z_3$$

is a 1-dimensional subspace disjoint from Z_2. This proves the result.   □

Now comparing (2.9.4) and (2.9.5) we deduce the assertion stated in the paragraph before (2.9.5) from the basic properties of the projective plane of order 2 and its automorphism group.

2.10 Automorphisms of G_3^s

We start by calculating the outer automorphism group of G_3^s.

Lemma 2.10.1 *The outer automorphism group of G_3^s is an extension of the image of the central automorphisms (this image is isomorphic to Z_3) by $GL(Z_3) \cong L_3(2)$.*

Proof. By the results established in Section 2.9, we know that the subgroup in $\mathrm{Out}(G_3^s)$ generated by the images of the central automorphisms together with the image of $\bar{G}_3$ induces an extension of Z_3 by $GL(Z_3)$ on the set of classes of complements to Q_3 in G_3^s. Furthermore, the action of the whole automorphisms group $\bar{A}_3$ on the classes of complements does not get larger and the kernel of the action of $\bar{A}_3$ centralizes Z_3. A consideration of the chief factors of Y inside Q_3 shows that every automorphism of G_3^s normalizes each of Z_3, R_3, and T_3. In particular, such an automorphism cannot induce an outer automorphism of $G_3^s/Q_3 \cong Y \cong 3 \cdot S_6$. Hence it suffices to show that the identity is the only automorphism of G_3^s which centralizes both Y and Z_3. Let α be such an automorphism. Since $\mathcal{H}$, $\mathcal{H}^\diamond$ and U_4 are absolutely irreducible Y-modules and since

$$K_3 := [R_3, X] \cong \mathcal{H}^\diamond$$

is a complement to Z_3 in R_3, it is immediate from (2.8.8) that α centralizes Z_3, K_3, T_3/R_3, and Q_3/T_3. As in the paragraph before (2.8.6), put $Q_3^{(a)}$ to be the preimage in Q_3 of the Y-irreducible submodule in Q_3/T_3 corresponding to $a \in Z_3^{\#}$. Then by (2.8.6), $Q^{(a)}/R_3$ is an elementary abelian and since $X = O_3(Y)$ acts trivially on T_3/R_3 and is fixed-point freely on $Q^{(a)}/T_3$, we have the following direct sum decomposition into Y-submodules

$$Q_3^{(a)}/R_3 = L_3^{(a)}/R_3 \oplus T_3/R_3,$$

where $L_3^{(a)}$ is a subgroup of order 2^{15} in Q_3 such that $L_3^{(a)} \cap T_3 = R_3$. Since $L_3^{(a)}/R_3 \cong \mathcal{H}$ is an absolutely irreducible Y-module, the above direct sum decomposition shows that α centralizes $Q_3^{(a)}/R_3$. Since Q_3 is generated by the $Q_3^{(a)}$'s taken for all $a \in Z_3^{\#}$, we conclude that α centralizes the whole of Q_3/R_3. The triviality of the actions of α on $L_3^{(a)}$ and on T_3 will be proved separately.

For the case of $L_3^{(a)}$, assuming that a is taken to be the generator z of Z_1, we reduce the calculations into G_1. In fact, $Q_1/Z_1 \cong \bar{\Lambda}$ is the Leech lattice modulo 2. By (1.8.6), $\bar{\Lambda}$, considered as an M_{24}-module, contains two faithful composition factors, isomorphic to C_{11} and to C_{11}^*, respectively. By (1.10.6) and its dual version, this implies that $\bar{\Lambda}$, considered as a module for the hexacode group Y (the latter being a complement to $O_2(G_{123})$ in G_{123}) besides the trivial factors and the ones isomorphic to U_4, contains one irreducible factor isomorphic to $\mathcal{H}$ and one isomorphic to $\mathcal{H}^\diamond$. This shows that

$$Q_3^{(z)} = Q_1 T_3.$$

Put $D_3^{(z)} = [X, Q_1]$. It is easy to check using (1.8.4) or otherwise that the restriction of the Co_1-invariant quadratic form $q_{\bar{\Lambda}}$ to $D_3^{(z)}/Z_1$ is non-singular, therefore D_3 is an extraspecial group of order 2^{13}. This and the order consideration shows that

$$L_3^{(z)} = Z_3 D_3^{(z)} \cong Z_3 * D_3^{(z)}$$

(where the latter expression stands for the central product of Z_3 and $D_3^{(z)}$ over Z_1). The automorphism α normalizes $D_3^{(z)}$ and commutes with the action of Y on $D_3^{(z)}$. We claim that Y has trivial centralizer in the automorphism group of $D_3^{(z)}$. Since Y acts fixed-point freely on $D_3^{(z)}/Z_1$, it is sufficient to show that $C_{\mathrm{Out}(D_3^{(z)})}(Y)$ is trivial. The action of Y on $D_3^{(z)}/Z_1$ preserves an irreducible submodule (isomorphic to) $\mathcal{H}^\diamond$ such that the corresponding factor-module is (isomorphic to) $\mathcal{H}$. Since $\mathcal{H}^\diamond$ is not self-dual, it must be totally isotropic with respect to $q_{\bar{\Lambda}}$. This means that $D_3^{(z)}$ is of plus type and that the stabilizer of $\mathcal{H}^\diamond$ in $\mathrm{Out}(D_3^{(z)}) \cong O_{12}^+(2)$ is isomorphic to

$$(\mathcal{H}^\diamond \wedge \mathcal{H}^\diamond) : GL(\mathcal{H}^\diamond).$$

To prove the triviality of the centralizer it is now sufficient to show that Y acts fixed-point freely on $\mathcal{H}^\diamond \wedge \mathcal{H}^\diamond$. The vectors of the latter module are indexed by the symplectic forms on the dual space $\mathcal{H}$ of $\mathcal{H}^\diamond$ (cf. lemma 1.4.3 in [Iv04]). The action of Y on $\mathcal{H}$ is irreducible and $\mathcal{H}$ is not self-dual, therefore no such form is stabilized by Y. Thus the triviality of the action of α on $L_3^{(z)}$ has been established.

Finally, let us turn to T_3. As in the proof of (2.9.5) let W be the preimage in T_3 of an irreducible Y-submodule in T_3/R_3. Then

$$W = C_W(X) \oplus [X, W],$$

and $[X, W] = K_3$ is already known to be centralized by α. By the proof of (2.9.5), $C_W(X)$ is the direct sum of $[Y, C_W(X)]$ and a 2-dimensional complement to $[Y, C_W(X)] \cap Z_3$ in Z_3. The complement is certainly centralized by α, while $[Y, C_W(X)]$ is isomorphic to $V_{1\backslash 4}$ and therefore Y has trivial centralizer in $GL([Y, C_W(X)])$. This completes the proof. $\square$

Corollary 2.10.2 *Exactly one of the following two possibilities occur:*

(i) $\bar{G}_3 = \operatorname{Aut}(G_3^s)$ *and* $\bar{G}_3/\bar{G}_3^s \cong 2^3.L_3(2)$;
(ii) $\bar{G}_3/\bar{G}_3^s \cong L_3(2)$.

Proof. The result is immediate from (2.8.2) and (2.10.2) since the action of $GL(Z_3)$ on Z_3 is obviously irreducible. $\square$

Notice that if the option (i) in (2.10.2) were to occur, a group G_3 satisfying ($\mathcal{M}4$) could not possibly exist. This option will be eliminated in the next section.

2.11 $L_3(2)$-amalgam

By (2.8.1), the pair of natural homomorphisms

$$\psi_1 : G_{13} \to \operatorname{Out}(G_3^s) \text{ and } \psi_2 : G_{23} \to \operatorname{Out}(G_3^s)$$

define a faithful completion ψ of the amalgam

$$\{G_{13}/G_3^s, G_{23}/G_3^s\}$$

isomorphic to the amalgam $\mathcal{L}$ of maximal parabolics in $GL(Z_3) \cong L_3(2)$. Therefore, $\bar{G}_3/\bar{G}_3^s$ is a generating completion of $\mathcal{L}$. From (2.8.2) we know that $\bar{G}_3/\bar{G}_3^s$ possesses $L_3(2)$ as a factor group. The condition ($\mathcal{M}4$) requires the isomorphism $G_3/G_3^s \cong L_3(2)$ and hence for G_3 to exist, the quotient $\bar{G}_3/\bar{G}_3^s$ must be precisely $L_3(2)$. By (2.10.2), the only other possibility for this quotient is to be an extension of 2^3 by $L_3(2)$. In fact such an extension is never a generating completion of $\mathcal{L}$. This important property of completions of $\mathcal{L}$ (known as $L_3(2)$-lemma) was originally proved by Sergey Shpectorov in [Sh88] (see also Section 8.3 in [ISh02]). Here we suggest an alternative proof of the $L_3(2)$-lemma based on analysis of abelian covers of the incidence graph of the order 2 projective plane.

Let $L = GL(Z_3) \cong L_3(2)$, let L_1 and L_2 be the stabilizers in L of Z_1 and Z_2, respectively, and let $L_{12} = L_1 \cap L_2$. Then

$$L_1 \cong L_2 \cong S_4, \quad L_{12} \cong D_8,$$

and $\mathcal{L} = \{L_1, L_2\}$ the *amalgam of maximal parabolics in* $L_3(2)$.

With a faithful generating completion

$$\psi^{(X)} : \mathcal{L} \to X$$

we associate the coset graph $\Theta^{(X)}$ whose vertex-set $V(\Theta^{(X)})$ is formed by the (left) cosets in X of $\psi^{(X)}(L_1)$ and of $\psi^{(X)}(L_2)$ and whose edge-set $E(\Theta^{(X)})$ is formed by the pairs of cosets having non-empty intersections. Then $\Theta^{(X)}$ is bipartite of valency

$$3 = [L_1 : L_{12}] = [L_2 : L_{12}]$$

and X acts on $\Theta^{(X)}$ by left translations with L_{12} being the stabilizer of an edge and with L_1, L_2 being the stabilizers of the vertices incident to this edge.

The group L is of course a faithful generating completion of $\mathcal{L}$, and $\Theta^{(L)}$ (denoted simply by Θ) is the incidence graph of the projective plane of order 2 associated with Z_3 (in particular $|V(\Theta)| = 14$ and $|E(\Theta)| = 21$).

Let

$$\psi^{(U)} : \mathcal{L} \to U$$

be the universal completion of $\mathcal{L}$. Then U is the free amalgamated product of L_1 and L_2 over L_{12}, and $\Theta^{(U)}$ is the cubic tree. If

$$\eta^{(L)} : U \to L$$

is the homomorphism of completion groups (defined via $\eta^{(L)} : \psi^{(U)}(l) \mapsto \psi^{(L)}(l)$ for every $l \in \mathcal{L}$), then the kernel $K^{(L)}$ of $\eta^{(L)}$ is the fundamental group of Θ which is a free group of rank 8, the latter being the number of fundamental cycles in Θ

$$8 = |E(\Theta)| - |V(\Theta)| + 1$$

(cf. [Serr77], [Vi94], [Ve98]).

Consider two further generating completions

$$\psi^{(A)} : \mathcal{L} \to A \text{ and } \psi^{(E)} : \mathcal{L} \to E$$

defined as follows. If $K^{(A)}$ and $K^{(E)}$ are the kernels of the completion homomorphisms of U onto A and E, respectively, then $K^{(A)}$ is the commutator subgroup of $K^{(L)}$, and $K^{(E)}$ is the subgroup in $K^{(L)}$ generated by all the commutators and the squares. This means that

$$K^{(L)} \geq K^{(E)} \geq K^{(A)},$$

$K^{(L)}/K^{(A)}$ is the largest abelian factor group of $K^{(L)}$, and $K^{(L)}/K^{(E)}$ is the largest factor group of $K^{(L)}$ which is an elementary abelian 2-group. Since $K^{(L)}$ is a free group of rank 8, $K^{(L)}/K^{(A)}$ is a free abelian group of rank 8 and $K^{(L)}/K^{(E)}$ is an elementary abelian group of order 2^8. By definition, A is the largest generating completion of $\mathcal{L}$ which is an extension by $L_3(2)$ of an abelian group, while E is the largest such completion which is an extension by $L_3(2)$ of an elementary abelian 2-group.

Combining the above paragraph with the discussions at the beginning of the section we have the following result:

Lemma 2.11.1 *The quotient* $\bar{G}_3/\bar{G}_3^s$ *is isomorphic to a factor group of* $E \cong 2^8.L_3(2)$ *and* $O_2(E) \cong K^{(L)}/K^{(E)}$. $\qquad\square$

Since $\bar{G}_3/\bar{G}_3^s$ is a subgroup in $\mathrm{Out}(G_3^s) \cong 2^3.L_3(2)$, in order to establish the isomorphism $\bar{G}_3/\bar{G}_3^s \cong L_3(2)$ all we need is the following proposition:

Proposition 2.11.2 *The quotient* $K^{(L)}/K^{(E)}$ *is the 8-dimensional irreducible Steinberg module* Σ_8 *for* $L \cong L_3(2)$.

Proof. The Hurewicz homomorphism (Prop. 4.21 in [Vi94]) establishes an isomorphism between $K^{(L)}/K^{(A)}$ and the first homology group $H_1(\Theta)$ of the graph Θ. This isomorphism commutes with the action of $U/K^{(L)} \cong L_3(2)$. The homology group $H_1(\Theta)$ can be defined in the following way. Let $E_{\mathbb{Z}}$ be the free abelian group on the set of arcs (which are the ordered edges) of Θ modulo the identification $(x, y) = -(y, x)$. Let $V_{\mathbb{Z}}$ be the free abelian group on the set of vertices of Θ, and let

$$\partial : E_{\mathbb{Z}} \to V_{\mathbb{Z}}$$

be the homomorphism defined by $\partial : (x, y) \mapsto y - x$. The following results are standard:

 (i) the kernel $\ker(\partial)$ of ∂ is (isomorphic to) the homology group $H_1(\Theta)$;
 (ii) the sum h_T of arcs over a cycle T in Θ is an element of $\ker(\partial)$;
 (iii) the elements h_T taken for all fundamental cycles T freely generate
      $H_1(\Theta) \cong \ker(\partial)$.

By the above, $K^{(E)}/K^{(L)}$ is canonically isomorphic to the subspace $\mathcal{T}$ in the space of $GF(2)$-valued functions on the edge-set of Θ generated by the cycles. We are going to identify $\mathcal{T}$ with the 8-dimension Steiberg module Σ_8 of $GL(Z_3)$. The module Σ_8 is well known to be a direct summand of the tensor product $Z_3 \otimes Z_3^*$ (where Z_3^* is the dual space of Z_3). The tensor product is of course 9-dimensional and its other direct summand is trivial 1-dimensional. Since $\mathcal{T}$ is known to be 8-dimensional, in order to establish the identification it is sufficient to associate with every pair (a, b^*) (where $a \in Z_3 \setminus \{0\}$, $b^* \in Z_3^* \setminus \{0\}$) a cycle $\tau(a, b^*)$ from $\mathcal{T}$ and to justify the bilinearity of the resulting mapping $(a, b^*) \mapsto \tau(a, b^*)$. To simplify the notation we consider a and b^* as 1- and 2-dimensional subspaces in Z_3 and also as 2- and 1-dimensional subspaces in Z_3^*. Define $\tau(a, b^*)$ to be the subgraph in Θ induced by

$$(Z_3 \setminus (b^* \cup a)) \cup (Z_3^* \setminus (a \cup b^*)).$$

It is an easy combinatorial exercise to check that $\tau(a, b^*)$ is always a cycle, whose length is 8 or 6 depending on whether or not $a \in b^*$, and also that the mapping

$$(a, b^*) \mapsto \tau(a, b^*)$$

is indeed bilinear. $\qquad\square$

Lemma 2.11.3 $\bar{G}_3/\bar{G}_3^s \cong L_3(2)$.

Proof. This is a consequence of (2.11.1) and (2.11.2). $\qquad\square$

2.12 Constructing G_3

Let $\mathcal{A}_3 = \{G_{13}, G_{23}\}$ be the subamalgam in $\{G_1, G_2\}$, where the latter is as in (2.6.2). Then by the definition, G_3 is a faithful generating completion group of $\mathcal{A}_3$. In this section we construct G_3 as a quotient of the $\widetilde{G}_3$ where

$$\widetilde{\psi}_3 : \mathcal{A}_3 \to \widetilde{G}_3$$

is the universal completion of $\mathcal{A}_3$. First we deduce necessary and sufficient conditions for the kernel of a homomorphism of $\widetilde{G}_3$ onto G_3 and then prove that subject to these conditions the kernel exists and is unique.

Lemma 2.12.1 *Let us identify $\mathcal{A}_3$ with its image in $\widetilde{G}_3$ under the universal completion $\widetilde{\psi}_3$. Then the kernel $\widetilde{K}_3$ of the completion homomorphism η_3 : $\widetilde{G}_3 \to G_3$ is a complement to Z_3 in $C_{\widetilde{G}_3}(G_3^s)$.*

Proof. By the definition, G_3^s is normal both in G_{13} and G_{23}, so it is normal in $\widetilde{G}_3$. Of course, $\widetilde{K}_3$ is also normal in $\widetilde{G}_3$ and since $\widetilde{\psi}_3$ is faithful

$$\widetilde{K}_3 \cap G_3^s = 1,$$

which shows that $\widetilde{K}_3 \leq C_{\widetilde{G}_3}(G_3^s)$ and that $\widetilde{K}_3$ is disjoint from Z_3. On the other hand

$$\widetilde{\psi}_3 : \mathcal{A}_3 \to \mathrm{Aut}(G_3^s)$$

is a (non-faithful) completion of $\mathcal{A}$ having $\bar{G}_3$ as the completion group and Z_3 as the kernel. The kernel of the completion homomorphism

$$\bar{\eta}_3 : \widetilde{G}_3 \rightarrow \bar{G}_3$$

is $C_{\widetilde{G}_3}(G_3^s)$. The condition ($\mathcal{M}4$) requires the equality $C_{G_3}(G_3^s) = Z_3$, and therefore $\bar{\eta}_3$ is the composition of η_3 and the natural homomorphism of G_3 onto $\bar{G}_3$ with kernel Z_3. Therefore $C_{\widetilde{G}_3}(G_3^s) = \widetilde{K}_3 Z_3$. $\qquad\square$

The next lemma shows that G_3 exists and is unique.

Lemma 2.12.2 *In terms of* (2.12.1) *there exists a unique normal subgroup* $\widetilde{K}_3$ *in* $\widetilde{G}_3$ *such that* $C_{\widetilde{G}_3}(G_3^s) = Z_3 \times \widetilde{K}_3$.

Proof. We claim that the quotient $C_{\widetilde{G}_3}(G_3^s)/Z_3$ is isomorphic to the fundamental group of the incidence graph Θ of the projective plane of order 2 associated with Z_3. Consider the coset graph $\widetilde{\Theta}$ associated with the completion $\widetilde{\psi}_3$. Since G_3^s is the largest subgroup in the edge stabilizer G_{123} which is normal in the stabilizers (G_{13} and G_{23}) of the vertices incident to the edge, we conclude that G_3^s is the kernel of the action of $\widetilde{G}_3$ on $\widetilde{\Theta}$. Therefore, the image $\widetilde{G}_3/G_3^s$ of this action is a completion of the amalgam

$$\mathcal{L} = \{G_{13}/G_3^s, G_{23}/G_3^s\}$$

of maximal parabobolics in $L_3(2)$. Since the completion $\widetilde{\psi}_3$ is universal, $\widetilde{\Theta}$ is a tree. Since a completion is universal if and only if the corresponding coset graph is a tree, we conclude that $\widetilde{G}_3/G_3^s$ is the universal completion group of $\mathcal{L}$ and hence the kernel of the completion homomorphism

$$\varphi_3 : \widetilde{G}_3/G_3^s \rightarrow L_3(2) \cong \bar{G}_3/\bar{G}_3^s$$

is the fundamental group of Θ. On the other hand, the natural homomorphism

$$\widetilde{\varphi}_3 : \widetilde{G}_3 \rightarrow \bar{G}_3/\bar{G}_3^s \cong L_3(2)$$

has $G_3^s C_{\widetilde{G}_3}(G_3^s)$ as the kernel and $\widetilde{\varphi}_3$ is the composition of the homomorphism $\widetilde{G}_3 \rightarrow \mathrm{Aut}(\widetilde{\Theta})$ with kernel G_3^s and φ_3. Since $G_3^s \cap C_{\widetilde{G}_3}(G_3^s) = Z_3$ the claim follows.

The cohomological dimension of a free group is zero and therefore the above paragraph implies that

$$C_{\widetilde{G}_3}(G_3^s) \cong Z_3 \times K^{(L)}$$

(here $K^{(L)}$ is the fundamental group of Θ, which is a free group of rank 8 by the paragraph before (2.11.1)). Thus Z_3 is indeed complemented in $C_{\widetilde{G}_3}(G_3^s)$, but we still have to justify the existence of a $\widetilde{G}_3$-invariant complement.

Let N be the smallest normal subgroup in $C_{\widetilde{G}_3}(G_3^s)$ such that the corresponding quotient is an elementary abelian 2-group. It is clear that N is normal in $\widetilde{G}_3$ and because of the above direct product decomposition, the quotient

$$V := C_{\widetilde{G}_3}(G_3^s)/N$$

is elementary abelian of order 2^{11}. The action of $G_3^s C_{\widetilde{G}_3}(G_3^s)$ on V is trivial and therefore V is a module for $\widetilde{G}_3/(G_3^s C_{\widetilde{G}_3}(G_3^s)) \cong L_3(2)$. Furthermore, by (2.11.2) U is an extension of Z_3 by the 8-dimensional Steinberg module Σ_8. Since the Steinberg module is projective, the extensions splits into a direct sum. Thus, there exists a unique $\widetilde{G}_3$-invariant complement to Z_3 in U and hence there is a unique $\widetilde{G}_3$-normal complement to Z_3 in $C_{\widetilde{G}_3}(G_3^s)$. $\qquad\square$

Now we a ready to prove the main result of the chapter.

Theorem 2.12.3 *The Monster amalgam $\mathcal{M}$ exists and is unique up to isomorphism.*

Proof. Let

$$\psi_3 : \mathcal{A}_3 \to G_3$$

be the faithful completion of the amalgam $\mathcal{A}_3 = \{G_{13}, G_{23}\}$ such that the kernel $\widetilde{K}_3$ of the completion homomorphism $\eta_3 : \widetilde{G}_3 \to G_3$ is the unique $\widetilde{G}_3$-normal complement to Z_3 in $C_{\widetilde{G}_3}(G_3^s)$ as in (2.12.2). Then $C_{G_3}(G_3^s) = Z_3$ and by (2.11.3) we have $G_3/G_3^s \cong L_3(2)$. If $\mathcal{A}_3$ is identified with its image under ψ_3, then

$$G_3 \cap (G_1 \cup G_2) = G_{13} \cup G_{23}$$

and an amalgam $\mathcal{M}$ satisfying conditions $(\mathcal{M}1)$ to $(\mathcal{M}4)$ exists. The uniqueness of $\mathcal{M}$ follows from (2.6.2) and (2.12.2). $\qquad\square$

2.13 G_3 contains $L_3(2)$

In this section we analyse further the structure of G_3 showing in particular that it splits over G_3^s. We follow the notation adopted in the previous sections. In particular, Y denotes a complement to $Q_3 = O_3(G_3^s)$ in G_3^s, so that $Y \cong 3 \cdot S_6$ is the hexacode group.

Let D^+ be a Sylow 3-subgroup in Y and let D^- be a subgroup of order 15 in Y. Then for $\varepsilon = +$ or $-$, the subgroup D^ε preserves on U_4 a quadratic form of type ε where the associated bilinear form is the one preserved by Y. Furthermore, D^ε is a maximal odd order subgroup in Y subject to this property.

Lemma 2.13.1 *The following assertions hold:*

(i) $C_{G_3}(D^\varepsilon) = Z(D^\varepsilon) \times L^\varepsilon$, *where* L^ε *is an extension of* Z_3 *by* $L_3(2)$;
(ii) L^- *splits over* Z_3, *so that* $L^- \cong 2^3 : L_3(2)$;
(iii) L^+ *does not split over* Z_3, *so that* $L^+ \cong 2^3 \cdot L_3(2)$ (*the unique non-split extension by* $L_3(2)$ *of its natural module*);
(iv) G_3 *splits over* G_3^s.

Proof. First, $C_{Q_3}(D^\varepsilon) = Z_3$ because of (2.8.8) and the fact that D^ε acts fixed-point freely on each of U_4, $\mathcal{H}$ and $\mathcal{H}^\diamond$. Second, $C_Y(D^\varepsilon) = Z(D^\varepsilon)$ by the basic properties of S_6 and $C_{G_3}(D^\varepsilon)G_3^s = G_3$ by Frattini argument. Finally, the Schur multiplier of $L_3(2)$ has order 2, which gives the direct product structure in (i). In view of Gaschütz's theorem, the assertions (ii) and (iii) can be deduced from the structure of $C_{D_{123}}(D^\varepsilon)$ and we suggest the reader to perform the necessary calculations. Finally, (iv) is immediate from (ii). $\qquad\square$

By the above lemma, G_3 could have been constructed as a semidirect product of G_3^s and an $L_3(2)$-complement to Z_3 in L^-. On the one hand, this would save us from dealing with the universal completion of the amalgam $\{G_{13}, G_{23}\}$; on the other hand, this would require the identification of

$$\{C_{G_{13}}(D^-)/D^-, C_{G_{23}}(D^-)/D^-\}$$

with the amalgam of maximal parabolics in the semidirect product $2^3 : L_3(2)$. I did not find an easy way to accomplish this identification. The fact that L^- splits over Z_3 is not so crucial for the alternative approach. We could have used L^+ instead and construct G_3 as a partial semidirect product over Z_3.

2.14 Essentials

In this section we state a proposition which accumulates the essential properties of the Monster amalgam $\mathcal{M}$ established during its construction and uniqueness proof. We follow the notation introduced prior and within the defining conditions of $\mathcal{M}$ in Section 2.1, in addition G_i^s denotes $C_{G_i}(Z_i)$ for $i = 2$ and 3. Recall that in Section 1.3 the anti-permutation and anti-heart modules were defined as M_{24}-sets

$$W_{24} = \{(u, v^*) \mid u \in \mathcal{C}_{12}, v^* \in \mathcal{C}_{12}^*\} \text{ and}$$
$$\mathcal{A}(M_{24}) = \{(u, v^*) \mid u \in \mathcal{C}_{11}, v^* \in \mathcal{C}_{11}^*\};$$

and subject to the notational convention the addition rule can be written as

$$(u_1, v_1^*) + (u_2, v_2^*) = (u_1 + u_2, v_1^* + v_2^* + (u_1 \cap u_2)^*)$$

in both the cases. The unique non-trivial M_{24}-invariant quadratic form on $\mathcal{A}(M_{24})$ is

$$q_h = (u, v^*) = (\frac{1}{4}|u| + |u \cap v|) \bmod 2.$$

If $M(\mathcal{S})$ is the sextet stabilizer in M_{24}, then $\mathcal{H} = O_2(M(\mathcal{S})) \cong 2^6$ is the hexacode module for the hexacode group $Y \cong 3 \cdot S_6$ (the latter being a complement to $\mathcal{H}$ in $M(\mathcal{S})$).

Proposition 2.14.1 *Let* $\mathcal{M} = \{G_1, G_2, G_3\}$ *be the unique Monster amalgam. Then:*

(I) *if* $\widetilde{G}_1$ *is the subdirect product of the standard holomorph*
   $G_1^{(s)} \sim 2_+^{1+24}.Co_1$ *and of* Co_0 *over homomorphisms onto* Co_1, *then* G_1
   is the twisted holomorph which is the quotient of $\widetilde{G}_1$ *over the central*
   order 2 subgroup which is contained in neither of the subdirect factors;

   (a) *the minimal complex representation of* G_1, *considered as a*
       representation of $\widetilde{G}_1$, *is the tensor product of the* 2^{12}-*dimensional*
       representation where the acting group is $G_1^{(s)}$ *and of the*
       24-dimensional representation on $\Lambda \otimes \mathbb{C}$ *where the acting group is*
       Co_0;

   (b) $G_{12} = N_{G_1}(Z_2)$, $Q_1 \leq G_{12}$ *and* G_{12}/Q_1 *is the monomial subgroup*
       $F \cong 2^{11} : M_{24}$ *in* $G_1/Q_1 \cong Co_1$ *and the quotient* Q_1/Z_1 *is the*
       anti-permutation module for the M_{24}-*complements in* F;

   (c) $G_{13} = N_{G_1}(Z_3)$, $Q_1 \leq G_{13}$ *and* G_{13}/Q_1 *is the three bases*
       subgroup $T \cong 2^{4+12} \cdot (3 \cdot S_6 \times S_3)$ *in* $G_1/Q_1 \cong Co_1$;

(II) Q_2 *has order* 2^{35} *with centre* Z_2, *the preimage* R_2 *of* $Z(Q_2/Z_2)$ *in* Q_2
   has order 2^{13}, $G_2^s/Q_2 \cong M_{24}$ *and* $G_2/Q_2 \cong M_{24} \times S_3$;

   (d) $Z_2 = Z(G_2^s)$, R_2, *as a module for* G_2/Q_2, *is the direct sum of* Z_2
       and $\mathcal{C}_{11}^*$, *while* Q_2/R_2 *is isomorphic to the tensor product of* Z_2 *and*
       a module isomorphic to $\mathcal{C}_{11}$;

   (e) Q_2 *is a product of three subgroups* $Q_2^{(a)}$, $a \in Z_2^\#$ *of order* 2^{24} *each*
       with pairwise intersection coinciding with the triple intersection
       which is R_2 *where* $Q_2^{(z)} = C_{Q_1}(Z_2)$;

   (f) G_2^s *is a partial semidirect product over* R_2 *of* Q_2 *and a group* M^δ
       which is the direct product of Z_2 *with the unique non-split extension*
       $\mathcal{C}_{11}^* \cdot M_{24}$;

   (g) *if* $(\mathcal{P}, \mathcal{B})$ *is the Steiner system of type* $S(5, 8, 24)$ *on which* M^δ *acts*
       as the automorphism group with kernel R_2, *then there are*
       M^δ-*module isomorphisms*

$$\psi^{(a)} : \mathcal{A}(M_{24}) \to Q_2^{(a)}/Z_2$$

such that $\psi^{(a)}(0, v^*)$ is independent of $a \in Z_2^{\#}$ and is denoted simply by $\psi(v^*)$ and the following equalities hold for $\{a, b, c\} = Z_2^{\#}$ and for all $u, u_1, u_2 \in \mathcal{C}_{11}, v^*, v_1^*, v_2^* \in \mathcal{C}_{11}^*$

$$\psi^{(a)}(u, v^*)\psi^{(b)}(u, v^*) = \psi^{(c)}(u, v^*),$$

$$[\psi^{(a)}(u_1, v_1^*), \psi^{(b)}(u_2, v_2^*)] = \psi((u_1 \cap u_2)^*);$$

(h) *for a permutation π of $Z_2^{\#}$ let π denote the automorphism of G_2^s/Z_2 which centralizes M^{δ}/Z_2 and maps $\psi^{(a)}(u, v^*)$ onto $\psi^{(\pi(a))}(u, v^*)$ for every possible $a, u,$ and v^*, and let μ be an automorphism centralizing Q_2/Z_2 and inducing an outer automorphism of M^{δ}/Z_2 centralizing M^{δ}/R_2, then $\mathrm{Aut}(G_2^s/Z_2)$ is the semidirect product of G_2^s/Z_2 and*

$$A = \langle \pi, \mu \mid \pi \in Sym(Z_2^{\#}) \rangle \cong S_3 \times 2,$$

while G_2/Z_2 is the semidirect product of G_2^s/Z_2 and

$$S_t = \{\pi \mu^{t(\pi)} \mid \pi \in Sym(Z_2^{\#})\} \cong S_3,$$

where $t(\pi) = 0$ if the permutation π is even and $t(\pi) = 1$ otherwise;

(i) *if $\widehat{G}_2^s$ is the largest perfect central extension of G_2^s such that (subject to the hat convention) $\widehat{R}_2$ is elementary abelian and for $a \in Z_2^{\#}$ the squaring map $Q_2^{(a)}/Z_2 \to \widehat{Z}_2$ is the image under $\psi^{(a)}$ of the M_{24}-invariant non-singular quadratic form q_h on $\mathcal{A}(M_{24})$, then $\widehat{Z}_2$ is of order 2^4 acted on faithfully by A with an element of order 3 acting fixed-point freely;*

(j) *G_2^s is the quotient of $\widehat{G}_2^s$ over a subgroup of order 2^2 which is the minimal S_t-invariant subgroup containing the order 2 kernel of the homomorphism of $\widehat{G}_2^s$ onto the preimage of G_{12} in $\widetilde{G}_1$;*

(k) *all the automorphisms of G_2 are inner;*

(l) *define N^{δ} to be the semidirect product of the alternating group A_4 of degree four and the non-split extension $\mathcal{C}_{12}^* \cdot M_{24}$ which acts on A_4 as a transposition (with $\mathcal{C}_{11}^* \cdot M_{24}$ being the kernel), then G_2 is a partial semidirect product over $R_2 = O_2(N^{\delta})$ of Q_2 and N^{δ}, so that M^{δ} is the kernel of the unique homomorphism of N^{δ} onto S_3 and $Z_2 = O_2(A_4)$;*

(m) *$G_{12} = C_{G_2}(Z_1) = N_{G_2}(Q_2^{(z)})$, while $G_{23} = N_{G_2}(Z_3)$ is the partial semidirect product of Q_2 and the stabilizer in N^{δ} of the sextet $\mathcal{S}(\mathcal{T})$;*

(III) *G_3 is the quotient of the universal completion group $\widetilde{G}_3$ of the amalgam $\{G_{13}, G_{23}\}$ over the unique normal complement to Z_3 in $C_{\widetilde{G}_3}(G_3^s)$;*

(o) Q_3 has order 2^{39}, G_3^s is a semidirect product of Q_3 and the
hexacode subgroup $Y \cong 3 \cdot S_6$; $G_3/Q_3 \cong 3 \cdot S_6 \times L_3(2)$ and Q_3 is
not complemented in G_3, although G_3 does contain an
$L_3(2)$-subgroup;

(p) Q_3 possesses a unique G_3-chief series

$$Z_3 < R_3 < T_3 < Q_3,$$

and the following are isomorphisms of Y-modules where $\mathcal{H}^\circ$ is the
dual of the hexacode module $\mathcal{H}$ and U_4 is a natural symplectic
module for $Y/O_3(Y) \cong Sp_4(2)$

$$R_3 \cong Z_3 \oplus \mathcal{H}^\circ, \quad T_3/R_3 \cong Z_3^* \otimes U_4, \quad Q_3/T_3 \cong Z_3 \otimes \mathcal{H};$$

(q) the automorphism group of G_3^s is a semidirect product of G_3^s/Z_3
and the affine group $2^3 : L_3(2)$ of Z_3; there are precisely 64 classes
of complements to Q_3 in G_3^s which are transitively permuted by
$\mathrm{Out}(G_3^s)$ with stabilizer isomorphic to the Frobenius group of order
21;

(r) if D^+ and D^- are subgroups in Y of order 3^3 and $3 \cdot 5$, respectively,
then $C_{G_3}(D^+) \cong 2^3 \cdot L_3(2)$ (non-split) while
$C_{G_3}(D^-) \cong 2^3 : L_3(2)$ (split extension).

Proof. The isomorphism type of G_1 as in (I) was established in (2.6.2),
while (a) follows from (2.6.1). The assertions (b) and (c) are direct consequences of the definitions of $\mathcal{M}$ in Section 2.1, of the monomial subgroup in
Section 1.8 and of the three bases subgroup in Section 1.11. The assertions (II),
(d) and (e) are extractions from (2.3.1), (2.3.2), and (2.3.4). The structure of
M^δ is given in (2.3.3). Since $G_2^s = Q_2 M^\delta$ and $Q_2 \cap M^\delta = R_2$, the group G_2^s
is indeed a partial semidirect product of Q_2 and M^δ over R_2 which is (f). The
assertion (g) is precisely (2.4.1). The information on automorphisms of G_2^s/Z_2
in (h) is taken from (2.4.4) and 92.4.5). The assertions (i), (j), (k), and (l) summarize the results in Section 2.5, while (m) is a restatement of the definition of
$\mathcal{M}$. Finally, (III) follow from the results in Section 2.12; (o) is (2.8.8); (p) is
(2.9.4); (q) and (r) are (2.13.1 (iv) and (iii)).   □

It should be possible to produce a rather explicit description of G_3 by (a)
presenting Q_3 as a product of seven copies of $C_{Q_1}(Z_3)$ indexed by the elements of $Z_3^\#$, (b) exploring the commutator and product relations in (2.14.1
(g)), and (c) making use of the structure of the three bases subgroup revealed
in Section 1.11. Since there is no immediate need for such a description, we
will reserve this project for a later consideration.

3

196 883-representation of $\mathcal{M}$

In this chapter we show that the Monster amalgam $\mathcal{M}$ (whose existence and uniqueness were established in the previous chapter) possesses a faithful completion into a general linear group of an m-dimensional real vector space, where m is the famous number 196 883 known as the degree of the smallest faithful representation of the Monster group and also as the linear coefficient of the modular invariant $J(q)$, reduced by one. First we construct a faithful representation of the subamalgam $\{G_1, G_2\}$ in $\mathcal{M}$ of the required degree. Thus we construct a pair $\varphi = (\varphi_1, \varphi_2)$ of representations

$$\varphi_1 : G_1 \to GL_m(\mathbb{R}), \quad \varphi_2 : G_2 \to GL_m(\mathbb{R})$$

which are *conformal* in the sense that their restrictions to G_{12} coincide. We carry on the construction over complex numbers and as a byproduct we observe that 196 883 is the smallest degree of a non-trivial representation in characteristic zero of the universal completion group of $\{G_1, G_2\}$ (which is the free product of G_1 and G_2 amalgamated over G_{12}). The representations φ_1 and φ_2 have three and four irreducible constituents, respectively, while their (equal) restrictions to G_{12} have six such constituents. Since

$$3 + 4 - 6 = 1,$$

by Thompson's criterion (which is a specialization of Goldschmidt's theorem), we conclude that φ is unique up to simultaneous conjugation by an element of $GL(\mathbb{C})$. More importantly, φ extends to a representation of the whole amalgam $\mathcal{M}$. The crucial property of φ which enables us to extend is the equality

$$C_{\langle \varphi(G_{13}), \varphi(G_{23}) \rangle}(\varphi(G_3^s)) = \varphi(Z_3).$$

By the defining properties of $\mathcal{M}$ this implies that

$$\varphi(\mathcal{M}) := \{\varphi_1(G_1), \varphi_2(G_2), \langle \varphi_1(G_{13}), \varphi_2(G_{23}) \rangle_{GL_m(\mathbb{C})}\}$$

is isomorphic to $\mathcal{M}$. Since $\varphi(\mathcal{M})$ is a subamalgam in $GL_m(\mathbb{C})$, this provides us with a faithful completion of $\mathcal{M}$. Eventually in this volume we show that there exists a unique faithful generating completion of $\mathcal{M}$ where the completion group is the Monster. This means that the Monster can be redefined as the image of the minimal non-trivial complex representation of the amalgam $\{G_1, G_2\}$.

3.1 Representing $\{G_1, G_2\}$

The goal of this section is to construct a pair of monomorphisms

$$\varphi_1 : G_1 \to GL(\Pi), \quad \varphi_2 : G_2 \to GL(\Pi)$$

(where Π is a complex vector space) such that $\varphi_1(g) = \varphi_2(g)$ for every $g \in G_{12} = G_1 \cap G_2$.

We build up the vector space Π and the pair (φ_1, φ_2) inductively. On the ith step we produce a triple

$$(l(i), \varphi(i), \Pi(i)),$$

where $l(i) \in \{1, 2\}$ and $\varphi(i) : G_{l(i)} \to GL(\Pi(i))$ is a faithful representation. The restriction of $\varphi(i)$ to G_{12} will be denoted by $\varphi^{12}(i)$. We start by putting $l(1) = 1$ and taking $\varphi(1)$ to be the minimal faithful representation of G_1. By (2.2.3 (iii)) and (2.6.2) we know that the degree of $\varphi(1)$ is $24 \cdot 2^{12}$. On the ith step we put $l(i) = 3 - l(i-1)$ and take

$$\varphi(i) : G_{l(i)} \to GL(\Pi(i))$$

to be a minimal representation of $G_{l(i)}$ whose restriction to G_{12} possesses $\varphi^{12}(i-1)$ as a direct summand. As soon as we reach the equality $\Pi(i) = \Pi(i-1)$ the construction is completed with $\varphi_{l(i-1)} = \varphi(i-1)$ and $\varphi_{l(i)} = \varphi(i)$. It might be helpful to consult the diagram presented before (3.1.22) to understand the individual steps of the construction.

We may notice that our strategy is essentially the one adopted in [Iv04] for constructing a 1333-dimensional representation of a rank 2 amalgam from the fourth Janko group J_4. For the rank 2 amalgam under consideration we accomplish the construction in six steps ending up with a 196 883-dimensional space. A closer look at the construction shows that in the considered situation we obtain the minimal faithful complex representation of $\{G_1, G_2\}$.

By the universality property of the induced representations, $\varphi(i)$ is always a suitable constituent of the representation induced from $\varphi^{12}(i-1)$ to $G_{l(i)}$. The following easy observation might help to understand better our construction procedure.

Lemma 3.1.1 *In the above terms suppose that*

$$\varphi^{12}(i) = \varphi^{12}(i-1) \oplus \eta,$$

where η *is an irreducible representation of* G_{12}. *Then*

$$\varphi(i+1) = \varphi(i-1) \oplus \psi,$$

where ψ *is an irreducible representation of* $G_{l(i+1)}$.

Proof. According to our strategy, the restriction of ψ to G_{12} contains η as an irreducible constituent. Since we are dealing with complex representations of finite groups, all the modules are semi-simple, and therefore η appears as a constituent of an irreducible constituent of $\varphi(i+1)$ restricted to G_{12}. □

In order to make our notation more meaningful and construction independent, for $l \in \{1, 2, 12\}$ and an integer n by Π_n (or Π_n^l) we will denote an n-dimensional complex vector space on which G_l acts via a representation φ_n^l. The precise form of the representation and the relevant value of l will be clear from the context.

3.1.1 $\Pi(1) = \Pi_{24 \cdot 2^{12}}$, $\Pi(2) = \Pi_{3 \cdot 24 \cdot 2^{11}}$

As has been announced, we start with the unique minimal faithful representation of $G_1 = G_1^{(t)}$ of degree $24 \cdot 2^{12}$

$$\varphi(1) = \varphi_{24 \cdot 2^{12}}^1 : G_1 \to GL(\Pi_{24 \cdot 2^{12}}).$$

By the proof of (2.2.3), $\varphi_{24 \cdot 2^{12}}^1$ is the tensor product of a 2^{12}-dimensional representation of $\widetilde{G}_1$ (with the acting group $G_1^{(s)}$) and a 24-dimensional representation of $\widetilde{G}_1$ (where the acting group is Co_0). We follow the notation for elements and subgroups in G_{12} introduced in Section 2.3. In particular, x, y, and z are the non-identity elements of $Z_2 = Z(Q_2) = Z(G_3^s)$, where z is the generator of $Z_1 = Z(G_1)$.

Lemma 3.1.2 *Let* $\Pi(1) = \Pi_{24 \cdot 2^{12}}$. *Then:*

(i) $\Pi(1) = C_{\Pi(1)}(y) \oplus C_{\Pi(1)}(x)$, *where each of the two direct summands is* $24 \cdot 2^{11}$-*dimensional*;

(ii) G_2^s *stabilizes each of the direct summands in* (i), *while elements from* $G_{12} \setminus G_2^s$ *permute them;*

(iii) *for $a = x$ or y the representation of G_2^s on $C_{\Pi(1)}(a)$ is induced from a non-trivial linear representation ρ of the normalizer N in G_2^s of an index 2 subgroup S in $Q_2^{(a)}$ such that $N / Q_2^{(a)} \cong M_{23}$;*

(iv) *the subgroup S in* (iii) *is unique up to conjugation in G_2^s.*

Proof. The result follows from (2.6.1) in view of the discussion prior to that lemma. The uniqueness statement in (iv) is by (1.8.7). $\qquad\square$

It appears appropriate to recall some notation introduced in Section 1.8: $\bar\lambda_4 = \sum_{p \in \mathcal{P}} 2p + 2\Lambda$ is in $\bar\Lambda_4$ and stabilized by the subgroup $F \cong \mathcal{C}_{11} : M_{24}$ of Co_1; $\bar\lambda_2 = (3p - \sum_{r \in \mathcal{P} \setminus \{p\}} r) + 2\Lambda$, $\bar\lambda_3 = (5p + \sum_{r \in \mathcal{P} \setminus \{p\}} r) + 2\Lambda$ (for some $p \in \mathcal{P}$) are elements from $\bar\Lambda_2$ and $\bar\Lambda_3$, respectively, such that $\bar\lambda_2 + \bar\lambda_3 = \bar\lambda_4$. For $\bar\mu \in \bar\Lambda$, put

$$\bar\mu^\perp = \{\bar\lambda \mid \bar\lambda \in \bar\Lambda, (\bar\mu, \bar\lambda)_{\bar\Lambda} = 0\}$$

which is the orthogonal complement of $\bar\mu$ (or rather of $\langle\bar\mu\rangle$) in $\bar\Lambda$ with respect to the Co_1-invariant bilinear form $(\ ,\ )_{\bar\Lambda}$.

Define a pair

$$S_2^{(z)} = \chi^{-1}(\bar\lambda_2^\perp) \text{ and } S_3^{(z)} = \chi^{-1}(\bar\lambda_3^\perp),$$

of index 2 subgroups in Q_1. Since the sum of $\bar\lambda_2$, $\bar\lambda_3$, and $\bar\lambda_4$ is zero and since $Q_2^{(z)} = \chi^{-1}(\bar\lambda_4^\perp)$, the intersections $S_2^{(z)} \cap Q_2^{(z)}$ and $S_3^{(z)} \cap Q_2^{(z)}$ are the same subgroup of index 2 in $Q_2^{(z)}$ which we denote by $S^{(z)}$.

Let $N^{(z)}$ be the normalizer of $S^{(z)}$ in G_2^s; let $N_2^{(z)}$ and $N_3^{(z)}$ be the normalizers in G_{12} of $S_2^{(z)}$ and $S_3^{(z)}$, respectively. By (1.8.7) the images in $G_{12}/Q_1 \cong F$ of the above-defined three normalizers is the same M_{23}-subgroup. Therefore

$$N^{(z)}/S^{(z)} \cong N_2^{(z)}/S_2^{(z)} \cong N_3^{(z)}/S_3^{(z)} \cong M_{23} \times 2.$$

Let $\psi^{(z)}$, $\psi_2^{(z)}$, and $\psi_3^{(z)}$ be representations of G_2^s, G_{12}, and G_{12} induced from non-trivial linear presentations of $N^{(z)}$, $N_2^{(z)}$, and $N_3^{(z)}$, whose kernels contain $S^{(z)}$, $S_2^{(z)}$, and $S_3^{(z)}$, respectively. Since $M_{23} \times 2$ has a unique non-trivial cyclic factor group (whose order is 2), each of the three representations is uniquely determined. Finally, for $a = x$ or y define $X^{(a)}$ to be the conjugate of $X^{(z)}$ under an element of G_2 which conjugates z onto a (here $X^{(z)}$ is one of the above-defined subgroups and representations).

Lemma 3.1.3 *The restriction of $\varphi(1)$ to G_2^s is $\psi^{(x)} \oplus \psi^{(y)}$. The restriction of $\varphi(1)$ to G_{12} is the representation induced from $\psi^{(x)}$ to G_{12}.*

Proof. In terms introduced before the lemma, the subgroup S in (3.1.2 (iii)) is (a conjugate of) $S^{(a)}$ and the normalizer N in that lemma is $N^{(a)}$. This observation immediately gives the result.   □

According to our strategy, we need to find a minimal representation $\varphi(2)$ of G_2 where the restriction to G_{12} contains as a direct summand the restriction to G_{12} of $\varphi(1)$. Because of (3.1.3) and the universality property of induced representations, $\varphi(2)$ must be a constituent of the representation induced from $\psi^{(a)}$ to G_2, where $a \in Z_2^{\#}$. Since the action of G_2 on $Z_2^{\#}$ is transitive, the choice of a is irrelevant.

Lemma 3.1.4 *Let ψ_2 and ψ_3 be the representations of G_2 induced from $\psi_2^{(z)}$ and $\psi_3^{(z)}$, respectively. Then:*

(i) *ψ_2 and ψ_3 are non-isomorphic irreducible;*
(ii) *$\psi_2 \oplus \psi_3$ is the representation induced from $\psi^{(z)}$ to G_2;*
(iii) *for $i = 2$ and 3 the restriction of ψ_i to G_{12} is the direct sum of $\psi_i^{(z)}$ and of the restriction of $\varphi(1)$ to G_{12}.*

Proof. Since $\psi_2^{(z)}$ and $\psi_3^{(z)}$ are non-isomorphic, so are ψ_2 and ψ_3. The restriction to G_2^s of either ψ_2 or ψ_3 is the direct sum of $\psi^{(z)}$, $\psi^{(x)}$, and $\psi^{(y)}$. Since each of the three summands is G_2^s-irreducible and since G_2 permutes them transitively, the irreducibility in (i) follows. Now (ii) follows from (i) and its proof in view of the dimension comparison. By a standard property of induced representations, the restriction of ψ_i to G_{12} is the direct sum of $\psi_i^{(z)}$ and the representation induced from $\psi^{(x)}$ to G_{12}. Hence (iii) follows from (3.1.3).   □

As an immediate consequence of the above lemma, we obtain the following:

Lemma 3.1.5 *$\varphi(2) = \psi_i$ for $i = 2$ or 3.*   □

The choice for $\varphi(2)$ between ψ_2 and ψ_3 will be finalized in the next subsection.

3.1.2 $\Pi(3) = \Pi(1) \oplus \Pi_{|\bar{\Lambda}_2|}$

In view of (3.1.4) at this stage we need to find a minimal representation ψ of G_1, where the restriction to G_{12} contains as a direct summand $\psi_i^{(z)}$ for $i = 2$ or 3 and put $\varphi(3) = \varphi(1) \oplus \psi$. By (3.1.1), ψ must be irreducible representation of G_1, since both $\psi_2^{(z)}$ and $\psi_3^{(z)}$ are irreducible representations of G_{12}.

Lemma 3.1.6 *Let ψ be a minimal representation of G_1, whose restriction to G_{12} contains $\psi_i^{(z)}$ as a direct summand for $i = 2$ or 3. Then:*

(i) ψ *is induced from a non-trivial linear representation ρ of the normalizer in G_1 of $S_i = \chi^{-1}(\bar{\lambda}_i^\perp)$ with $S_i \leq \ker(\rho)$;*

(ii) *for a given $i \in \{2, 3\}$, the representation ψ is uniquely determined;*

(iii) *the rank of ψ is*

$$|\bar{\Lambda}_2| = 98\,280 \;\; or \;\; |\bar{\Lambda}_3| = 189\,017\,000$$

for $i = 2$ or 3, respectively.

Proof. In view of the remark prior to the lemma, we assume that ψ is irreducible. Since Z_1 is a normal subgroup of G_1 contained in (in fact equal to) the kernel of $\psi_i^{(z)}$, we conclude that Z_1 is in the kernel of ψ. Since the whole of Q_1 is not in the kernel, $\ker(\psi) = Z_1$ and $\psi(Q_1) \cong \bar{\Lambda}$. Decomposing $\mathrm{Im}(\psi)$ into the eigenspaces of Q_1, immediately gives (i). In fact, the eigenspaces of Q_1 in $\mathrm{Im}(\psi_i^{(z)})$ are 1-dimensional and S_i is the kernel of one of them. The normalizer of S_i in $Co_1 \cong G_1/Q_1$ is the ith Conway group Co_i which is known to be simple. Therefore, each of

$$N_{G_1}(S_2)/S_2 \cong Co_2 \times 2 \text{ and } N_{G_1}(S_3)/S_3 \cong Co_3 \times 2$$

contains a unique non-trivial cyclic factor group (of order 2), which gives (ii). Now (iii) is a consequence of (i) and the known parameters of the Leech lattice. $\qquad\square$

According to our strategy to minimize the degree on every step, we put

$$\varphi(3) = \varphi(1) \oplus \varphi^1_{|\bar{\Lambda}_2|},$$

where the latter summand is the representation of G_1 induced from the unique non-trivial linear representation of $N_{G_1}(S_2) \cong Q_1.Co_2$. This forces $\varphi(2)$ to be

$$\varphi^2_{3\cdot24\cdot2^{11}} := \psi_2$$

so that the choice between the possibilities given by (3.1.4) has been finalized.

For further references we need the following statement:

Lemma 3.1.7 *Let $\bar{\mu} \in \bar{\Lambda}_2^2$, let $\Pi(\bar{\mu})$ be the Q_1-eigenspace of dimension 1 in $\Pi^1_{|\bar{\Lambda}_2|}$ corresponding to $\bar{\mu}$ in the sense that*

$$C_{Q_1}(\Pi(\bar{\mu})) = \chi^{-1}(\bar{\mu}^\perp).$$

Let $q \in \chi^{-1}(\bar{\mu})$ and let g be an element of G_1. Then:

(i) *g normalizes $\Pi(\bar{\mu})$ if and only if g stabilizes $\bar{\mu}$;*

(ii) *if g normalizes $\Pi(\bar{\mu})$, then it centralizes $\Pi(\bar{\mu})$ whenever g centralizes q and inverts $\Pi(\bar{\mu})$ otherwise.*

Proof. The assertion (i) follows directly from the definition of $\Pi_{|\bar{\Lambda}_2|}$ as an induced module, while (ii) follows from the fact that $C_{Q_1}(\Pi(\bar{\mu})) \cong Co_2 \times 2$ has a unique subgroup of index 2, which is the kernel of the actions on $\Pi(\bar{\mu})$ and on $\chi^{-1}(\bar{\mu})$.   □

3.1.3 $\Pi(4) = \Pi(2) \oplus \Pi_{2^6 \cdot 759} \oplus \cdots$

By a general property of induced representations, the irreducible constituents of $\varphi^{12}_{|\bar{\Lambda}_2|}$ correspond to the orbits of G_{12} on $\bar{\Lambda}_2$. The orbits are well known and can be deduced directly from (1.8.5).

Lemma 3.1.8 *The group G_{12} acts on $\bar{\Lambda}_2$ with three orbits $\bar{\Lambda}_2^3$, $\bar{\Lambda}_2^2$, and $\bar{\Lambda}_2^4$, where:*

(i) *$\bar{\Lambda}_2^3$ contains $\bar{\lambda}_2 = (3p - \sum_{r \in \mathcal{P} \setminus p} r) + 2\Lambda$ and has length $24 \cdot 2^{11}$;*
(ii) *$\bar{\Lambda}_2^2$ contains $\sum_{p \in \mathcal{O}} 2p + 2\Lambda$ (where $\mathcal{O}$ is an octad) and has length $2^6 \cdot 759$;*
(iii) *$\bar{\Lambda}_2^4$ contains $(4p + 4q) + 2\Lambda$ (where p and q are distinct elements of $\mathcal{P}$) and has length $2 \cdot \binom{24}{2} = 2 \cdot 276$.*   □

Therefore

$$\varphi^{12}_{|\bar{\Lambda}_2|} = \varphi^{12}_{|\bar{\Lambda}_2^3|} \oplus \varphi^{12}_{|\bar{\Lambda}_2^2|} \oplus \varphi^{12}_{|\bar{\Lambda}_2^4|}$$

and every direct summand is irreducible. The way we have produced $\varphi^1_{|\bar{\Lambda}_2|}$ immediately shows that $\varphi^{12}_{|\bar{\Lambda}_2^3|} = \psi_2^{(z)}$. Therefore, $\varphi(2)$ is the lifting of $\varphi^{12}_{|\bar{\Lambda}_2^3|}$ and our nearest goal is to 'lift' the other two direct summands to representation of G_2, keeping the degree as small as possible.

We are going to show that $\varphi^{12}_{|\bar{\Lambda}_2^2|}$ can be lifted to a representation of G_2 without increasing the degree.

It follows from the construction of $\Pi_{|\bar{\Lambda}_2|}$ as an induced module that the Q_1-eigenspaces in $\Pi_{|\bar{\Lambda}_2^2|}$ are 1-dimensional indexed by the elements of $\bar{\Lambda}_2^2$. An element of $\bar{\Lambda}_2^2$ is described by a pair $(\mathcal{O}, \sigma)$, where $\mathcal{O}$ is an octad and

$$\sigma : \mathcal{O} = s_1 \cup s_2$$

is a partition of $\mathcal{O}$ into a pair of (disjoint) even subsets, so that

$$\bar{\lambda}(\mathcal{O}, \sigma) = \sum_{p \in s_1} 2p - \sum_{q \in s_2} 2q + 2\Lambda = \sum_{p \in s_2} 2p - \sum_{q \in s_1} 2q + 2\Lambda.$$

This reconfirms that $|\bar{\Lambda}_2^2|$ is 759 (the number of octads) times 2^6 (the number of partitions of an 8-set into a pair of even subsets). If $\Pi(b, \sigma)$ is the egenspace corresponding to $\bar{\lambda}(\mathcal{O}, \sigma)$, then

$$C_{Q_1}(\Pi(\mathcal{O}, \sigma)) = \chi^{-1}(\bar{\lambda}(\mathcal{O}, \sigma)^{\perp}).$$

Since $\bar{\Lambda}_2^2 \subseteq \bar{\Lambda}^{(23)} = \bar{\lambda}_4^{\perp}$, the subgroup Z_2 centralizes every Q_1-eigenspace in $\Pi_{|\bar{\Lambda}_2^2|}$ and hence $R_2 = \chi^{-1}(\bar{\Lambda}^{(12)})$ acts on $\Pi_{|\bar{\Lambda}_2^2|}$ as

$$\bar{R}_2 = R_2/Z_2 \cong \mathcal{C}_{11}^*$$

(compare (2.3.1 (vii))).

Before decomposing $\Pi_{|\bar{\Lambda}_2^2|}$ into R_2-eigenspaces recall a few basic facts about the actions the octad stabilizer on the Golay code and Todd modules (cf. Lemma 3.8.4 in [Iv99]).

Lemma 3.1.9 *Let* $\mathcal{O}$ *be an octad, let* $M(\mathcal{O})$ *be the stabilizer of* $\mathcal{O}$ *in* $M \cong M_{24}$, *let* $Q_{\mathcal{O}} = O_2(M(\mathcal{O})) \cong 2^4$ *be the elementwise stabilizer of* $\mathcal{O}$, *and let* $K_{\mathcal{O}} \cong A_8 \cong L_4(2) \cong \Omega_6^+(2)$ *be a complement to* $Q_{\mathcal{O}}$ *in* $M(\mathcal{O})$. *Then:*

(i) *if* X *is* $\mathcal{C}_{11}$ *or* $\mathcal{C}_{11}^*$, *then* X *is a uniserial* $M(\mathcal{Q})$-module with the composition series

$$1 < C_X(Q_{\mathcal{O}}) < [X, Q_{\mathcal{O}}] < X;$$

(ii) $C_{\mathcal{C}_{11}^*}(Q_{\mathcal{O}}) \cong \mathcal{C}_{11}/[\mathcal{C}_{11}, Q_{\mathcal{O}}] \cong Q_{\mathcal{O}} \wedge Q_{\mathcal{O}}$ *is the natural 6-dimensional orthogonal module for* $K_{\mathcal{O}}$;

(iii) $[\mathcal{C}_{11}^*, Q_{\mathcal{O}}]/C_{\mathcal{C}_{11}^*}(Q_{\mathcal{O}}) \cong Q_{\mathcal{O}}$;

(iv) *if* $v \subseteq \mathcal{P}$, *then* $v^* \in [\mathcal{C}_{11}^*, Q_{\mathcal{O}}]$ *if and only if* $|v \cap \mathcal{O}|$ *is even, in particular the codimension of* $[\mathcal{C}_{11}^*, Q_{\mathcal{O}}]$ *in* $\mathcal{C}_{11}^*$ *is one;*

(v) *if* $u \in \mathcal{C}_{11}$, *then* $u \in [\mathcal{C}_{11}, Q_{\mathcal{O}}]$ *if and only if* u *is the image of an octad equal to or disjoint from* $\mathcal{O}$.

$\square$

If U_6 is the natural 6-dimensional orthogonal module of $K_{\mathcal{O}}$ and q_o is the non-singular invariant quadratic form on U_6 with the associated bilinear form f_o, then the elements of U_6 can be identified with the partitions of $\mathcal{O}$ into pairs of even subsets, so that for $\sigma : \mathcal{O} = s_1 \cup s_2$ and $\tau : \mathcal{O} = t_1 \cup t_2$ we have

$$q_o(\sigma) = \frac{1}{2}|s_1| \bmod 2, \quad f_o(\sigma, \tau) = |s_1 \cap t_1| \bmod 2.$$

Lemma 3.1.10 $(R_2 \cap C_{Q_1}(\Pi(\mathcal{O}, \sigma))/Z_2 = [\bar{R}_2, Q_{\mathcal{O}}]$.

Proof. Let (u, v^*) be an element of W_{24} such that $\omega(u, v^*) \in \bar{\Lambda}^{(12)}$. Then $\chi^{-1}(\omega(u, v^*)) \in R_2$, $u \in \{\emptyset, \mathcal{P}\}$ and $v^* \in C_{11}^*$. By (1.3.3) and (1.8.4), the equality

$$(\omega(u, v^*), \bar{\lambda}(\mathcal{O}, \sigma))_{\bar{\Lambda}} = 0$$

holds if and only if $|v \cap \mathcal{O}|$ is even. Hence the result follows from (3.1.9 (v)). $\square$

The above lemma together with a dimension comparison gives the following:

Lemma 3.1.11 *The R_2-eigenspaces in $\Pi_{|\bar{\Lambda}_2^2|}$ are 2^6-dimensinal indexed by the octads. The eigenspace $\Pi_{\mathcal{O}}$ corresponding to $\mathcal{O}$ is the sum of Q_1-eigenspaces $\Pi(\mathcal{O}, \sigma)$ taken for all even partitions σ of $\mathcal{O}$. If $G_{12}(\mathcal{O})$ is the stabilizer of $\mathcal{O}$ in G_{12}, then $\varphi_{|\bar{\Lambda}_2^2|}^{12}$ is the 2^6-dimensional representation $\rho_{\mathcal{O}}$ of $G_{12}(\mathcal{O})$ on $\Pi_{\mathcal{O}}$ induced to G_{12}.* $\square$

By the last assertion in (3.1.11) in order to accomplish our current task of lifting $\varphi_{|\bar{\Lambda}_2^2|}^{12}$ preserving the degree, it suffices to show that $\rho_{\mathcal{O}}$ is the restriction to $G_{12}(\mathcal{O})$ of a $G_2(\mathcal{O})$-representation. Towards this end we calculate the kernel and the image of $G_2^s(\mathcal{O})$ under $\rho_{\mathcal{O}}$ (denoting by M the quotient $G_2^s/Q_2 \cong M_{24}$).

Lemma 3.1.12 *Let $G_2^s(\mathcal{O})$ be the stabilizer of $\mathcal{O}$ in G_2^s (so that $Q_2 \leq G_2^s(\mathcal{O})$ and $G_2^s(\mathcal{O})/Q_2 = M(\mathcal{O})$), and let $Q_{\mathcal{O}} = O_2(M(\mathcal{O})) \cong 2^4$. Put $C_{\mathcal{O}} = C_{G_2^s(\mathcal{O})}(\Pi_{\mathcal{O}})$. Then:*

(i) $C_{\mathcal{O}}$ *contains Z_2 and $(C_{\mathcal{O}} \cap R_2)/Z_2 = [\bar{R}_2, Q_{\mathcal{O}}]$;*

(ii) $(C_{\mathcal{O}} \cap Q_2^{(a)})R_2/R_2 = [Q_2^{(a)}/R_2, Q_{\mathcal{O}}]$ *for every $a \in Z_2^{\#}$;*

(iii) $Q_2^{(x)}$ *permutes transitively the Q_1-eigenspaces contained in $\Pi_{\mathcal{O}}$;*

(iv) $C_{\mathcal{O}} Q_2/Q_2 = Q_{\mathcal{O}}$;

(v) $C_{\mathcal{O}}$ *is normal in $G_2(\mathcal{O})$.*

Proof. The assertion (i) follows from (3.1.10). In terms of Section 2.4 (subject to the identification of $\mathcal{O}$ with its image in C_{11}) if $\sigma : \mathcal{O} = s_1 \cup s_2$, then the element $\psi^{(z)}(\mathcal{O}, s_1^*)$ generates the image of $\chi^{-1}(\bar{\lambda}(\mathcal{O}, \sigma))$ in

$$\bar{Q}_2^{(z)} = Q_2^{(z)}/Z_2 = C_{Q_1}(Z_1)/Z_2.$$

Considering the $(\mathcal{O}, s_1^*)$s as elements of the anti-heart module $\mathcal{A}(M_{24})$, we obtain from (1.3.3) that for another element (u, v^*) of this module the equality

$$\beta_h((\mathcal{O}, s_1^*), (u, v^*)) = 0$$

holds if and only if u is the image in C_{11} of an octad equal to or disjoint from $\mathcal{O}$. By (3.1.9 (v)), this proves (ii) for case $a = z$. It is quite clear that for every even subset s_1 of $\mathcal{O}$ there exists a Golay set $u(\mathcal{O}, s_1)$ such that $\mathcal{O} \cap u(\mathcal{O}, s_1) = s_1$. By (2.4.1 (iii)), for such a set we have

$$\psi^{(x)}(u(\mathcal{O}, s_1^*), 0)\psi^{(z)}(\mathcal{O}, 0)\psi^{(x)}(u(\mathcal{O}, s_1^*), 0) = \psi^{(z)}(\mathcal{O}, s_1^*),$$

which proves both (ii) for the case $a \in \{x, y\}$ and (iii). We consider $\bar{G}_2^s$ as a partial semidirect product over $\bar{R}_2$ of $\bar{Q}_2$ and of the stabilizer $\bar{M}^\delta(\mathcal{O})$ of $\mathcal{O}$ in $\bar{M}^\delta \cong C_{11}^* \cdot M_{24}$. This viewpoint equips us with a well-defined action of $M(\mathcal{O}) \cong \bar{M}^\delta(\mathcal{O})/\bar{R}_2$ on the set of Q_1-eigenspaces contained in $\Pi_\mathcal{O}$. It is clear that the action of $Q_\mathcal{O}$ is trivial, while $K_\mathcal{O} \cong A_8$ acts faithfully. Furthermore

$$C_{G_2^s(\mathcal{O})}(\chi^{-1}(\mathcal{O}, \emptyset))Q_2/Q_2 \geq Q_\mathcal{O}.$$

Since $Q_\mathcal{O}$ centralizes $Q_2^{(x)}/R_2$, because of the above inclusion the assertion (iv) follows from (iii). Since

$$G_2(\mathcal{O}) = N_{G_2}(G_2^s(\mathcal{O})) = N_{G_2}([\bar{R}_2, Q_\mathcal{O}])$$

and since (ii) respects the full symmetry between the elements of $Z_2^\#$, we have $G_2(\mathcal{O})G_2^s = G_2$ and (v) follows. □

Lemma 3.1.13 *In terms of* (3.1.12), *put*

$$I^\mathcal{O} = G_2^s(\mathcal{O})/C_\mathcal{O}, \quad Q^\mathcal{O} = Q_2/(C_\mathcal{O} \cap Q_2) \ and \ R^\mathcal{O} = R_2/(C_\mathcal{O} \cap R_2).$$

Then:

(i) $R^\mathcal{O}$ *is of order* 2 *contained in the centre of* $I_\mathcal{O}$;
(ii) $Q^\mathcal{O} = O_2(I^\mathcal{O})$, $Q^\mathcal{O}/R^\mathcal{O}$ *is elementary abelian of order* 2^{12} *and* $I^\mathcal{O}/Q^\mathcal{O} \cong K_\mathcal{O}$;
(iii) $Q^\mathcal{O} \cong 2_+^{1+12}$;
(iv) $I^\mathcal{O} \cong 2_+^{1+12}.A_8$ *and* $G_2(\mathcal{O})/C_\mathcal{O} \cong 2_+^{1+12}.(A_8 \times S_3)$.

Proof. Since R_2 is normal in G_2^s, (i) follows from (3.1.12 (i)) and (3.1.9 (vi)). Since Q_2/R_2 is elementary abelian by (2.3.4), (ii) follows from (3.1.12 (ii), (iv)) and (3.1.9 (ii)). In order to prove (iv) it is sufficient to show that the squaring map

$$\pi : Q_2/R_2 \to \bar{R}_2$$

induces on $Q^\mathcal{O}/R^\mathcal{O}$ a non-singular quadratic form $q_\mathcal{O}$ of plus type taking values in $R^\mathcal{O} \cong 2$. Let U_6 be the natural orthogonal module of $K_\mathcal{O} \cong I^\mathcal{O}/Q^\mathcal{O}$ whose elements are identified with the partitions of $\mathcal{O}$ into pairs of even subsets, and let q_o be the non-singular $K_\mathcal{O}$-invariant quadratic form on U_6. For $a \in Z_2^\#$, put

$$Q_\mathcal{O}^{(a)} = Q_2^{(a)}/((Q_2^{(a)} \cap C_\mathcal{O})R_2).$$

Then by (3.1.12 (ii)) and (3.1.9 (ii)) there is a $I^\mathcal{O}/Q^\mathcal{O}$-isomorphism

$$\eta^{(a)} : U_6 \to Q_\mathcal{O}^{(a)}$$

of vector spaces. Furthermore, if $\sigma : \mathcal{O} = s_1 \cup s_2$, then $\eta^{(a)}(\sigma)$ coincides with the image in Q_2/R_2 of $\psi^{(a)}(u(\mathcal{O}, s_1), 0)$ (cf. Section 2.4), where $u(\mathcal{O}, s_1)$ is a Golay set which intersects $\mathcal{O}$ in s_1. The commutator relation in (2.4.1 (iii)) shows that

$$\pi(\psi^{(a)}(u_1, 0)\psi^{(b)}(u_2, 0)) = (1 - \delta_{a,b}) \, \psi(0, (u_1 \cap u_2)^*)Z_2.$$

Therefore, the form on $Q^\mathcal{O}/R^\mathcal{O}$ induced by π is defined by

$$q_\mathcal{O}(\eta^{(a)}(\sigma)\eta^{(b)}(\tau)) = (1 - \delta_{a,b}) \, (|s_1 \cap t_1| \bmod 2),$$

for σ as above and $\tau : \mathcal{O} = t_1 \cup t_2$. It is easy to check that $q_\mathcal{O}$ is indeed a non-singular quadratic form of plus type with $Q_\mathcal{O}^{(a)}$ being a maximal totally isotropic subspace. Thus (iii) is proved. Finally, (iv) follows from (ii), (iii), and (3.1.12 (v)). $\square$

By (3.1.13 (iv)), the image $I^\mathcal{O}$ of $G_2^s(\mathcal{O})$ under the representation $\rho_\mathcal{O}$ is an extraspecial extension from the class

$$\mathcal{E}(A_8, U_6 \oplus U_6, q_\mathcal{O}).$$

Since $2^6 = \dim(\Pi_\mathcal{O})$ is the degree of the minimal faithful representation of $Q^\mathcal{O} \cong 2_+^{1+12}$, we conclude that

$$G_{12}(\mathcal{O})/C_\mathcal{O} \cong 2_+^{1+12}.(A_8 \times 2)$$

is a standard holomorph. Since $G_{12}(\mathcal{O})$ contains a Sylow 2-subgroup of $G_2(\mathcal{O})$, this implies (by Gaschütz' theorem) that $G_2(\mathcal{O})/C_\mathcal{O}$ is also a standard holomorph. Hence $G_2(\mathcal{O})$ possesses a 2^6-dimensional representation $\rho_\mathcal{O}^2$ whose kernel is $C_\mathcal{O}$. Because of the uniqueness of the minimal faithful representation of $Q^\mathcal{O}$, the restriction of $\rho_\mathcal{O}^2$ to $G_{12}(b)$ can be assumed to coincide with $\rho_\mathcal{O}$. By (3.1.11) this means that $\varphi_{|\bar{\Lambda}_2^2|}^{12}$ indeed can be lifted without increasing of the degree.

Lemma 3.1.14 *There is a unique representation $\varphi_{|\bar{\Lambda}_2^2|}^2 = \varphi_{2^6 \cdot 759}^2$ of G_2 whose restriction to G_{12} coincides with $\varphi_{|\bar{\Lambda}_2^2|}^{12}$.* $\square$

3.1.4 $\Pi(4) = \Pi(2) \oplus \Pi_{26 \cdot 759} \oplus \Pi_{3 \cdot 276}$

In the case of $\varphi_{|\bar{\Lambda}_2^4|}^{12}$ we construct a lifting to a G_2-representation increasing the degree by a half. The representation will be constructed by inducing from a 3-dimensional representation of the stabilizer in G_2 of a pair from $\mathcal{P}$. Before proceeding to the construction we introduce some notation.

As above $(\mathcal{P}, \mathcal{B})$ denotes the Steiner system of type $S(5, 8, 24)$ on which G_2 induces the natural action of M_{24} with kernel $C_{G_2}(\bar{R}_2) \cong Q_2 : S_3$. Let $\{p, q\}$ be a *pair*, which is a 2-element subset of $\mathcal{P}$. For a subgroup X of G_2, let $X(\{p, q\})$ denote the stabilizer of $\{p, q\}$ in X. Since $G_2 / G_2^s \cong S_3$, $G_2^s = Q_2 M^\delta$, where $M^\delta \cap Q_2 = R_2$ and $M^\delta / R_2 \cong M_{24}$ (cf. Sections 2.3, 2.4 and 2.5), we have the following

$$Q_2(\{p, q\}) = Q_2, \ M^\delta(\{p, q\})/R_2 \cong M_{22} : 2,$$
$$G_2^s(\{p, q\}) = Q_2 M^\delta(\{p, q\}), \ G_2(\{p, q\}) = C_{G_2}(\bar{R}_2) M^\delta(\{p, q\}).$$

Let $a \in Z_2^\#$. It is a standard property of $M_{24} \cong M^\delta / R_2$ (cf. Section 2.15 in [Iv99]) that under the action of $M^\delta(\{p, q\})/R_2 \cong M_{22} : 2$ the irreducible Golay code module $\mathcal{C}_{11} \cong Q_2^{(a)}/R_2$ is uniserial possessing a unique codimension 1 irreducible submodule formed by the images in $\mathcal{C}_{11}$ of the Golay sets evenly intersecting the pair $\{p, q\}$. Therefore, $Q_2^{(a)}$ possesses a unique index 2 subgroup containing R_2 and normalized by $M^\delta(\{p, q\})$. We denote this unique subgroup by $Q_{\{p,q\}}^{(a)}$ and put

$$Q_{\{p,q\}} = Q_{\{p,q\}}^{(x)} Q_{\{p,q\}}^{(y)} Q_{\{p,q\}}^{(z)},$$

so that $Q_2 / Q_{\{p,q\}} \cong 2^2$ and $C_{G_2}(\bar{R}_2)/Q_{\{p,q\}} \cong S_4$. As in (3.1.7) for $\bar{\mu} \in \bar{\Lambda}_2$ let $\Pi(\bar{\mu})$ denote the Q_1-eigenspace in $\Pi_{|\bar{\Lambda}_2|}^1$ such that

$$C_{Q_1}(\Pi(\bar{\mu})) = \chi^{-1}(\bar{\mu}^\perp).$$

A typical element of $\bar{\Lambda}_2^4$ is described by a pair from $\mathcal{P}$ and a sign, so that

$$\bar{\Lambda}_2^4 = \{\bar{\lambda}(\{p, q\}, \zeta) \mid \{p, q\} \in \binom{\mathcal{P}}{2}, \zeta = \pm 1\},$$

where

$$\bar{\lambda}(\{p, q\}, \zeta) = \zeta_p 4p + \zeta_q 4q + 2\Lambda$$

for $\zeta_p, \zeta_q = \pm 1$ with $\zeta_p \zeta_q = \zeta$.

Lemma 3.1.15 *If* $\Pi(\{p, q\}) = \Pi(\bar{\lambda}(\{p, q\}, 1)) \oplus \Pi(\bar{\lambda}(\{p, q\}, -1))$, *then the following assertions hold:*

(i) $C_{Q_2}(\Pi(\{p,q\})) = Q_{\{p,q\}}$;

(ii) $C_{M^\delta}(\Pi(\{p,q\})) = M^\delta(\{p,q\})$;

(iii) $Q_2^{(z)}$ *acts on* $\Pi(\{p,q\})$ *by* ± 1-*scalar operators with kernel* $Q_{\{p,q\}}^{(z)}$;

(iv) *if* $a \in Z_2^\# \setminus \{z\}$ *and* $q \in Q_2^{(a)} \setminus Q_{\{p,q\}}^{(a)}$, *then* q *acts on* $\Pi(\{p,q\})$
with one eigenvalue 1 *and one eigenvalue* -1.

Proof. The assertions are easy to deduce from the known action of $F \cong C_{11} : M_{24}$ on $\bar{\Lambda}$ making use of (3.1.7). In fact, both $\bar{\lambda}(\{p,q\}, 1)$ and $\bar{\lambda}(\{p,q\}, -1)$ are contained in $\bar{\Lambda}^{(12)} = \chi(R_2)$; by (2.3.3 (iii)), R_2, considered as an M^δ-module, is isomorphic to the direct sum of C_{11}^* and a 2-dimensional trivial module, while by (2.3.1 (viii)), $Q_2^{(x)}$ acts on R_2 via transvections with centre x. □

Lemma 3.1.16 *The following assertions hold:*

(i) $\varphi_{|\bar{\Lambda}_2^4|}^{12}$ *is induced from the representation of* $G_{12}(\{p,q\})$, *whose dimension is 2 and the kernel is* $Q_{\{p,q\}} M^\delta(\{p,q\})$;

(ii) *under the action of* G_2^s *the module* $\Pi_{|\bar{\Lambda}_2^4|}$ *splits into the direct sum*

$$\Pi_{|\bar{\Lambda}_2^4|} = \Sigma^{(x)} \oplus \Sigma^{(y)},$$

where $\Sigma^{(a)} = C_{\Pi_{|\bar{\Lambda}_2^4|}}(Q_2^{(a)})$;

(iii) *the representation* $\xi^{(x)}$ *of* G_2^s *on* $\Sigma^{(x)}$ *is induced from the linear representation of* $G_2^s(\{p,q\})$ *with kernel* $Q_2^{(x)} Q_{\{p,q\}}^{(y)} M^\delta(\{p,q\})$.

Proof. The assertions follow from (3.1.15) in view of the equality

$$2 \cdot 276 = \dim(\Pi_{|\bar{\Lambda}_2^4|}) = \dim(\Pi(\{p,q\})) \cdot [G_2^s : G_2^s(\{p,q\})]. \qquad \square$$

Since 276 is the length of the shortest M_{24}-orbit on the set of hyperplanes in $C_{11} \cong Q_2^{(x)}/R_2$, it is immediate from (3.1.16) that $\xi^{(x)}$ is a 276-dimensional irreducible representation of G_2^s and that $\varphi_{|\bar{\Lambda}_2^4|}^{12}$ is induced from $\xi^{(x)}$ to G_{12}. Let ξ be an irreducible representation of G_2 such that the restriction of ξ to G_{12} contains $\varphi_{|\bar{\Lambda}_2^4|}^{12}$ as an irreducible constituent. By the above, ξ is a constituent of the representation induced from $\xi^{(x)}$ to the whole of G_2. The proof of the following lemma is analogous to that of (3.1.3) and can be deduced from elementary properties of the representations of $S_4 \cong G_2(\{p,q\})/Q_{\{p,q\}} M^\delta(\{p,q\})$.

Lemma 3.1.17 *In terms of* (3.1.16), *let* ϑ *be the representation induced to* G_2 *from the representation* $\xi^{(x)}$ *of* G_2^s. *Then:*

(i) $\vartheta = \xi_1 \oplus \xi_2$ *where* ξ_1 *and* ξ_2 *are irreducible of dimension* $3 \cdot 276$ *each;*

(ii) *for* $i \in \{1, 2\}$, *the restriction of* ξ_i *to* G_{12} *is the direct sum of* $\varphi^{12}_{|\tilde{\Lambda}^4_2|}$ *and a* 276-*dimensional irreducible representation* ρ_i;

(iii) ρ_1 *and* ρ_2 *are not isomorphic and each of them is induced from a linear representation of* $G_{12}(\{p, q\})$ *whose kernel contains* $Q_{\{p,q\}} Q_2^{(z)} M^\delta(\{p, q\})$ *and does not contain* Q_2;

(iv) *the kernel of* ρ_1 *contains* Q_1, *while that of* ρ_2 *does not.* $\qquad\square$

By (3.1.17), according to our strategy we put

$$\varphi(4) = \varphi(2) \oplus \varphi^2_{2^6.759} \oplus \xi_i$$

for $i = 1$ or 2. In the next subsection we show that the choice $i = 1$ leads to the minimal representation of the amalgam $\{G_1, G_2\}$.

3.1.5 $\Pi(5) = \Pi(3) \oplus \Pi_{299}$

By (3.1.1), (3.1.14), (3.1.17), and the paragraph at the end of the previous subsection, $\Pi(5) = \Pi(3) \oplus \psi$, where ψ is an irreducible representation of G_1, whose restriction to G_{12} contains ρ_i for $i \in \{1, 2\}$ as an irreducible constituent.

By (3.1.17 (iv)) under the representation ρ_2 the subgroup Q_1 acts by ± 1-scalar operators with kernel $Q_2^{(z)}$. Since the normalizer of $Q_2^{(z)}$ in G_1 is precisely G_{12}, the choice $i = 2$ requires ψ to have degree at least

$$276 \cdot [G_1 : G_{12}]$$

which is much higher than our target of 196 883. On the other hand, Q_1 is in the kernel of ρ_1 and since Q_1 is normal in G_1 the choice $i = 1$ forces G_1 to act under ψ via its factor group $G_1/Q_1 \cong Co_1$. Therefore, in this case we should be able to find ψ among the irreducible representations of Co_1 (whose character table can be found in [CCNPW]).

Let $\widehat{\varphi}_{24}$ be the representation of the universal perfect central extension $\widehat{G}_1$ on G_1 on $\Lambda \otimes \mathbb{R}$ with the acting group Co_0. Then it can be deduced from the character table of Co_0 that

$$\widehat{\varphi}_{24} \otimes \widehat{\varphi}_{24} = \varphi^1_1 \oplus \varphi^1_{276} \oplus \varphi^1_{299},$$

where on the right-hand side we have the first, second, and third smallest degree representations of Co_1. Furthermore, φ^1_{276} and $\varphi^1_1 \oplus \varphi^1_{299}$ are the exterior and symmetric squares of $\widehat{\varphi}_{24}$, respectively.

Given a representation of G_1 with kernel Q_1, in order to restrict it to G_{12} it is sufficient to describe the action of $F = E : M \cong C_{11} : M_{24}$ on the corresponding module. The following result follows from the known action of $F_0 \cong \mathcal{C}_{12}$: M_{24} on Λ and the decomposition of $\widehat{\varphi}_{24} \otimes \widehat{\varphi}_{24}$ into Co_1-irreducibles.

Lemma 3.1.18 *Let* φ_{276}^1 *and* φ_{299}^1 *be the second and third minimal representations of* Co_1. *Let* $\{p, q\}$ *be a pair from* $\mathcal{P}$, *let* $F(\{p, q\})$ *and* $M(\{p, q\})$ *be the stabilizers of* $\{p, q\}$ *in* F *and* M, *respectively (so that* $F(\{p, q\}) = E$: $M(\{p, q\})$ *and* $M(\{p, q\}) \cong M_{22} : 2$), *and let* $E_{\{p,q\}}$ *be the unique hyperplane in* E *stabilized by* $M(\{p, q\})$. *Then:*

(i) φ_{276}^1 *restricted to* F *is induced from a linear representation of* $F(\{p, q\})$ *whose kernel does not contain* $M(\{p, q\})$;

(ii) φ_{299}^1 *restricted to* F *is the direct sum* $\varphi_{23}^{12} \oplus \varphi_{276}^{12}$;

(iii) φ_{23}^{12} *has* E *in the kernel and its restriction to* $M \cong M_{24}$ *is the faithful irreducible constituent of the permutation module on* $\mathcal{P}$;

(iv) φ_{276}^{12} *is induced from the linear representation of* $F(\{p, q\})$ *with kernel* $E_{\{p,q\}} M(\{p, q\})$. $\qquad\square$

Comparing (3.1.17) and (3.1.18) we have the following lemma:

Lemma 3.1.19 *In terms of* (3.1.18) *the representation* ρ_1 *is isomorphic to* φ_{276}^{12} *and not isomorphic to the restriction of* φ_{276}^1 *to* F. $\qquad\square$

By the above lemma we put $\psi = \varphi_{299}^1$.

3.1.6 $\Pi(6) = \Pi(5) = \Pi(4) \oplus \Pi_{23}$

The final step in our construction can be reduced to the following elementary lemma.

Lemma 3.1.20 *Let* φ_{23}^2 *be the representation of* G_2 *whose kernel is* $C_{G_2}(\bar{R}_2)$ *and with respect to which* $M^\delta / R_2 \cong M_{24}$ *acts as on the faithful irreducible constituent of the permutation module on* $\mathcal{P}$. *Then the restriction of* φ_{23}^2 *to* G_{12} *is the representation* φ_{23}^{12} *as in* (3.1.18 (iii)). $\qquad\square$

Since φ_{23}^{12} can be lifted to the representation φ_{23}^2 of G_2 without any increase of the degree, the construction of a pair of conformal representations of G_1 and G_2 is completed and the resulting degree is

$$24 \cdot 2^{12} + |\bar{\Lambda}_2| + 299 = 196\,883$$

(recall that $|\bar{\Lambda}_2| = 98\,280$).

3.1.7 Rounding up

In this subsection we summarize the results of the construction accomplished in the previous subsections. The first proposition addresses the existence of the 196 883-dimensional representation of $\{G_1, G_2\}$ and specifies the irreducible constituents of the individual representations of G_1 and G_2.

Proposition 3.1.21 *Let* $\{G_1, G_2\}$ *be the subamalgam formed by the first two members in the Monster amalgam* $\mathcal{M}$. *Let* Π *be a complex vector space of dimension* 196 883. *Then there are representations*

$$\varphi_1 : G_1 \to GL(\Pi) \text{ and } \varphi_2 : G_2 \to GL(\Pi)$$

such that $\varphi_1(g) = \varphi_2(g)$ *for every* $g \in G_{12}$. *Furthermore*

$$\varphi_1 = \varphi^1_{24 \cdot 2^{12}} \oplus \varphi^1_{|\bar{\Lambda}_2|} \oplus \varphi^1_{299},$$
$$\varphi_2 = \varphi^2_{3 \cdot 24 \cdot 2^{11}} \oplus \varphi^2_{2^6 \cdot 759} \oplus \varphi^2_{3 \cdot 276} \oplus \varphi^2_{23},$$

where the direct summands are irreducible representations defined as follows (where $(\mathcal{P}, \mathcal{B})$ *is the Steiner system of type* $S(5, 8, 24)$ *acted on by* $M_{24} \cong G_2/C_{G_2}(\bar{R}_2)$, $p \in \mathcal{P}$, $b \in \mathcal{B}$ *and* $\{p, q\}$ *is a pair):*

(i) $\varphi^1_{24 \cdot 2^{12}}$ *is faithful and when considered as a representation of the universal perfect central extension* $\widehat{G}_1$ *of* G_1, *it is the tensor product of the 24-dimensional representation* $\widehat{\varphi}^1_{24}$ *of* $\widehat{G}_1$ *on* $\Lambda \otimes \mathbb{R}$ *with acting group* Co_0 *and of the* 2^{12}-*dimensional representation of* $\widehat{G}_1$ *where the acting group is the standard holomorph* $G_1^{(s)}$;

(ii) $\varphi^1_{|\bar{\Lambda}_2|}$ *has kernel* Z_1 *and it is induced from the unique non-trivial linear representation of the preimage in* G_1 *of the* Co_2-*subgroup of* $G_1/Q_1 \cong Co_1$;

(iii) φ^1_{299} *has kernel* Q_1 *with* $G_1/Q_1 \cong Co_1$ *acting as on the faithful irreducible constituent of the symmetric square of* $\Lambda \otimes \mathbb{R}$;

(iv) $\varphi^2_{3 \cdot 24 \cdot 2^{11}}$ *is faithful and it is induced from the unique non-trivial linear representation of the normalizer in* G_{12} *of* $\chi^{-1}(\bar{\lambda}_2^{\perp})$ *whose kernel contains* $\chi^{-1}(\bar{\lambda}_2^{\perp})$ *(where* $\bar{\lambda}_2 = (3p - \sum_{r \in \mathcal{P} \setminus \{p\}} r) + 2\Lambda$);

(v) $\varphi^2_{2^6 \cdot 759}$ *has kernel* Z_2 *and it is induced from a* 2^6-*dimensional representation of the stabilizer of* b *in* G_2 *(where the acting group is a standard holomorph* $2^{1+12}_+ .(A_8 \times S_3))$;

(vi) $\varphi^2_{3 \cdot 276}$ *has kernel* R_2 *and it is induced from a linear representation of the stabilizer of* $\{p, q\}$ *in* G_2 *with kernel* $Q_1 Q_{\{p,q\}} M^{\delta}(\{p, q\})$;

(vii) φ^2_{23} *has kernel* $C_{G_2}(\bar{R}_2)$ *and it is the irreducible faithful constituent of the permutation module of* $G_2/C_{G_2}(\bar{R}_2) \cong M_{24}$ *on* $\mathcal{P}$.

Proof. The validity of the proposition is justified by the success of the construction we have just accomplished. Thus it only remains to compare the descriptions of the irreducibles in (i) to (vii) with their original definitions. For (i) check the first paragraph in Subsection 3.1.1; for (ii) check (3.1.6) and its proof; for (iii) check the discussion at the beginning of Subsection 3.1.5; for (iv) check (3.1.4) and the paragraph before (3.1.3); for (v) check (3.1.10) and (3.1.12); for (vi) check (3.1.17) and (3.1.19); finally for (vii) it is just (3.1.20). $\qquad\square$

The way the irreducible constituents of φ_1 and φ_2 behave when restricted to G_{12} are shown on the following diagram which can also serve as an illustration for the above construction.

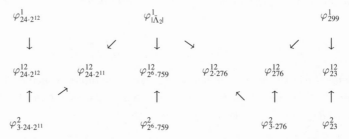

The next proposition deals with the minimality and uniqueness matters. Although they are not crucial for the mainstream of our construction and uniqueness proof for the Monster group, we believe the explicit statement is appropriate.

Proposition 3.1.22 *For a complex vector space* Π *let* ψ *be a homomorphism into* $GL(\Pi)$ *of the universal completion group of the amalgam* $\{G_1, G_2\}$ *whose image is a non-identity group, and let* ψ_1 *and* ψ_2 *be the restrictions of* ψ *to* G_1 *and* G_2, *respectively. Then:*

(i) ψ_1 *and* ψ_2 *are both faithful and*

$$\psi_1(G_1) \cap \psi_2(G_2) = \psi_1(G_{12}) = \psi_2(G_{12})$$

so that the pair (ψ_1, ψ_2) *defines a faithful completion of* $\{G_1, G_2\}$ *in* $GL(\Pi)$;

(ii) $\dim(\Pi) \geq 196\,883$;

(iii) *if the equality is attained in* (ii), *then there exists* $l \in GL(\Pi)$ *such that*

$$\psi_1(g_1) = l^{-1}\varphi_1(g_1)l \text{ and } \psi_2(g_2) = l^{-1}\varphi_2(g_2)l$$

for all $g_1 \in G_1$, $g_2 \in G_2$, *where* φ_1 *and* φ_2 *are as in* (3.1.21).

Proof. Let N_1 be a proper normal subgroup of G_1 (that is Z_1 or Q_1) and let N_2 be a proper normal subgroup of G_2 (that is Z_2, R_2, Q_2, G_2^s, $C_{G_2}(\bar{R}_2)$, $C_{G_2}(\bar{R}_2)'$, or G_2'). Then it is easy to check that for $\{i, j\} = \{1, 2\}$ the normal closer of $N_i \cap G_j$ in G_j intersected with G_i contains $N_i \cap G_j$ as a proper subgroup. Since the whole of G_i is also the normal closure of G_{12} in G_i, this means that whenever the kernel of ψ bites a non-identity element of the amalgam $\{G_1, G_2\}$ it swallows the whole amalgam. So to say, $\{G_1, G_2\}$ possesses no proper factor amalgams, which gives the faithfulness assertion in (i). The second assertion in (i) follows from the obvious fact that G_2 is not isomorphic to a subgroup in G_1. The assertion (ii) can be deduced by a somewhat careful analysis of our construction. In fact, if on any of the steps we were to go for a non-minimal lifting, the degree of the resulting representation would be greater than $196\,883$ (we leave this analysis to a dedicated reader). It also follows from our construction that in the minimal case the individual representations φ_1 and φ_2 along with their (equal) restrictions to G_{12} are determined uniquely up to the obvious equivalence. Since φ_1 and φ_2 have three and four irreducible constituents respectively, while the restriction has six such constituents, the assertion (iii) follows from the main result of [Th81] (which can be viewed as a version of Goldschmidt's theorem). Notice that the amalgam $\{G_1, G_2\}$ along with the J_4-amalgam $\{2^{10} : L_5(2), 2^{6+4\cdot2}.(L_4(2) \times 2)\}$ employed in [Iv04] were the motivation examples for [Th81]. $\qquad\square$

3.2 Incorporating G_3

Let $\varphi : G_1 \cup G_2 \rightarrow GL(\Pi)$ be the completion of $\{G_1, G_2\}$ such that the restrictions φ_1 and φ_2 of φ to G_1 and G_2 are representations from (3.1.21). We are going to extend φ to the whole of $\mathcal{M}$ by putting

$$G_3^\varphi = \langle \varphi(G_{13}), \varphi(G_{23}) \rangle_{GL(\Pi)}$$

and showing that G_3^φ is in fact isomorphic to G_3. Since φ is a faithful completion, by the proof of (2.12.3) in order to establish the isomorphism it suffices to check the *crucial equality*

$$C_{G_3^\varphi}(\varphi(G_3^s)) = \varphi(Z_3).$$

Towards this end we calculate the irreducible constituents of φ restricted to G_3^s and describe how the corresponding irreducible submodules in Π are permuted by $\varphi(G_{12})$ and $\varphi(G_{23})$. These data are given in the following proposition whose proof will be achieved in the next two sections:

Proposition 3.2.1 *The following assertions hold:*

(i) *the restriction of φ to G_3^s is the sum of 23 pairwise distinct irreducible representations;*

(ii) *the degrees of the irreducibles are 5, $2^6 \cdot 360$, 18, 240, and $24 \cdot 2^{10}$;*

(iii) *each of the former two degrees appears once, while each of the latter three appears 7 times;*

(iv) *the irreducible G_3^s-submodules in Π of any given dimension are transitively permuted by G_3^φ.*

3.3 Restricting to G_3^s

We calculate the restriction of φ to G_3^s in three stages. First we restrict to G_2^s, then to $G_2^s \cap G_3$, and finally to G_3^s. In this way we will also see the way $\varphi(G_{23})$ permutes the $\varphi(G_3^s)$-irreducible submodules in Π. We start by restricting to G_2^s, which is rather straightforward.

Lemma 3.3.1 *The following assertions hold:*

(i) *the representations φ_{23}^2 and $\varphi_{2^6 \cdot 759}^2$ restricted to G_2^s are irreducible;*

(ii) *$\varphi_{3 \cdot 24 \cdot 2^{11}}^2$ restricted to G_2^s is the sum of three irreducibles $\varphi_{24 \cdot 2^{11}}^{(a)}$, $a \in Z_2^\#$, where the kernel of $\varphi_{24 \cdot 2^{11}}^{(a)}$ is $\langle a \rangle$;*

(iii) *$\varphi_{3 \cdot 276}^2$ restricted to G_2^s is the sum of three irreducibles $\varphi_{276}^{(a)}$, $a \in Z_2^\#$, where the kernel of $\varphi_{276}^{(a)}$ is $Q_2^{(a)}$;*

(iv) *the subgroup $\varphi(G_2)$ permutes transitively the $\varphi(G_2^s)$-irreducible submodules in Π of any given dimension.*

Proof. The result is immediate from the definitions of the irreducible constituents of φ_2 (compare either (3.1.21) or the original definitions). $\square$

Next we study the restriction to $G_2^s \cap G_3$ and to G_3^s of the representations φ_{23}^2, $\varphi_{2^6 \cdot 759}^2$, $\varphi_{24 \cdot 2^{11}}^{(a)}$, and $\varphi_{276}^{(a)}$. By (3.3.1 (iv)), it is sufficient to consider the case $a = z$, so that we have just four representations to restrict. The following lemma is an easy application of Clifford's theory and the fact that $G_2^s \cap G_3$ contains the subgroup Q_2 which is normal in G_2^s.

Lemma 3.3.2 *Let ψ be a representation of G_2^s and suppose that ψ restricted to Q_2 is the sum of pairwise distinct irreducibles indexed by a set $\mathcal{I}$ acted* *on by* $G_2/Q_2 \cong M_{24}$. *Then the irreducible constituents of* ψ *restricted to* $G_2^s \cap G_3$ *are pairwise distinct indexed by the orbits on* $\mathcal{I}$ *of the sextet stabilizer* $(G_2^s \cap G_3)/Q_2 \cong 2^6 : 3 \cdot S_6$. $\qquad\square$

Since G_3^s is a normal subgroup of index four in $G_2^s \cap G_3$, we have the following lemma (which is another easy consequence of Clifford's theory):

Lemma 3.3.3 *Let* ψ *be an irreducible representation of* $G_2^2 \cap G_3$ *and let* η *be the restriction of* ψ *to* G_3^s. *Then* η *has* k *irreducible constituents of the same degree, where* $k = 1, 2,$ *or* 4. $\qquad\square$

It is appropriate to refresh some notations related to G_3 (cf. Section 2.8). Let $\mathcal{T}$ be a 4-subset in $\mathcal{P}$ and $\mathcal{S} = \mathcal{S}(\mathcal{T})$ be the sextet in $(\mathcal{P}, \mathcal{B})$ determined by $\mathcal{T}$. Let

$$\bar{\lambda} = \sum_{p \in \mathcal{P}} 2p + 2\Lambda \text{ and } \bar{\mu} = \sum_{q \in \mathcal{T}} 4q + 2\Lambda$$

be elements of $\bar{\Lambda}$ such that $Z_3 = \chi^{-1}(\langle \bar{\lambda}, \bar{\mu} \rangle)$. For a subgroup X whose action on $\mathcal{P}$ is defined we denote by $X(\mathcal{S})$ the stabilizer of $\mathcal{S}$ in X, so that $G_2(\mathcal{S})$ contains $C_{G_2}(\bar{R}_2)$ and $G_2(\mathcal{S})/C_{G_2}(\bar{R}_2) \cong M^\delta(\mathcal{S})/R_2 \cong 2^6 : 3 \cdot S_6$. For $a \in Z_2^\#$ put

$$Q_{\mathcal{S}}^{(a)} = C_{Q^{(a)}}(Z_3) \text{ and } Q_{\mathcal{S}} = Q_{\mathcal{S}}^{(x)} Q_{\mathcal{S}}^{(y)} Q_{\mathcal{S}}^{(z)}.$$

Then $Q_{\mathcal{S}}^{(a)}/R_2$ is the unique hyperplane in $Q_2^{(a)}/R_2 \cong \mathcal{C}_{11}$ stabilized by $M^\delta(\mathcal{S})$, while

$$G_2^s \cap G_3 = Q_2 M^\delta(\mathcal{S}) \text{ and } G_3^s = Q_{\mathcal{S}} M^\delta(\mathcal{S}).$$

Lemma 3.3.4 *The restriction of* φ_{23}^2 *to either* $G_2^s \cap G_3$ *or* G_3^s *is the sum of two irreducibles of degrees 5 and 18.*

Proof. Under φ_{23}^2 the subgroup G_2^s acts by $G_2^s/Q_2 \cong M_{24}$. Hence result follows from (3.3.2) and elementary properties of the permutation module of M_{24} on $\mathcal{P}$. $\qquad\square$

Lemma 3.3.5 *The restriction of* $\varphi_{276}^{(z)}$ *to* $G_2^s \cap G_3$ *is the sum of two irreducibles of degrees 36 and 240. When further restricted to* G_3^s *the 36-constituent splits into two distinct irreducibles of degree 18 while the 240-constituent stays irreducible.*

Proof. By (3.1.17 (iv)) the kernel of φ_{276}^{12} is Q_1 and G_{12} acts under this representation via its factor group $F = EM \cong \mathcal{C}_{11} : M_{24}$. Furthermore, the action is induced from the linear representation of $F(\pi)$ with kernel $E_\pi M(\pi)$

where π is a pair. Let E_S be the hyperplane in E stabilized by $M(S) \cong 2^6$: $3 \cdot S_6$. Then $G_2^s \cap G_3$ acts under φ_{276}^{12} via $EM(S)$, while G_3^s acts via $E_S M(S)$. Since $M(S)$ acts on the pairs with two orbits of lengths 36 (the pairs inside tetrads) and 240 (the pairs across the tetrads), the irreducibles of $G_2^s \cap G_3$ can be deduced from (3.3.2). To calculate the irreducibles of G_3^s notice that whenever π_1 and π_2 are pairs, the equality

$$E_{\pi_1} + E_{\pi_2} = E_S$$

holds if and only if π_1 and π_2 partitions of a tetrad from S. This shows that the irreducibles of E_S inside the 240-constituent of $G_2^s \cap G_3$ are pairwise distinct and hence this constituent stays irreducible under G_3^s. On the other hand, the E_S-eigenspaces in the 36-submodule are 2-dimensional indexed by the partitions of the tetrads into pairs. Therefore, the 36-constituent restricted to G_3^s is induced from a 2-dimensional representation η of the stabilizer in $E_S M(S)$ of such a partition. The acting group of η is an elementary abelian of order 4 and hence η is the direct sum of two distinct linear representations and the result follows. $\qquad\square$

Lemma 3.3.6 *The representation* $\varphi_{24 \cdot 2^{11}}^{(z)}$ *stays irreducible under its restriction to* $G_2^s \cap G_3$, *although being restricted to* G_3^s *the representation splits into two distinct irreducibles of equal degrees.*

Proof. It can be seen from the consideration of $\varphi_{24 \cdot 2^{11}}^{(z)}$ as a constituent of $\varphi_{|\bar{\Lambda}_2|}^1$ restricted to G_2^s that

$$(Q_1 \cap G_2^s)/Z_1 = Q_2^{(z)}/Z_1 \cong \bar{\Lambda}^{(23)} = \bar{\lambda}^\perp$$

acts under $\varphi_{24 \cdot 2^{11}}^{(z)}$ with 1-dimensional eigenspaces indexed by the elements of $\bar{\Lambda}_2^3$, so that $\bar{\lambda}^\perp \cap \bar{\lambda}_2^\perp$ (where $\bar{\lambda}_2 = 3p - \sum_{r \in \mathcal{P} \setminus \{p\}} r$) is one of the kernels.
The group

$$G_2^s/Q_2^{(z)} \cong G_{12}/Q_1 \cong F = EM \cong C_{11} : M_{24}$$

permutes these eigenspaces as it does the elements of $\bar{\Lambda}_2^3$, that is acting on the cosets of

$$M(p) \cong M_{23}.$$

Since the sextet stabilizer acts transitively on $\mathcal{P}$, the irreducibility under the restriction to $G_2^s \cap G_3$ follows from (3.3.2). Turning to the restriction to G_3^s notice that

$$(Q_1 \cap G_3^s) = \bar{\lambda}^\perp \cap \bar{\mu}^\perp.$$

We claim that the kernels of the linear representations of $(Q_1 \cap G_3^s)$ involved in $\varphi_{24 \cdot 2^{11}}^{(z)}$ are still pairwise distinct. The claim is immediate from the fact that $\bar{\lambda}_2$ is the only element of $\bar{\Lambda}_2^3$ contained in $\langle \bar{\lambda}, \bar{\mu}, \bar{\lambda}_2 \rangle$. On the other hand

$$G_3^s/(Q_1 \cap G_3^s) \cong E_{\mathcal{S}} : M(\mathcal{S})$$

is easily seen to act on $\bar{\Lambda}_2^3$ intransitively with two orbits of equal lengths. Therefore, under G_3^s we have two distinct irreducibles of equal degrees. □

In order to restrict $\varphi_{2^6 \cdot 759}^2$ we require the following:

Lemma 3.3.7 *Let* $\mathcal{O}$ *be an octad, let* $\mathcal{S}$ *be a sextet with tetrads* w_i, $1 \leq i \leq 6$, *let*

$$v = \max_{1 \leq i \leq 6} |\mathcal{O} \cap w_i|,$$

let $M(\mathcal{O})$ *and* $M(\mathcal{S})$ *be the stabilizers of* $\mathcal{O}$ *and* $\mathcal{S}$ *in* $M \cong M_{24}$, *and let* $K_{\mathcal{O}} = O_2(M(\mathcal{O})) \cong 2^4$. *Let* $E_{\mathcal{S}}$ *be the unique hyperplane in* $E \cong \mathcal{C}_{11}$ *stabilized by* $M(\mathcal{S})$, *and let* $E_{\mathcal{O}} = [E, K_{\mathcal{O}}] \cong 2^5$. *Then the following assertions hold:*

(i) $v \in \{4, 3, 2\}$ *and* v *determines the* M-*orbit containing the pair* $(\mathcal{O}, \mathcal{S})$;
(ii) *if* $v = 4$, *then* $[M(\mathcal{S}) : M(\mathcal{S}) \cap M(\mathcal{O})] = 15$ *and* $E_{\mathcal{O}} \leq E_{\mathcal{S}}$;
(iii) *if* $v = 3$, *then* $[M(\mathcal{S}) : M(\mathcal{S}) \cap M(\mathcal{O})] = 384$ *and* $E_{\mathcal{O}} \nleq E_{\mathcal{S}}$;
(iv) *if* $v = 2$, *then* $[M(\mathcal{S}) : M(\mathcal{S}) \cap M(\mathcal{O})] = 360$ *and* $E_{\mathcal{O}} \nleq E_{\mathcal{S}}$.

Proof. The assertion (i) and the length of the M-orbit of $(\mathcal{O}, \mathcal{S})$ as a function of v are well known (cf. Lemma 2.14.1 in [Iv99]). The hyperplane $E_{\mathcal{S}}$ is formed by the images in E of the Golay sets evenly intersecting every tetrad in $\mathcal{S}$, while $E_{\mathcal{O}}$ is formed by the images of the Golay sets disjoint from $\mathcal{O}$. These Golay sets include $\emptyset$, $\mathcal{P} \setminus \mathcal{B}$ and the cosets of the hyperplanes in the 4-dimensional $GF(2)$-space θ defined on $\mathcal{P} \setminus \mathcal{O}$. If $v = 4$, then we can assume that $\mathcal{O} = w_1 \cup w_2$ in which case the tetrads w_i for $3 \leq i \leq 6$ are the cosets of a 2-subspace t in θ. Since t intersects evenly every hyperplane in θ, in this case $E_{\mathcal{O}}$ is indeed contained in $E_{\mathcal{S}}$. To show that $E_{\mathcal{O}}$ is not contained in $E_{\mathcal{S}}$ when $v = 3$ or 2 we apply the following construction. Let $\mathcal{O}_1$ be an octad disjoint from $\mathcal{O}$ (so that $\mathcal{O}_1$ is a hyperplane in θ). Let u be a $(4 - v)$-subset in $\mathcal{P} \setminus \mathcal{O}$ intersecting $\mathcal{O}_1$ in a single element, let v be a v-subset in $\mathcal{O}$, let $w = u \cup v$, and let $\mathcal{S}$ be the sextet determined by w. Then by the construction, the image of $\mathcal{O}_1$ in E belongs to $E_{\mathcal{O}}$ but not to $E_{\mathcal{S}}$. □

Lemma 3.3.8 *The following assertions hold:*

(i) *the representation* $\varphi_{2^6 \cdot 759}^2$ *restricted to* $G_2^s \cap G_3$ *is the sum of three irreducibles of degrees* $2^6 \cdot 15 = 960$, $2^6 \cdot 360$ *and* $2^6 \cdot 384 = 24 \cdot 2^{10}$;

(ii) *the* $(2^6 \cdot 360)$*- and* $(2^6 \cdot 384)$*-constituents restricted to* G_3^s *stay irreducible;*

(iii) *the* $(2^6 \cdot 15)$*-constituents when restricted to* G_3^s *does one of the following:*

    (a) *stays irreducible of degree* 960;

    (b) *splits into two irreducibles of degree* 480;

    (c) *splits into four irreducibles of degree* 240.

Proof. It follows from the definition of $\varphi^2_{2^6 \cdot 759}$ (cf. Subsection 3.1.3) that its restriction to Q_2 is the sum of 759 irreducibles $\eta_{\mathcal{O}}$ of degree 2^6, each indexed by the octads in $\mathcal{B}$. These irreducibles are pairwise distinct (as they have different kernels) and $G_2^s/Q_2 \cong M_{24}$ permutes them as it permutes the octads. The assertion (i) follows from (3.3.2) together with the description the $M(\mathcal{S})$-orbits on $\mathcal{B}$ given in (3.3.7). If $C_{\mathcal{O}} \cap Q_2$ is the kernel of $\eta_{\mathcal{O}}$ and $a \in Z_2^\#$, then by (3.1.12 (ii)) $(C_{\mathcal{O}} R_2 \cap Q_2^{(a)})/R_2$ is the 5-dimensional subspace $E_{\mathcal{O}}$ in $Q_2^{(a)}/R_2 = E \cong C_{11}$. On the other hand, $(G_3^s \cap Q_2^{(a)})/R_2$ is the hyperplane $E_{\mathcal{S}}$ in E. Suppose first that $\nu = 3$ or 2. Then by (3.3.7 (iii), (iv)), $E_{\mathcal{O}}$ is not contained in $E_{\mathcal{S}}$ and hence for every $a \in Z_2^\#$ we have the equality

$$(G_3^s \cap Q_2^{(a)})/(C_{\mathcal{O}} \cap Q_2^{(a)}) = Q_2^{(a)}/(C_{\mathcal{O}} \cap Q_2^{(a)}).$$

Therefore, for $\nu = 3$ and 2 the acting group of $\eta_{\mathcal{O}}$ does not change after restricting to $G_3^s \cap Q_2$; in particular, the restricted representation is still irreducible and (ii) follows. If $\nu = 4$, then the situation is different (compare (3.3.7 (ii))); the above argument does not work and in (iii) we just list the possibilities given in (3.3.3). It should be possible to stretch the efforts here and to show that in fact the possibility (iii) (c) occurs. But in the next section this will be deduced with no effort at all. $\qquad\square$

3.4 Permuting the $\varphi(G_3^s)$-irreducibles

For an irreducible constituent φ_l^γ of the restriction of φ to a group H (which will be clear from the context) by Π_l^γ we denote a $\varphi(H)$-irreducible submodule in Π on which the constituent φ_l^γ is realized (in all the cases to be considered the submodule will be uniquely determined by the degree l together with the superscript γ).

Lemma 3.4.1 *Let* Π_l^2 *be a* $\varphi(G_2)$*-irreducible submodule in* Π *and let* $g \in G_1 \setminus G_{12}$. *Then* Π_l^2 *is not normalized by* $\varphi(g)$.

Proof. The subgroup G_{12} is maximal in G_1 (since G_{12} contains Q_1 and $G_{12}/Q_1 \cong C_{11} : M_{24}$ is maximal in Co_1 [CCNPW]). Hence the result follows from the irreducibility the image of φ. □

Lemma 3.4.2 *The following assertions hold:*

(i) *for every $\varphi(G_2)$-irreducible submodule Π_l^2, the subgroup $\varphi(G_{23})$ permutes transitively the $\varphi(G_3^s)$-irreducibles of any given dimension contained in Π_l^2;*

(ii) *if $g \in G_{13} \setminus G_{123}$, then $\varphi(g)$ maps a $(24 \cdot 2^{10})$-dimensional irreducible submodule from $\Pi_{3 \cdot 24 \cdot 2^{11}}^2$ onto such a submodule contained in $\Pi_{2^6 \cdot 759}^2$;*

(iii) *if $g \in G_{13} \setminus G_{123}$, then $\varphi(g)$ maps a 240-dimensional irreducible submodule from $\Pi_{3 \cdot 276}^2$ onto such a submodule contained in $\Pi_{2^6 \cdot 759}^2$;*

(iv) *(3.2.1) holds.*

Proof. By (3.3.4), (3.3.5), (3.3.6), (3.3.8), and their proofs the group $\varphi(G_2^s \cap G_3)$ permutes transitively the $\varphi(G_3^s)$-irreducibles of any given dimension inside every irreducible submodule of $\varphi(G_2^s)$ in Π (except possibly for $\Pi_{2^6 \cdot 759}^2$). Furthermore, $\varphi(G_2)$ permutes transitively $\Pi_{276}^{(a)}$'s as well as $\Pi_{24 \cdot 2^{11}}^{(a)}$'s. Since $G_2 = G_2^s G_{23}$, (i) follows (except for the case $l = 2^6 \cdot 759$ to be proved later). If $g \in G_{13} \setminus G_{123}$, then $\varphi(g)$ normalizes $\varphi(G_3^s)$ and it maps every $\varphi(G_3^s)$-irreducible submodule in Π onto a similar submodule preserving the dimension. On the other hand, by (3.4.1) in every Π_l^2 there is at least one $\varphi(G_3^s)$-irreducible submodule which is mapped by $\varphi(g)$ onto a submodule outside Π_l^2. In order to establish (ii) and (iii) it suffices to compare the dimensions (notice that $\varphi(g)$ must map the 18-dimensional irreducible contained in Π_{23}^2 onto a similar submodule in Π_{276}^2). Thus (3.3.7 (iii) (c)) must hold and φ restricted to G_3^s has seven irreducible constituents of degree 240. Three of them involved in $\varphi_{3 \cdot 276}^2$ are easily seen to have pairwise different kernels. Hence all the seven of the 240-constituents must be pairwise distinct (since seven is a prime number). In particular, those four from $\varphi_{2^6 \cdot 759}^2$ are distinct and (i) holds for $l = 2^6 \cdot 759$ as well. Thus the proof of (3.2.1) is completed and (iii) is established. □

Now we are ready to describe the irreducible constituents of the representation of G_3^φ in $GL(\Pi)$ (cf. (3.4.3) below).

The diagram below shows the fusions patten between the irreducible constituents of the representations φ_2 and φ_3 and their restrictions to G_{23}.

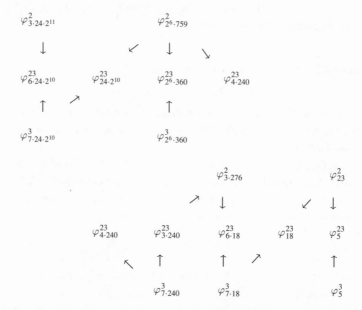

Lemma 3.4.3 *Let φ_3 be the natural representation of*

$$G_3^\varphi = \langle \varphi(G_{13}), \varphi(G_{23}) \rangle_{GL(\Pi)}$$

in $GL(\Pi)$. Then:

(i) *φ_3 is the sum of five irreducibles φ_l^3 for $l = 7 \cdot 24 \cdot 2^{10}$, $2^6 \cdot 360$, $7 \cdot 240$,*
    $7 \cdot 18$, 5 and their restrictions to G_3^s have kernels 1, Z_3, R_3, T_3, and
    $O_{2,3}(G_3^s)$, respectively;
(ii) *if $l = 2^6 \cdot 360$ or 5, then φ_l^3 stays irreducible when restricted to G_3^s;*
(iii) *if $l = 7 \cdot k$ where $k = 24 \cdot 2^{10}$, 240, or 18, then φ_l^3 restricted to G_3^s is the*
     sum of seven distinct irreducibles of degree k;
(iv) *G_3^φ induces on the set of irreducible constituents of the restriction of φ to*
     G_3^s an action of $GL(Z_3) \cong L_3(2)$, furthermore:
     (a) *G_3^φ permutes the constituents in $\varphi_{7 \cdot 240}^3$ as 1-subspaces in Z_3;*
     (b) *G_3^φ permutes the constituents in $\varphi_{7 \cdot k}^3$ for $k = 24 \cdot 2^{10}$ and 18 as*
         2-subspaces in Z_3.

Proof. The degrees of the irreducibilities in (i) as well as (ii) and (iii) follow from (3.4.2) and its proof, while the kernels in (i) are easy to calculate from those of φ_n^2s considering intersections and/or normal closures. Since the amalgams $\{\varphi(G_{13}), \varphi(G_{23})\}$ and $\{G_{13}, G_{23}\}$ are isomorphic

$$G_3^\varphi/\varphi(G_3^s)C_{G_3^\varphi}(\varphi(G_3^s)) \cong GL(Z_3) \cong L_3(2)$$

by (2.11.3). The latter group possesses only two transitive actions of degree 7 which are the actions on 1- and 2-subspaces in Z_3 and hence (iii) follows. $\square$

3.5 G_3^φ is isomorphic to G_3

It has been already mentioned that $G_3^\varphi/\varphi(G_3^s)C_{G_3^\varphi}(\varphi(G_3^s))$ is isomorphic to $\bar{G}_3/\bar{G}_3^s \cong L_3(2)$ by (2.11.3). Therefore, in order to achieve the objection of the section it suffices to check the crucial equality

$$C_{G_3^\varphi}(\varphi(G_3^s)) = \varphi(Z_3).$$

The equality will be deduced from the fact that $C_{G_3^\varphi}(G_3^s)$ does not involve the Steinberg module of $L_3(2)$. The argument runs as follows.

Let C_E be the centralizer of $\varphi(G_3^s)$ in the endomorphism ring of Π and let C_I be the centralizer of $\varphi(G_3^s)$ in $GL(\Pi)$ (so that C_I consists of the invertible elements of C_E). By (3.2.1) and Schur's lemma, C_E is 23-dimensional naturally identified with the space of $\mathbb{R}$-valued functions on the set of irreducible constituents of $\varphi(G_3^s)$ and C_I consists of such functions with the full support (of size 23). The subgroup $\varphi(Z_3)$ is certainly contained in C_I and it consists of the (± 1)-valued functions whose (-1)-values are on a complement of a line in the set of $\varphi(G_3^s)$-irreducible constituents inside $\varphi_{7\cdot 24\cdot 2^{10}}^3$ (this set is identified with the set of non-zero vectors in the dual of Z_3 via (3.4.3 (iv) (b))).

The group $G_3^\varphi/\varphi(G_3^\varphi)$ is a faithful completion of the amalgam

$$\{\varphi(G_{13})/\varphi(G_3^s), \varphi(G_{23})/\varphi(G_3^s)\}$$

of maximal parabolics in $L := GL(Z_3)$ and it possesses a homomorphism onto L whose kernel

$$C_{G_3^\varphi}(G_3^s)/\varphi(Z_3) \leq C_I/\varphi(Z_3)$$

is an abelian group. Arguing as in Section 2.11 we conclude that there is an L-homomorphism

$$\nu : H_1(\Theta) \to C_I/\varphi(Z_3)$$

whose image is $C_{G_3^\varphi}(G_3^s)/\varphi(Z_3)$. Here $H_1(\Theta)$ is the homology group of the incidence graph Θ of the projective plane of Z_3. The crucial equality is a consequence of the following:

Lemma 3.5.1 *Every L-homomorphism* $\nu : H_1(\Theta) \to C_I/\varphi(Z_3)$ *has a trivial image.*

Proof. The homology group $H_1(\Theta)$ is spanned by a set of 28 infinite cyclic subgroups indexed by the minimal (length 6 undirected) cycles in Θ. Let Σ be such a cycle, let $I(\Sigma)$ be the corresponding infinite cyclic subgroup in $H_1(\Theta)$, let $N(\Sigma)$ be the setwise stabilizer of Σ in $L \cong L_3(2)$, and let $O(\Sigma)$ be the subgroup in $N(\Sigma)$ preserving an orientation of Σ). Then $N(\Sigma) \cong D_6$, $O(\Sigma) \cong 3$ and by the definition $O(\Sigma)$ centralizes $I(\Sigma)$ while every element from $N(\Sigma) \backslash O(\Sigma)$ inverts $I(\Sigma)$. It follows from the description of C_E as a sum of certain permutation modules of L that whenever an element of $C_I/\varphi(Z_3)$ is centralized by $O(\Sigma)$ it is centralized by the whole of $N(\Sigma)$ (due to the fact that $O(\Sigma)$ and $N(\Sigma)$ have the same orbits on the set of subspaces in Z_3). This means that the image of $I(\Sigma)$ under ν is inverted and centralized by every element from $N(\Sigma) \backslash O(\Sigma)$ simultaneously. Hence $\mathrm{Im}(\nu)$ is of (multiplicative) exponent 2 and it is an L-submodule in the largest $GF(2)$-submodule C_B in C_I. The structure of C_E as a permutation module immediately shows that C_B does not involve the (8-dimensional irreducible) Steinberg module which implies the triviality of the image of ν by (2.11.2) and its proof. $\square$

Thus G_3^φ and G_3 are indeed isomorphic and we have proved the principal result of the chapter.

Proposition 3.5.2 *There exists a faithful completion*

$$\varphi : \mathcal{M} \to GL(\Pi)$$

where $\mathcal{M}$ *is the Monster amalgam and* Π *is a real vector space of dimension* 196 883. *The restriction* φ_i *of* φ *to* G_i *for* $i = 1, 2,$ *and 3 are as described in* (3.1.21) *and* (3.4.3). $\square$

4

2-local geometries

Let G be the Monster group which, according to our definition, is the universal completion of the unique Monster amalgam $\mathcal{M}$. From the previous chapter we know that G is a faithful completion of $\mathcal{M}$ and that it possesses a 196 883-dimensional complex representation. In what follows $\mathcal{M}$ will be identified with the corresponding subamalgam in G, so that G_1, G_2, and G_3 are subgroups in G. In this chapter we start enlarging $\mathcal{M}$ by adjoining subgroups which are universal completions of the subamalgams formed by their intersections with the subamalgam already available. The well-known simple connectedness results for the geometries of spherical buildings will enable us to reconstruct all the maximal constrained 2-local subgroups of the Monster along with the subgroup $2^2 \cdot (^2E_6(2)) : S_3$. After a retour to the 196 883-dimensional G-module Π in the next chapter we continue the enlarging process making use of more recent simple connectedness results for geometries of the sporadic groups. This allows us to bring about the subgroups related to the Fischer groups culminating in reconstructing the non-split central extension of the Baby Monster group.

4.1 Singular subgroups

It seems appropriate to start by giving the names to a pair of conjugacy classes of involutions in G consistent with the standard notation for the Monster involutions adopted in [CCNPW]. Let t be an involution in Q_1 such that $\chi(t) \in \bar{\Lambda}_2$. The involutions in G conjugate to t will be called $2A$-*involutions*, while those conjugate to the generator z of $Z_1 = Z(G_1)$ will be called $2B$-*involutions*.

Lemma 4.1.1 $2A$- *and* $2B$-*involutions are not conjugate in* G.

Proof. It is sufficient to show that

$$\text{tr}(\varphi(z)) \neq \text{tr}(\varphi(t)),$$

where $\varphi : G \to GL(\Pi)$ is the representation constructed in Chapter 3. Since z is a central involution of G_1 contained in Q_1, it is immediate from (3.1.21) that

$$\text{tr}(\varphi(z)) = \text{tr}(\varphi^1_{24 \cdot 2^{12}}(z)) + \text{tr}(\varphi^1_{|\bar{\Lambda}_2|}(z)) + \text{tr}(\varphi^1_{299}(z))$$

$$= -24 \cdot 2^{12} + |\bar{\Lambda}_2| + 299 = 275.$$

On the other hand, $\text{tr}(\varphi^1_{24 \cdot 2^{12}}(t)) = 0$ since both z and tz are $2A$-involutions whose product is a $2B$-involution. It follows from the definition of $\varphi^1_{|\bar{\Lambda}_2|}$ (compare (3.1.7)) that t centralizes $\Pi(\bar{\mu})$ if $\chi(t) \in \bar{\mu}^\perp$ and negates $\Pi(\bar{\mu})$ otherwise. Since both $\bar{\mu}$ and $\chi(t)$ are elements from $\bar{\Lambda}_2$, some standard calculations in the Leech lattice taken modulo 2 show that

$$\text{tr}(\varphi^1_{|\bar{\Lambda}_2|}(t)) = 4072$$

(cf. Section 4.11 in [Iv99]). Finally, since t acts on Π_{299} by the identity operator, $\text{tr}(\varphi(t))$ is 4371 (which is a very important number in the Monster numerology). $\square$

The subgroups Z_1, Z_2, and Z_3 are $2B$-pure in the sense that all their nonidentity elements are $2B$-involutions. We are going to construct a further triple of $2B$-pure subgroups and to identify the subgroups in G generated by their respective normalizers in G_1, G_2, and G_3.

As above, $(\mathcal{P}, \mathcal{B})$ denotes the Steiner system of type $S(5, 8, 24)$ on which G_2 induces the natural action of $M \cong M_{24}$ with kernel $C_{G_2}(\bar{R}_2)$ and $\mathcal{T}$ denotes the 4-subset in $\mathcal{P}$ such that G_{23} is the stabilizer in G_2 of the sextet in which $\mathcal{T}$ is one of the tetrads. Let

$$\rho : \mathcal{C}^*_{11} \to \bar{R}_2$$

be the M-isomorphism of modules whose existence is guaranteed by (2.3.1 (vii)) and recall that M acts on the non-zero elements in $\mathcal{C}^*_{11}$ with two orbits: 276 pairs and 1771 sextets.

Lemma 4.1.2 *For a non-zero element w^* in $\mathcal{C}^*_{11}$, let $Z(w^*)$ denote the preimage in R_2 of the subgroup of order 2 in $\bar{R}_2$ generated by $\rho(w^*)$. Then either:*

(i) *w^* is a sextet, $Z(w^*)$ is a G_2-conjugate of Z_3, in particular $Z(w^*)$ is $2B$-pure, or*

(ii) *w^* is a pair and $Z(w^*)$ contains $2A$-involutions.*

Proof. The assertion (i) follows from the definition of Z_3 (in fact Z_3 is precisely $Z(v^*)$), while (ii) follows from (1.8.3), (1.8.7), and the defining property of $2A$-involutions in Q_1. □

By the above lemma, the image in $\bar{R}_2$ of a $2B$-pure subgroup containing Z_2 and contained in R_2 is a sextet-pure subspace in C_{11}^*. An important class of such subspaces is characterized by Lemma 3.3.8 in [Iv99] restated below. Notice, that whenever S_1 and S_2 are sextets refining the partition

$$\mathcal{P} = \mathcal{O} \cup (\mathcal{P} \setminus \mathcal{O})$$

for an octad $\mathcal{O}$, the sum of S_1 and S_2 in C_{11}^* is again a sextet if and only if every tetrad in S_1 intersects evenly with every tetrad in S_2.

Lemma 4.1.3 *Let U be a proper sextet-pure subspace in C_{11}^* such that all the sextets in $U^\#$ refine the partition of $\mathcal{P}$ into an octad $\mathcal{O}$ and its complement. Then one of the following holds:*

(i) $\dim(U) = 1$;
(ii) $\dim(U) = 3$ and $U^\#$ is formed by the sextets refining a trio containing $\mathcal{O}$;
(iii) $\dim(U) = 3$ and $U^\#$ is formed by
    the sextets S such that $\tau \in O_2(M(S))$ for an involution $\tau \in O_2(M(\mathcal{O}))$;
(iv) $\dim(U) = 2$ and
    U is the intersection of a subspace from (ii) and a subspace from (iii);

In any event, U is contained in the subspace $C_{C_{11}^*}(O_2(M(\mathcal{O})))$ (of dimension 6) and $N_M(U)/C_M(U) \cong GL(U)$. A subspace in (iv) is contained in the three subspace from (iii) and in a unique subspace from (ii). □

It is implicit in the above lemma that the subspaces in each of (i) to (iv) form a single M-orbit. The stabilizers in M of a clique from (ii) is the trio stabilizer $2^6 : (L_3(2) \times S_3)$ while the stabilizer of a clique from (iii) is the involution centralizer $2_+^{1+6} : L_3(2)$.

Let Z_4, $Z_5^{(t)}$, and $Z_5^{(l)}$ be the preimages in R_2 of the subgroups $\rho(U)$ for sextet-pure subspaces corresponding to the cases (iv), (ii), and (iii) in (4.1.3), respectively. If the octad $\mathcal{O}$ contains the tetrad T defining Z_3, then $Z_3 < Z_4$ and we can further assume that

$$Z_4 = Z_5^{(t)} \cap Z_5^{(l)}.$$

Lemma 4.1.4 *Let Y be one of the subgroups Z_4, $Z_5^{(t)}$, and $Z_5^{(l)}$. Then:*

(i) *Y is $2B$-pure of order 2^4, 2^5, and 2^5, respectively;*
(ii) *$G_i = G_i^s N_{G_i}(Y)$ for $i = 2$ and 3.*

Proof. The assertion (i) follows from (4.1.2). Since G_2 acts on the set $Z_2 \times \bar{R}_2$ as the direct product of $GL(Z_2)$ and M_{24}, the assertion (ii) for $i = 2$ follows. In order to prove this assertion for $i = 3$ notice that Y is contained in R_3. This can be seen by comparing of the remark after (4.1.3) and the definition of R_3 in the paragraph prior to (2.8.3). Next, Y is clearly the full preimage in R_3 of $Y/Z_3 < \bar{R}_3$, while G_3 induces on $Z_3 \times \bar{R}_3$ the direct product $GL(Z_3) \times 3 \cdot S_6$. $\square$

For $Y = Z_4$, $Z_5^{(t)}$, and $Z_5^{(l)}$, define $G(Y)$ to be the subgroup in G generated by the normalizers $N_{G_i}(Y)$ taken for $i = 1, 2, 3$; put

$$G_4 = G(Z_4), \ G_5^{(t)} = G(Z_5^{(t)}) \text{ and } G_5^{(l)} = G(Z_5^{(l)}),$$

and let $C(Y)$ denote $C_{G_{123}}(Y)$.

Lemma 4.1.5 *In terms introduced in the paragraph prior to the lemma the following assertions hold for each of the three choices for Y and for every $i \in \{1, 2, 3\}$:*

(i) $C_{G_i}(Y) = C(Y)$ *and* $N_{G_i}(Y)/C(Y)$ *is the stabilizer of Z_i in $GL(Y)$;*
(ii) $C_{G(Y)}(Y) = C(Y)$ *and* $G(Y)/C(Y) \cong GL(Y)$;
(iii) $G(Y) \cap G_i = N_{G_i}(Y) = N_{G(Y)}(Z_i)$.

Proof. Since $Z_1 < Z_2 < Z_3$, the first assertion in (i) is immediate. The second assertion we first justify for $i = 1$. Since Q_1 is extraspecial, it induces on Y the group generated by the transvections with centre z. It follows from the description of the maximal subgroups in $Co_1 \cong G_1/Q_1$ (cf. [CCNPW] and referencies therein) that $N_{G_1}(Z_5^{(t)})$ is the maximal subgroup of the shape

$$2^{2+12} : (A_8 \times S_3)$$

which induces on the 4-dimensional $GF(2)$-space $Z_5^{(t)}/Z_1$ its full automorphism group $L_4(2) \cong A_8$. Therefore, the assertion holds for $Y = Z_5^{(t)}$. Since $Z_4 < Z_5^{(t)}$, it holds for $Y = Z_4$ as well. It is easy to deduce from the last assertion in (4.1.3) together with (4.1.4 (ii)) that $N_{G_1}(Z_5^{(l)})$ does not stabilize Z_4 and the proof of (i) is completed for $i = 1$. Now in order to extend (i) for the cases $i = 2$ and $i = 3$ it is sufficient to apply (4.1.4 (ii)). Put

$$\mathcal{A}(Y) = \{N_{G_i}(Y)/C(Y) \mid 1 \le i < \log_2(|Y|)\}.$$

Then by (i) and induction on the order of Y we observe that $\mathcal{A}(Y)$ is the amalgam of maximal parabolic subgroups in $GL(Y)$ associated with the natural projective geometry of Y. Since the projective geometry is simply connected, $G(Y)/C(Y)$ is the universal and the only completion of $\mathcal{A}(Y)$ isomorphic to $GL(Y)$, and (ii) holds. The assertion (iii) is a direct consequence of (i) and (ii). $\square$

The subgroups Z_1, Z_2, Z_3, Z_4, $Z_5^{(t)}$, and $Z_5^{(l)}$ are *singular* in the sense that if Y is one of them, then Y is contained in $Q_1^{(u)}$ for every $u \in Y$ where $Q_1^{(u)}$ denotes Q_1 conjugated by an element from $N_G(Y)$ which conjugates z onto u. This terminology was used in [MSh02] with attribution of its introducing to J. Thompson. When knowing that G_1 is the centralizer of z in G, we can redefine $Q_1^{(u)}$ simply as $O_2(C_G(u))$. The above six subgroups in fact represent up to conjugation all the singular subgroups in the Monster. The subgroups G_1, G_2, G_3, and $G_5^{(t)}$ are maximal 2-locals in the Monster and

$$G_5^{(t)} \sim 2^{5+10+20}.(S_3 \times L_5(2)).$$

The subgroup G_4 is the normalizer of Z_4 in $G_5^{(t)}$ (having index 31), while $G_5^{(l)}$ is contained in a $2^{10+16}.\Omega_{10}^+(2)$-subgroup to be constructed in Section 4.3.

4.2 Tilde geometry

Define $\mathcal{G}(G)$ to be the coset geometry of G associated with the subamalgam

$$\mathcal{A}(G) = \{G_i \mid 1 \leq i \leq 5\},$$

in G, where G_5 stays for $G_5^{(t)}$. This means that the elements of type i in $\mathcal{G}(G)$ are the right cosets of G_i and two elements of type i and j are incident if their intersection is a coset of $G_i \cap G_j$. The geometry $\mathcal{G}(G)$ possesses the diagram

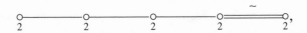

that is to say, $\mathcal{G}(G)$ is a *tilde geometry* of rank 5. The rightmost edge on the diagram symbolizes the triple cover of the generalized quadrangle of order $(2, 2)$ (the automorphism group of the triple cover is the hexacode group $3 \cdot S_6$). The group G acting on $\mathcal{G}(G)$ by right translations induces a flag-transitive action (which in fact is the full automorphism group of the geometry).

The classification of the flag-transitive tilde geometries are given in [Iv99] and [ISh02]. Besides $\mathcal{G}(G)$ there is only one further flag-transitive tilde geometry of rank 5, whose automorphism group is the non-split extension $3^{35} \cdot Sp_{10}(2)$. Within the simple connectedness proof for the tilde geometry of the Monster in [Iv99], we referred to certain properties of a graph associated with the Monster and some of these properties are still unpublished. One of the purposes of the present volume is to present a self-contained treatment of Monster and of its tilde geometry.

The Monster amalgam $\mathcal{M}$ is formed by the parabolics of the truncation of $\mathcal{G}(G)$ by the elements of types 4 and 5. The content of the previous section amounts to reconstruction of the complete set of maximal parabolics from the truncation. An alternative reconstruction in terms of the collinearity graph of $\mathcal{G}(G)$ (the points and lines are the cosets of G_1 and G_2 respectively) can be found in Section 5.2 in [Iv99].

4.3 $2^{10+16} . \Omega^+_{10}(2)$-subgroup

In this section we locate in G an elementary abelian subgroup Z_{10} of order 2^{10}. By declaring the $2B$- and $2A$-involutions in Z_{10} to be singular and non-singular, respectively, we define on Z_{10} a non-singular orthogonal form κ, whose Witt index is 5 (that is κ is of plus type). The subgroups $Z_5^{(t)}$ and $Z_5^{(l)}$ are maximal totally isotropic subspaces in Z_{10} with respect to κ (it goes without saying that κ is preserved by the normalizer of Z_{10} in G.) If we put

$$\mathcal{Y} = \{G_1, G_2, G_3, G_5^{(t)}, G_5^{(l)}\},$$

then $C_Y(Z_{10})$ is independent of the choice of $Y \in \mathcal{Y}$. This independent centralizer (contained in G_{123} and denoted below by Q_{10}) will be proved to be a special group of the shape 2^{10+16}. The amalgam

$$\mathcal{A}(G_{10}) = \{N_Y(Z_{10})/Q_{10} \mid Y \in \mathcal{Y}\}$$

is that of maximal parabolics in the simple commutator subgroup of the automorphism group of the orthogonal space (Z_{10}, κ). Since the corresponding orthogonal polar space is simply connected, we conclude that the subgroup G_{10} in G generated by the normalizers $N_Y(Z_{10})$ taken for all $Y \in \mathcal{Y}$ satisfies $G_{10}/G_{10}^s \cong \Omega^+_{10}(2)$. It is a classical and well-known fact (cf. Proposition 12.49 in [Pa94]) that the latter group is the universal and the only faithful completion of the amalgam $\mathcal{A}(G_{10})$. This gives us the desired 2-local subgroup

$$G_{10} \sim 2^{10+16} . \Omega^+_{10}(2).$$

We start by reviewing some known facts about the centralizer in Co_1 of a central involution and its action on $\bar{\Lambda}$. Let σ be a involution of $G_1/Q_1 \cong Co_1$ contained in $E = O_2(F)$, where

$$F = G_{12}/Q_1 \cong \mathcal{C}_{11} : M_{24}$$

is the monomial subgroup, and suppose that σ corresponds to an octad $\mathcal{O}$ of $(\mathcal{P}, \mathcal{B})$. This means that σ is contained in the 759-orbit of $M \cong M_{24}$ on $E \cong$

C_{11}. The following lemma is a standard result from the Conway–Leech theory (cf. Lemma 4.8.6 in [Iv99]).

Lemma 4.3.1 *In the above terms, the following assertions hold, where $D = C_{G_1/Q_1}(\sigma)$ and $\widehat{D} = D/O_2(D)$:*

(i) $\widehat{D} \cong \Omega_8^+(2)$, *$D$ is 2-constrained, $O_2(D) \cong 2_+^{1+8}$, and D does not split over $O_2(D)$;*

(ii) *$\bar{\Lambda}$ possesses a unique D-composition series*

$$1 < [\sigma, \bar{\Lambda}] < C_{\bar{\Lambda}}(\sigma) < \bar{\Lambda};$$

(iii) *$O_2(D)/\langle\sigma\rangle$, $[\sigma, \bar{\Lambda}]$, and $C_{\bar{\Lambda}}(\sigma)/[\sigma, \bar{\Lambda}]$ are pairwise non-isomorphic self-dual faithful 8-dimensional $GF(2)$-modules for $\widehat{D} \cong \Omega_8^+(2)$ (they are permuted by the triality outer automorphism of $\widehat{D}$);*

(iv) *$[\sigma, \bar{\Lambda}]$ and $\bar{\Lambda}/C_{\bar{\Lambda}}(\sigma)$ are isomorphic $\widehat{D}$-modules;*

(iv) *if $U = \chi^{-1}([\sigma, \bar{\Lambda}])$, then U is an elementary abelian (of order 2^9) and the image in $\mathrm{Aut}(U)$ of the preimage of D in G_1 is an extension of $\bar{\Lambda}/C_{\bar{\Lambda}}(\sigma) \cong 2^8$ by $\widehat{D} \cong \Omega_8^+(2)$.* □

The subgroup Z_{10} we are after satisfies the following

$$Z_1 < Z_{10}, \ \chi(Z_{10} \cap Q_1) = [\sigma, \bar{\Lambda}] \text{ and } Z_{10}Q_1/Q_1 = \langle\sigma\rangle.$$

In order to show that such Z_{10} exists and to deduce its properties it appears convenient to adopt the G_2-viewpoint.

For $\mathcal{O}$ being the octad in $(\mathcal{P}, \mathcal{B})$, let $\bar{\Sigma}$ be the subgroup in $\bar{Q}_2 = Q_2/Z_2$ generated by the elements $\psi^{(a)}(\mathcal{O}, 0)$ taken for all $a \in Z_2^{\#}$ and let Σ be the preimage of $\bar{\Sigma}$ in Q_2. The element $(\mathcal{O}, 0)$ is isotropic with respect the the M-invariant quadratic form q_r on $\mathcal{A}(M_{24})$ which determines the squaring map on $Q_2^{(a)}$ (cf. (2.14.1 (g) and (i))), hence Σ is the elementary abelian of order 2^4. Since the automorphisms of $\bar{G}_2^s$ from S_t permute $\psi^{(a)}(\mathcal{O}, 0)$s among themselves (cf. (3.1.14 (h)), $\bar{\Sigma}$ is normalized by

$$N^\delta(\mathcal{O})/R_2 \cong 2^4 : A_8 \times S_3$$

(where N^δ is a subgroup of G_2 as in (2.14.1 (l)) such that $Q_2 \cap N^\delta = R_2$, $Q_2N^\delta = G_2$ and $N^\delta/M^\delta \cong S_3$, and $2^4 : A_8$ is the stabilizer of $\mathcal{O}$ in $M = M^\delta/R_2 \cong M_{24}$). Since a preimage of $\psi^{(a)}(\mathcal{O}, 0)$ acts on R_2 as the transvection whose centre is a and whose axis is $[R_2, O_2(N^\delta(\mathcal{O}))]$, we conclude that Σ is normalized by $N^\delta(\mathcal{O})$ and centralized by $M^\delta(\mathcal{O})$.

Lemma 4.3.2 *Put $Z_{10} = C_{R_2}(O_2(N^\delta(\mathcal{O})))\Sigma$. Then Z_{10} is elementary abelian of order 2^{10} normal in $G_2(\mathcal{O})$.*

Proof. The subgroup $C_{R_2}(O_2(N^\delta(\mathcal{O})))$ is the preimage in R_2 of $C_{\bar{R}_2}(O_2(M(\mathcal{O})))$, and by (3.1.9) the latter is generated by the elements $\psi(0, v^*)$ taken for all even subsets v of $\mathcal{O}$. The commutator relation in (2.14.1 (g)) shows that

$$[\psi^{(a)}(\mathcal{O}, 0), \psi^{(b)}(u, w^*)] = \psi(0, (\mathcal{O} \cap u)^*)$$

and since the Golay sets intersect evenly, the element on the right-hand side is contained in $C_{\bar{R}_2}(O_2(M(\mathcal{O})))$. Therefore

$$[\Sigma, Q_2] \leq C_{R_2}(O_2(N^\delta(\mathcal{O})))$$

and Z_{10} is normal in Q_2. Since $N^\delta(\mathcal{O})$ is proved to normalize Σ and since it obviously normalizes $C_{R_2}(O_2(N^\delta(\mathcal{O})))$, the normality of Z_{10} in $G_2(\mathcal{O}) = Q_2 N^\delta(\mathcal{O})$ follows. If $r \in Z_{10} \setminus Z_2$, then $rZ_2 = \psi^{(a)}(\mathcal{O}, v^*)$ for some $a \in Z_2^\#$ and for some even subset v of $\mathcal{O}$. Since $q_r(\mathcal{O}, v^*) = 0$ and since q_r determines the squaring map on $Q_2^{(a)}$, the element r is an involution and hence Z_{10} is elementary abelian as claimed. Finally

$$|Z_{10}| = |Z_2| \cdot |C_{\bar{R}_2}(O_2(M(\mathcal{O})))| \cdot |\bar{\Sigma}| = 2^2 \cdot 2^6 \cdot 2^2 = 2^{10}.$$

$\square$

Lemma 4.3.3 *Let κ be a $GF(2)$-valued form on Z_{10} which is zero on the $2B$-involutions and one on the $2A$-involutions. Then κ is a non-singular quadratic form of plus type.*

Proof. By the proof of (4.3.2) we have

$$Z_{10}/Z_2 = \{\psi^{(a)}(u, v^*) \mid a \in Z_2^\#, u \in \{0, \mathcal{O}\}, v \subseteq \mathcal{O}, |v| = 0 \bmod 2\}.$$

Let us calculate the numbers of $2A$- and $2B$-involutions in Z_{10}. The subgroup Z_2 is $2B$-pure; the four preimages of $\psi(0, v^*)$ in Z_{10} all have the same type which is $2B$ or $2A$ depending on whether or not $|v|$ is divisible by 4. Let us discuss the preimages of $\psi^{(z)}(\mathcal{O}, 0)$. If r is one of the preimages, then the remaining three are rz, rx, and $ry = rxz$. Since the preimages are contained in Q_1 which is extraspecial with centre $\langle z \rangle$, the elements r and rz are conjugate. Since $\chi(x) = \sum_{p \in \mathcal{P}} 2p + 2\Lambda$ by (1.8.3), up to renaming, we have

$$\chi(r) = \sum_{p \in \mathcal{O}} 2p + 2\Lambda, \quad \chi(rx) = \sum_{p \notin \mathcal{O}} 2p + 2\Lambda,$$

which means that r is of type $2A$, while rx is of type $2B$. If v is an even subset of $\mathcal{O}$, then $\psi^{(z)}(\mathcal{O}, 0)$ and $\psi^{(z)}(\mathcal{O}, v^*)$ are conjugate via an element $\psi^{(x)}(u, 0)$ where u is a Golay set intersecting $\mathcal{O}$ either in v or in $\mathcal{O} \setminus v$. Since the action of G_2 on $Z_2^{\#}$ is also transitive, among the four preimages of any $\psi^{(a)}(\mathcal{O}, v^*)$ two are of type $2A$ and two are of type $2B$. Thus the numbers of $2B$- and $2A$-elements in Z_{10} are respectively

$$3 + 35 \cdot 4 + 3 \cdot 2^7 = 527 \text{ and } 28 \cdot 4 + 3 \cdot 2^7 = 496,$$

which are the familiar numbers of non-zero singular and non-singular vectors in a 10-dimensional non-singular orthogonal space of plus type.

The verification that κ is in fact a quadratic form can be achieved by more or less direct calculations in the Leech lattice modulo two (or rather in the anti-permutation module for M_{24}) making use of (2.14.1 (g)). We believe that this can be left to the reader who might prefer a more conceptional proof anyway. One possibility is to consider the *octad sublattice* in Λ formed by the vectors whose support is contained in $\mathcal{O}$ (cf. Section 25 in [A94]). Alternatively, we are just about to see that $N_{G_1}(Z_{10})$ and $N_{G_2}(Z_{10})$ induce on Z_{10} actions isomorphic to

$$2_+^{1+12} : (A_8 \times S_3) \text{ and } 2^8 : \Omega_8^+(2),$$

respectively. It is not be difficult to show that κ being a non-trivial function invariant under both the actions must be of the required type. □

Lemma 4.3.4 *The following assertions hold:*

(i) $Z_5^{(t)}$ and $Z_5^{(l)}$ are contained in Z_{10} and they are maximal totally isotropic subspaces with respect to κ;

(ii) *for $i = 1$ and 2, the action on Z_{10} induced by the normalizer $N_{G_i}(Z_{10})$ is contained in the simple commutator subgroup*

$$\Omega(Z_{10}, \kappa) \cong \Omega_{10}^+(2)$$

of the automorphism group of the orthogonal space (Z_{10}, κ).

Proof. Since $Z_5^{(t)}$ and $Z_5^{(l)}$ are contained in $C_{R_2}(O_2(N^\delta(\mathcal{O})))$ by the penultimate sentence in (4.1.3), they are contained in Z_{10} by (4.3.2). Since these subgroups are $2B$-pure of order 2^5 each, (i) follows. By the last sentence in (4.1.3), $Z_5^{(t)}$ and $Z_5^{(l)}$ are conjugate neither in G_1 nor in G_2. Since $Z_5^{(t)} \cap Z_5^{(l)} = Z_4$, and Z_4 is 4-dimensional, $Z_5^{(t)}$ and $Z_5^{(l)}$ represent different types in the $D_5(2)$-geometry associated with (Z_{10}, κ). Hence the action of $N_{G_i}(Z_{10})$ is contained in the automorphism group of that geometry which is precisely $\Omega(Z_{10}, \kappa)$. □

Lemma 4.3.5 *The following assertions hold:*

(i) $N_{G_2}(Z_{10}) = G_2(\mathcal{O})$ *and* $C_{G_2}(Z_{10})$ *is the centralizer in* G_2 *of the subspace* $\Pi_{\mathcal{O}}$ *(the subgroup* $C_{\mathcal{O}}$ *in (3.1.12));*

(ii) $C_{G_2}(Z_{10})$ *is a special group* 2^{10+16} *in which* Z_{10} *is both the centre and the commutator subgroup;*

(iii) $N_{G_2}(Z_{10})/C_{G_2}(Z_{10}) \cong 2_+^{1+12} : (A_8 \times S_3)$ *is the stabilizer of* Z_2 *in* $\Omega(Z_{10}, \kappa)$.

Proof. Since $M(\mathcal{O})$ is the stabilizer of

$$\bar{\Sigma} = \langle \psi^{(a)}(\mathcal{O}, 0) \mid a \in Z_2^{\#} \rangle$$

in $M \cong M_{24}$ and since Z_{10} is normalized by $G_2(\mathcal{O})$ by (4.3.2), the first assertion in (i) follows. Since a preimage of $\psi^{(a)}(\mathcal{O}, 0)$ in Q_2 acts on R_2 as the transvection with centre a and with axis $[R_2, O_2(N^\delta(\mathcal{O}))]$, this axis is precisely the intersection of $C_{G_2}(Z_{10})$ with R_2. If $b \in Z_2^{\#} \setminus \{a\}$ and $\mathcal{O}'$ is an octad, then $\psi^{(a)}(\mathcal{O}, 0)$ and $\psi^{(b)}(\mathcal{O}', 0)$ commute if and only if $\mathcal{O}'$ is either equal to or disjoint from $\mathcal{O}$. Since for every such $\mathcal{O}'$ the element $\psi^{(b)}(\mathcal{O}', 0)$ also commutes with $\psi(0, v^*)$ for every even subset v of $\mathcal{O}$, in view of (3.1.9) we conclude that

$$C_{G_2}(Z_{10}) Q_2^{(a)} / R_2 = [O_2(N^\delta(\mathcal{O})), Q_2^{(a)}/R_2].$$

Finally, since $O_2(M^\delta(\mathcal{O}))$ and R_2 induce on Z_{10} the same action

$$C_{G_2}(Z_{10}) Q_2 / Q_2 = O_2(M(\mathcal{O}))$$

which proves (i) and (ii). Now (iii) is by the order considerations and the fact that in a Lie type group, parabolic subgroups split over their radicals. □

Lemma 4.3.6 *Let* $\widehat{\sigma}$ *be an element from* $(Q_2^{(x)} \cap Z_{10}) \setminus Z_2$ *such that* $\widehat{\sigma} Z_2 = \psi^{(x)}(\mathcal{O}, 0)$ *and let* σ *be the image of* $\widehat{\sigma}$ *in* $G_1/Q_1 \cong Co_1$. *Then:*

(i) σ *is a central involution in* G_1/Q_1 *contained in* $E = O_2(F)$, *where* $F = E : M$ *is the monomial subgroup and* $C_M(\sigma) = M(\mathcal{O})$;

(ii) $\chi(Z_{10} \cap Q_1) = [\sigma, \bar{\Lambda}]$;

(iii) $N_{G_1}(Z_{10})/C_{G_1}(Z_{10}) \cong 2^8 : \Omega_8^+(2)$ *is the stabilizer of* Z_1.

Proof. Since E the image of $Q_2^{(x)}$ in the monomial subgroup and M is the image of M^δ, (i) follows. The second assertion is easy to deduce from the proof of (4.3.2). Now (iii) follows from (i) and (4.3.1 (v)). □

Proposition 4.3.7 *Let* G_{10} *be generated by* $N_{G_1}(Z_{10})$ *and* $N_{G_2}(Z_{10})$. *Then*

$$G_{10} \sim 2^{10+16}.\Omega_{10}^+(2).$$

Proof. If Q_{10} denotes $C_{G_{12}}(Z_{10})$, then $Q_{10} = C_{G_i}(Z_{10})$ for $i = 1$ and 2. By (4.3.5 (ii)) Q_{10} is a special group of the shape 2^{10+16} normal in G_{10}. We claim that the quotient G_{10}/Q_{10} is a completion of the amalgam of maximal parabolics in $\Omega(Z_{10}, \kappa) \cong \Omega_{10}^+(2)$. By (4.3.5 (iii)) and (4.3.6 (iii))

$$G_{10}/C_{G_{10}}(Z_{10}) \cong \Omega(Z_{10}, \kappa) \cong \Omega_{10}^+(2).$$

Since the $D_5(2)$-geometry is residually connected, the following assertion holds: for every $Y \in \{Z_3, Z_5^{(t)}, Z_5^{(l)}\}$ the subgroup

$$\langle N_{G_1}(Y) \cap N_{G_1}(Z_{10}), N_{G_2}(Y) \cap N_{G_2}(Z_{10}) \rangle$$

contains Q_{10} and induces on Z_{10} the stabilizer of Y in $\Omega(Z_{10}, \kappa)$. This implies that G_{10}/Q_{10} is a completion of

$$\mathcal{A}(\bar{G}_{10}) = \{N_H(Z_{10})/Q_{10} \mid H \in \{G_1, G_2, G_3, G_5^{(t)}, G_5^{(l)}\}\}$$

which is the amalgam of maximal parabolics in $\Omega(Z_{10}, \kappa) \cong \Omega_{10}^+(2)$. Since the latter group is the universal and only faithful completion of $\mathcal{A}(\bar{G}_{10})$, the result follows. $\qquad \Box$

The coset geometry corresponding to the subamalgam

$$\mathcal{H}(G) = \{G_1, G_2, G_3, G_5^{(t)}, G_{10}\}$$

corresponds to the maximal 2-local parabolic geometry introduced by M. Ronan and St. Smith in [RSm80].

4.4 $2^2 \cdot (^2E_6(2))$: S_3-subgroup

For $\mathcal{T}$ being the 4-subset in $\mathcal{P}$ whose sextet is stabilized by G_{23}, let $\mathcal{Q}$ be a 3-subset in $\mathcal{T}$. Put

$$\bar{Y}_2 = \langle 4p - 4q + 2\Lambda \mid p, q \in \mathcal{Q}, p \neq q \rangle,$$

so that $\bar{Y}_2$ is an order four subgroup in $\bar{\Lambda}$. The stabilizer of $\bar{Y}_2$ in Co_1 is the 222-triangle stabilizer isomorphic to $U_6(2) : S_3$ (cf. Lemma 4.8.10 in [Iv99]). Let Y_2 be a subgroup in Q_1 which maps isomorphically onto $\bar{Y}_2$ under the homomorphism $\chi : Q_1 \to \bar{\Lambda}$. The following result is a direct implication of the definitions.

Lemma 4.4.1 *The subgroup Y_2 is $2A$-pure of order 2^2 contained in Z_{10} and perpendicular to Z_3 with respect to the bilinear form on Z_{10} associated with the form κ.* $\qquad \Box$

In this section we show that the subgroup in G generated by

$$N_{G_1}(Y_2) \text{ and } N_{G_{10}}(Y_2)$$

has the shape $2^2 \cdot (^2E_6(2)) : S_3$.

Lemma 4.4.2 *The following assertions hold:*

(i) $N_{G_{10}}(Y_2)$ *contains* $Q_{10} \cong 2^{10+16}$ *and*

$$N_{G_{10}}(Y_2)/Q_{10} \cong (3 \times \Omega_8^-(2)) : 2$$

is the stabilizer in $\Omega(Z_{10}, \kappa)$ *of a minus line in* $\Omega(Z_{10}, \kappa)$;
(ii) C_3-*geometry with the diagram*

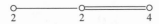

on which $N_{G_{10}}(Y_2)$ *acts as* $O_8^-(2)$ *is realized by the conjugates of* Z_1, Z_2,
and Z_3 *contained in the orthogonal complement of* Y_2 *in* Z_{10}, *so that*
$\{N_{G_{10}}(Y_2) \cap N_{G_{10}}(Z_i) \mid 1 \le i \le 3\}$ *is the amalgam of maximal*
parabolics associated with this action.

Proof. The subgroup Y_2 being $2A$-pure is a minus line in the orthogonal
space (Z_{10}, κ). By the construction, Y_2 is perpendicular to Z_3, therefore the
latter is a maximal totally isotropic subspace in $Y_2^\perp$, which is an orthogonal
8-dimensional subspace of minus type in (Z_{10}, κ).   □

Lemma 4.4.3 *The following assertions hold:*

(i) $(N_{G_1}(Y_2) \cap Q_1)/Y_2 \cong 2_+^{1+20}$; $N_{G_1}(Y_2)Q_1/Q_1 \cong U_6(2) : S_3$;
(ii) C_3-*geometry with the diagram*

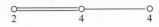

on which $N_{G_1}(Y_2)$ *induces the action of* $U_6(2) : S_3$ *is realized by the*
conjugates of Z_2, Z_3, *and* Z_{10} *which are contained in* Q_1, *contain* Z_1
and commute with Y_2, *so that* $\{N_{G_1}(Y_2) \cap N_{G_1}(Z_i) \mid i = 2, 3, 10\}$ *is the*
amalgam of maximal parabolics associated with this action.

Proof. The result follows from the structure of the 222-triangle stabilizer in
the Conway group and its action on $\bar{\Lambda}$. This is a standard part of the Conway–
Leech theory (cf. Section 4.10 in [Iv99]). Alternatively, we can just compare
the orders of the parabolics in the amalgam in (ii) against the orders of the
parabolic subgroups in $U_6(2) : S_3$.   □

Lemma 4.4.4 *The following assertions hold:*

(i) *the normalizer $N_{G_2}(Y_2)$ intersects Q_2 in a subgroup of index 4 in Q_2 and*

$$N_{G_2}(Y_2)Q_2/Q_2 \cong N^\delta(Q)Q_2/Q_2 \cong P\Sigma L_3(4) \times S_3;$$

(ii) *the amalgam $\{N_{G_2}(Y_2) \cap N_{G_2}(Z_i) \mid i = 1, 3, 10\}$ is the amalgam of maximal parabolics associated with the action of $N_{G_2}(Y_2)$ on the direct sum of the 3-element geometry and the projective plane of order 4.*

Proof. It is sufficient to notice that the octads containing Q together with the sextets containing Q in one tetrad form the projective plane of order 4. This is a very basic property of the $S(5, 8, 24)$-Steiner system. $\square$

It follows from (4.4.2), (4.4.3), and (4.4.4) that the amalgam

$$\mathcal{A}(^2G_2) := \{N_H(Y_2) \mid H = G_1, G_2, G_3, G_{10}\}$$

has kernel Y_2 and corresponds to the F_4-diagram

and the C_3-residues are buildings. By Tits' characterization [Tit82a] and the spherical buildings classification [Tit74], we have the following:

Proposition 4.4.5 *The subgroup $^2G_2$ in G, generated by the subamalgam $\mathcal{A}(^2G_2)$, is isomorphic to $2^2.(^2E_6(2)) : S_3$.* $\square$

4.5 Acting on the 196 883-module

To define an algebra on the 196 883-dimensional G-module Π we require some information about the action of G_{10} and G_4 on that module.

Lemma 4.5.1 *Let $Q_{10} = O_2(G_{10}) \cong 2^{10+16}$, and let $\mathcal{O}$ be the octad such that $G_{10} \cap G_2 = G_2(\mathcal{O})$. Then:*

(i) $C_\Pi(Q_{10}) \cap \Pi^2_{3 \cdot 24 \cdot 2^{11}} = 0$;

(ii) $C_\Pi(Q_{10}) \cap \Pi^2_{2^6 \cdot 759} = \Pi_{\mathcal{O}}$ *(the 2^6-dimensional $G_2(\mathcal{O})$-submodule in $\Pi^2_{2^6 \cdot 759}$);*

(iii) *if $\{p, q\} \subset \mathcal{P}$ and $a \in Z_2^\#$, then the 1-dimensional subspace $\Pi^{(a)}_{\{p,q\}}$ from $\Pi^2_{3 \cdot 276}$ is contained in $C_\Pi(Q_{10})$ if and only if $\{p, q\} \subset \mathcal{O}$, in particular*

$$\dim(C_\Pi(Q_{10}) \cap \Pi^2_{3 \cdot 276}) = 3 \cdot \binom{8}{2} = 84;$$

(iv) $C_\Pi(Q_{10}) \cap \Pi_{23}^2 = C_{\Pi_{23}^2}(O_2(M(\mathcal{O}))$ *is 8-dimensional isomorphic to the permutation module of* $M(\mathcal{O})/O_2(M(\mathcal{O})) \cong A_8$;

(v) $C_\Pi(Q_{10})$ *is the direct sum* $\Pi_1^{10} \oplus \Pi_{155}^{10}$ *of a 1-dimensional trivial and a 155-dimensional irreducible submodule for* $G_{10}/Q_{10} \cong \Omega_{10}^+(2)$, *where* Π_1^{10} *is contained in* Π_{23}^2.

Proof. Since Z_{10} contains Z_2 and the latter acts on $\Pi_{3\cdot24\cdot2^{11}}^2$ fixed-point freely, (i) follows. By (4.3.5 (i)), $Q_{10} = C_\mathcal{O}$ centralizes $\Pi_\mathcal{O}$. On the other hand, the Q_2-irreducibles in $\Pi_{2^6\cdot759}^2$ are indexed by the octads in $\mathcal{B}$, while

$$O_2(M(\mathcal{O})) = Q_{10}Q_2/Q_2 \cong 2^4$$

acts fixed-point freely on $\mathcal{B} \setminus \mathcal{O}$, hence (ii) follows. The subspace $\Pi_{\{p,q\}}^{(a)}$ in (iii) can be defined as the centralizer in $\Pi_{3\cdot276}^2$ of $Q_{p,q}Q_2^{(a)}$ (compare (3.1.17 (iii))). Since $Q_{p,q}^{(b)}/R_2$ is generated by the images of $\psi^{(b)}(u, v^*)$ taken for all the Golay sets evenly intersecting $\{p, q\}$, while $(Q_{10} \cap Q_2^{(b)})R_2/R_2$ is generated by the images of the elements $\psi^{(b)}(w, v^*)$ taken for all Golay sets w which are either disjoint from $\mathcal{O}$ or contained in $\mathcal{O}$, (iii) follows. Since G_2 acts on Π_{23}^2 via its factor group $M \cong M_{24}$, and Q_{10} acts on this module via $O_2(M(\mathcal{O})) \cong O_2(2^4 : A_8) \cong 2^4$, we have (iv). Acting on $\mathcal{P}$ the latter group has nine orbits which are the elements of $\mathcal{O}$ together with $\mathcal{P} \setminus \mathcal{O}$. Since Π_{23}^2 is the quotient of the permutation module of M on $\mathcal{P}$ over the 1-dimensional trivial submodule, (iv) follows. The assertions (i) to (iv) and easy counting show that the dimension of $C_\Pi(Q_{10})$ is 156. Since the latter number is one plus the dimension of the minimal non-trivial irreducible representation of $G_{10}/Q_{10} \cong \Omega_{10}^+(2)$ [CCNPW] (and since $C_\Pi(Q_{10})$ is not centralized by the whole of G_{10}), (v) follows. $\square$

Lemma 4.5.2 *The following assertions hold:*

(i) $C_\Pi(Z_4) < C_\Pi(Z_3)$;

(ii) $[\Pi, Z_4]$ *is the direct sum of* 15 *subspaces*

$$C_\Pi(Y) \cap [\Pi, Z_4]$$

of dimension $24 \cdot 2^9$ *each, taken for all the hyperplanes* Y *in* Z_4 *and* $G_4/C_{G_4}(Z_4) \cong GL(Z_4)$ *permutes the direct summands doubly transitively.*

Proof. The result follows from the inclusion $Z_3 < Z_4$ and the fact that the intersection

$$[\Pi, Z_3] \cap C_\Pi(Z_2)$$

has dimension $24 \cdot 2^{10}$ (compare the diagram at the end of Section 3.4). $\square$

5

Griess algebra

Let G be the Monster group which, according to our definition, is the universal completion of the Monster amalgam $\mathcal{M}$. In Chapter 3 we have constructed a 196 883-dimensional G-module Π over the field $\mathbb{R}$ of real numbers such that the universal completion of $\mathcal{M}$ factorized through the representation φ : $G \to GL(\Pi)$ remains faithful. In this chapter we show that G preserves on Π a commutative non-associative multiplication famously known as Griess' algebra. We start with the space A_1 of G_1-invariant algebras on

$$C_\Pi(Z_1) = \Pi^1_{299} \oplus \Pi^1_{|\bar{\Lambda}_2|}$$

(where $G_1 \sim 2^{1+24}_+.Co_1$). It is not hard to deduce a rather explicit description of A_1 from the known properties of the Conway group Co_1. Let A_{12} be the space formed by the restrictions to

$$C_\Pi(Z_2) = \Pi^2_{23} \oplus \Pi^2_{3.276} \oplus \Pi^2_{2^6.759}$$

of the algebras from A_1 and let A_2 be the space of G_2-invariant algebras on $C_\Pi(Z_2)$ (where $G_2 \sim 2^{2+11+22}.(M_{24} \times S_3)$). We show that A_1 projects bijectively onto A_{12} and that A_{12} intersects A_2 in a 1-dimensional subspace. Let $A^{(z)}$ denote the 1-dimensional subspace in A_1 formed by the algebras whose restrictions to $C_\Pi(Z_2)$ are G_2-invariant. Alternatively $A^{(z)}$ can be defined as the set of algebras in A_1 whose restrictions to $C_\Pi(Q_{10})$ are G_{10}-invariant (where $G_{10} \sim 2^{10+16}.\Omega^+_{10}(2)$ and $Q_{10} = O_2(G_{10})$). Next we make use of the subgroups $G_3 \sim (2^{3+6+12+18}.(3 \cdot S_3 \times L_3(2))$ and $G_4 \sim [2^{39}].(S_3 \times L_4(2))$. Since $G_i = \langle G_1 \cap G_i, \ G_2 \cap G_i \rangle$ for $i = 3$ and $i = 4$, every algebra from $A^{(z)}$ is G_i-invariant when restricted to $C_\Pi(Z_i)$. On the other hand, Π possesses the following decomposition into a sum of G_i-submodules

$$\Pi = C_\Pi(Z_i) \bigoplus_{Y \in Z_i^* \setminus \{0\}} (C_\Pi(Y) \cap [\Pi, Z_i]),$$

121

where $Z_i^* \setminus \{0\}$ is identified with the set of hyperplanes in Z_i and

$$\dim (C_\Pi(Y) \cap [\Pi, Z_i]) = 24 \cdot 2^{13-i}$$

for $Y \in Z_i^* \setminus \{0\}$. The action of G_i on Z_i^* is the natural doubly transitive action of $GL(Z_i) \cong GL_i(2)$ of degree $2^i - 1$. This means that whenever

$$u_1 \in C_\Pi(Y_1) \text{ and } u_2 \in C_\Pi(Y_2) \text{ for } Y_1, Y_2 \in Z_i^* \setminus \{0\}$$

there is an element $g_i \in G_i$ such that $g_i(Y_1) \cap g_i(Y_2) \geq Z_1$ (equivalently, such that $g_i(u_1), g_i(u_2) \in C_\Pi(Z_1)$). This observation enables us to expand an algebra multiplication from $A^{(z)}$ onto the whole of Π so that the expanded multiplication is G_i-invariant. Since the required element g_i can always be found inside $G_3 \cap G_4$, the expansion is invariant under the whole of $G = \langle G_3, G_4 \rangle$. This shows that the space of G-invariant algebras on Π is 1-dimensional. In this way we achieve a compromise between the abstraction of Norton's observation and the explicitness of Griess' construction.

5.1 Norton's observation

In early stages of studying the Monster, when the existence of the 196 883-dimensional representation φ was still a conjecture, Simon Norton calculating with the hypothetical character of degree 196 883 had observed that $(\varphi(G), \Pi)$ is a *Norton pair* according to the following definition:

Definition 5.1.1 *Let H be a group and let U be an irreducible H-module over the field $\mathbb{R}$ of real numbers. Then (H, U) is a* Norton pair *if:*

(i) *the trivial irreducible H-module appears in the symmetric part $S^2(U)$ of the tensor square of U, and*

(ii) *U appears with multiplicity one in its tensor square and is contained in $S^2(U)$.*

If (H, U) is a Norton pair, then by (5.1.1 (i)) H preserves on U a non-zero symmetric inner product $(\ ,\ )_U$, while by (5.1.1 (ii)) it preserves a non-zero commutative algebra product $\star$. Since U is absolutely irreducible, $(\ ,\ )_U$ is unique up to rescaling and so is $\star$ because of the 'multiplicity one' condition in (5.1.1 (ii)).

Lemma 5.1.2 *Let (H, U) be a group-representation pair and suppose that (5.1.1 (i)) holds. Then (H, U) is a Norton pair if and only if the trivial irreducible H-module appears with multiplicity one in the tensor cube of U and is contained in the symmetric part $S^3(U)$ of the tensor cube.*

Proof. The condition in the lemma is equivalent to the assertion that H preserves on H a unique (up to rescaling) trilinear form. The dimension of the space $T_H(U)$ of H-invariant trilinear forms on U is equal to the multiplicity of the trivial irreducible character in the tensor cube of U. By (5.1.1 (i)), U is isomorphic to its dual module U^* and therefore

$$U \otimes U \otimes U \cong (U \otimes U) \otimes U^*.$$

Since U is irreducible, the multiplicity of the trivial irreducible module on the right-hand side is equal to the multiplicity of U in its tensor square U. The latter multiplicity is by definition the dimension of the space $A_H(U)$ of H-invariant algebra products on U. Therefore, $T_H(U)$ and $A_H(U)$ are isomorphic. An explicit isomorphism assigns to $\star \in A_H(U)$ the form

$$(u, v, w) \mapsto ((u \star v), w)_U.$$

In any event, $A_H(U)$ is 1-dimensional if and only if $T_H(U)$ is such. Suppose now that $\dim(A_H(T)) = \dim(T_H(U)) = 1$. Then the symmetric group S_3 permuting the factors of the tensor cube acts on $T_H(U)$ by scalar multiplications. Therefore, the forms in $T_H(U)$ are either all symmetric or all antisymmetric. The above correspondence between $A_H(T)$ and $T_H(U)$ immediately shows that in the respective cases the products in $A_H(U)$ are commutative and anti-commutative. This completes the proof. $\qquad\square$

In [Gri82a] the (Monster-invariant to be) multiplication $\star$ (nowadays commonly known as *Griess algebra*) was selected among the algebras invariant under $\varphi(G_1)$. After that an additional automorphism of this algebra (the famous element σ) was constructed which together with $\varphi(G_1)$ generates the Monster.

When constructing φ in Chapter 3, we did not use any invariant form or algebra. On the other hand, having φ constructed, we can recover the inner product $(\, , \,)_\Pi$ and the algebra multiplication $\star$ via their $\varphi(G)$-invariance. The original construction of $\star$ in [Gri82a] was rather involved (as Paul Erdös once pointed out 'the first proof can be as complicated as it likes'). Later on, Griess' construction was modified by J. Conway [C84], J. Tits [Tit82b], [Tit83], [Tit84] and M. Aschbacher [A94]. We are hoping to expose a new angle from which the construction and its modifications can be viewed.

5.2 3-dimensional S_4-algebras

Let $S \cong S_4$ be the symmetric group of degree 4, let V_3 be a real vector space of dimension 3, and let

$$\theta : S \to GL(V_3)$$

be a faithful irreducible representation. Decomposing V_3 into $O_2(S)$-eigenspaces we observe that θ is induced from a non-trivial linear representation σ of a Sylow 2-subgroup $D \cong D_8$ of S. The kernel of σ must be one of the two subgroups of index 2 in D which are not normal in the whole of S. One of these subgroups (denoted by D^+) is elementary abelian generated by a pair of commuting transpositions, while the other one (denoted by D^-) is cyclic of order 4. For $\epsilon \in \{+, -\}$, let σ^ϵ be the linear representation of D whose kernel is D^ϵ, let θ^ϵ denote the S-representation induced from σ^ϵ, and let V_3^ϵ be the 3-dimensional complex vector space affording θ^ϵ. Let a^ϵ, b^ϵ and c^ϵ be eigenvectors of $O_2(S)$ forming a basis of V_3^ϵ stable under an element t of order 3 in S inducing the cycle $(a^\epsilon \, b^\epsilon \, c^\epsilon)$. A non-identity element from $O_2(S)$ centralizes one of the basis vectors negating the remaining two. There is a transposition s in S which normalizes the subgroup $\langle t \rangle$ and permutes a^ϵ and b^ϵ. Then s centralizes c^+ and negates c^-.

We are interested in the S-invariant bilinear multiplications $\star_\epsilon$ on V_3^ϵ. The product $a^\epsilon \star_\epsilon b^\epsilon$ is centralized by the element $e \in O_2(S)$ which negates both a^ϵ and b^ϵ. Since $C_{V_3}(e) = \{\lambda c^\epsilon \mid \lambda \in \mathbb{R}\}$, assuming that the product is non-zero and properly scaled we may assume that

$$a^\epsilon \star_\epsilon b^\epsilon = c^\epsilon.$$

Acting by t and by s we obtain

$$b^\epsilon \star c^\epsilon = a^\epsilon, \quad c^\epsilon \star a^\epsilon = b^\epsilon \text{ and } b^\epsilon \star_\epsilon a^\epsilon = \epsilon c^\epsilon.$$

The multiplication thus defined and expanded on the whole of V_3^ϵ via distributivity is clearly S-invariant and in fact it is the only such S-invariant up to rescaling.

The difference between the algebras

$$(V_3^+, \star_+) \text{ and } (V_3^-, \star_-)$$

can be demonstrated by calculating the idempotents. Let $v^\epsilon = \alpha a^\epsilon + \beta b^\epsilon + \gamma c^\epsilon$ be a vector in V_3^ϵ where α, β, and γ are parameters. The condition for v^ϵ to be an idempotent is the equality $v^\epsilon \star_\epsilon v^\epsilon = v^\epsilon$ which reduces to the following system of equations

$$\beta\gamma + \epsilon\beta\gamma = \alpha, \ \alpha\gamma + \epsilon\alpha\gamma = \beta, \ \alpha\beta + \epsilon\alpha\beta = \gamma.$$

If $\epsilon = -$, then $\alpha = \beta = \gamma = 0$ is the only solution, while for $\epsilon = +$ there are four non-zero solutions

$$\pm \frac{1}{2}a^+ \pm \frac{1}{2}b^+ \pm \frac{1}{2}c^+,$$

where the number of minus signs must be even. Now it is easily seen that S is in fact the full automorphism group of $(V_3^+, \star_+)$. On the other hand, $(V_3^-, \star_-)$ has an infinite automorphism group, because it is in fact isomorphic to the 3-dimensional simple Lie algebra $sl_2(\mathbb{C})$ of traceless 2×2 matrices with respect to the commutator product. An identification can be achieved by assigning

$$a^- = \frac{1}{2}\begin{pmatrix} i & 0 \\ 0 & -i \end{pmatrix}, \ b^- = \frac{1}{2}\begin{pmatrix} 0 & 1 \\ -1 & 0 \end{pmatrix}, \ c^- = \frac{1}{2}\begin{pmatrix} 0 & i \\ i & 0 \end{pmatrix},$$

where $i = \sqrt{-1}$. Conjugating by elements of $GL_2(\mathbb{C})$ we clearly obtain infinitely many automorphisms.

Proposition 5.2.1 *Let $S \cong S_4$ be the symmetric group of degree 4, let $D \cong D_8$ be a Sylow 2-subgroup of S, let $D^+ \cong 2^2$ and $D^- \cong 4$ be the index 2 subgroups in D different from $O_2(S)$. Let σ^ϵ be the linear representation of D with kernel D^ϵ and let*

$$\theta^\epsilon : S \to GL(V_3^\epsilon)$$

be the representation of S induced from σ^ϵ. Then:

  (i) *every 3-dimensional faithful S-module over $\mathbb{R}$ is isomorphic to V_3^ϵ for $\epsilon \in \{+, -\}$;*
 (ii) *V_3^+ is a permutation module for an S_3-subgroup in S, while V_3^- is strictly monomial for that subgroup;*
(iii) *there is an S-invariant non-zero algebra multiplication $\star_\epsilon$ on V_3^ϵ, which is unique up to rescaling, and:*
  (a) *$\star_+$ is commutative, there are exactly four non-zero idempotents and S is the full automorphism group of $\star_+$;*
  (b) *$\star_-$ is anti-commutative isomorphic to the Lie algebra $sl_2(\mathbb{C})$ and its automorphism group is infinite;*
 (iv) *(S, V_3^+) is a Norton pair.*

Proof. The assertions (i) to (iii) were established above in this section. The inner product with respect to which $(a^\epsilon, b^\epsilon, c^\epsilon)$ is an orthonormal basis is manifestly S_4-invariant (hence it is the unique one up to rescaling). If τ is a non-zero S-invariant trilinear form on V_3^+, then

$$u \mapsto \tau(a^+, b^+, u)$$

is a linear form stabilized by $O_2(S)$ and centralized by $C_{O_2(S)}(c^+)$. Applying the self-duality of V_3^+ performed by the invariant inner product and rescaling τ if necessary, we observe that

$$\tau(a^+, b^+, a^+) = \tau(a^+, b^+, b^+) = 0 \text{ and } \tau(a^+, b^+, c^+) = 1.$$

Since $\{a^+, b^+, c^+\}$ is stabilized and permuted naturally by the subgroup $\langle t, s \rangle \cong S_3$, the form τ is symmetric. □

5.3 Krein algebras

An important class of commutative algebras can be constructed from multiplicity-free permutation modules (cf. [Sc77], [CGS78] and Section 2.8 in [BI84]). Let Ω be a set transitively permuted by a group H and let V be the corresponding $\mathbb{R}$-permutation module. Considering V as the space of real-valued functions on Ω and identifying the elements of Ω with their characteristic functions we turn Ω into a basis of V. Then H preserves an inner product and an algebra multiplication on V with respect to which Ω is a basis consisting of pairwise orthogonal idempotents of norm 1

$$(a, b) = \delta_{a,b} \text{ and } a \star b = \delta_{ab}a \text{ for } a, b \in \Omega.$$

Suppose that the permutation module is *multiplicity-free*. This means that V possesses an orthogonal decomposition

$$V = V_0 \oplus V_1 \oplus ... \oplus V_{r-1}$$

into pairwise non-isomorphic irreducible H-modules, where r is the permutation rank, which is the number of H-orbits on $\Omega \times \Omega$. It is common to assume that V_0 is the trivial 1-dimensional module formed by the constant functions. Let π_i denote the (orthogonal) projection of V into V_i, so that $\pi_i(v) \in V_i$ and

$$v = \pi_0(v) + \pi_1(v) + ... + \pi_{r-1}(v)$$

for every $v \in V$.

The multiplication $\star_i$ on V_i defined by

$$u \star_i w = \pi_i(u \star w) \text{ for } u, w \in V_i$$

is clearly H-invariant and the following assertion is proved in Section 2.8 of [BI84].

Lemma 5.3.1 *The algebra* $(V_i, *_i)$ *is non-zero if and only if the diagonal Krein parameter* q^i_{ii} *is non-zero.* $\qquad\square$

We call $(V_i, \star_i)$ the ith Krein algebra of (H, Ω).

Lemma 5.3.2 *Let* (H, Ω) *be a multiplicity-free permutation representation* H, *and let* $(V_i, \star_i)$ *be the associated ith Krein algebra. Then for every* $\alpha \in \Omega$ *(considered as a vector from V) the element* $\pi_i(\alpha)$ *is either nilpotent or a scalar multiple of an idempotent.*

Proof. Let K be the stabilizer of α in H. Then the multiplicity-freeness condition together with the Frobenius reciprocity imply that

$$C_{V_i}(K) = \{\lambda \pi_i(\alpha) \mid \lambda \in \mathbb{R}\}.$$

Since $\pi_i(\alpha) \star_i \pi_i(\alpha)$ is centralized by K, we have

$$\pi_i(\alpha) \star_i \pi_i(\alpha) = \lambda \pi_i(\alpha)$$

for some λ. If $\lambda = 0$, then $\pi_i(\alpha)$ is nilpotent, otherwise $\frac{1}{\lambda}\pi_i(\alpha)$ is an idempotent. $\qquad\square$

The algebra $(V_3^+, \star_+)$ is a Krein algebra of the natural degree four permutation representation of S_4, and the non-zero idempotents are projections of the elements of the 4-set. Both $(V_3^+, \star_+)$ and $(V_3^-, \star_-)$ are Krein algebras of the permutation representation of S_4 on the cosets of a cyclic subgroup of order 3.

Let us mention some less trivial examples of Krein algebras. First, let W be a 10-dimensional $GF(2)$-space equipped with a non-singular quadratic form q of plus type. Let Ξ be a graph whose vertices are the vectors in W which are non-isotropic with respect to q, and let two such vectors be adjacent if they are not perpendicular with respect to the bilinear form associated with q. It is well known and easy to check (cf. Lemma 5.12.3 in [Iv99]) that Ξ is strongly regular on $n = 496$ vertices of valency $k = 240$ with $\lambda = 120$ and $\mu = 112$ (in the theory of distance-regular graphs the parameters λ and μ are better known under the names a_1 and c_2, respectively). The orthogonal group $O_{2n}^+(2)$ is the automorphism group of Ξ and we can define Ξ as a graph on the class of 3-transpositions in $O_{10}^+(2)$ with two transpositions being adjacent whenever they do not commute. Calculating by the formulas on p. 170 in [Bi93] we deduce the following:

Lemma 5.3.3 *The adjacency matrix of a strongly regular graph with parameters*

$$(n, k, \lambda, \mu) = (496, 240, 120, 112)$$

has three eigenvalues $p_0 = 240$, $p_1 = 16$, and $p_2 = -17$ with multiplicities $m_0 = 1$, $m_1 = 155$, and $m_2 = 340$, respectively. The Krein parameter q_{11}^1 is non-zero, equal to $1 - (2^4 - 1)^{-2}$. $\square$

In view of the above discussions and since $\Omega_{10}^+(2)$ possesses a unique 155-dimensional irreducible complex representation which can be lifted to a representation of $O_{2n}^+(2)$, (5.3.3) implies that the module Π_{155}^{10} for the group $G_{10}/Q_{10} \cong \Omega_{10}^+(2)$ as in (4.5.1) carries an invariant algebra multiplication. Because of the uniqueness the module is clearly self-dual. In Section 5.5 we will show that $(\Omega_{10}^+(2), \Pi_{155}^{10})$ is in fact a Norton pair.

The 196 883-dimensional representation of the Monster appears as an irreducible constituent of the multiplicity-free rank 9 permutation representation of the Monster acting by conjugation on its class of $2A$-involutions. Realization of the Griess algebra as the Krein algebra coming from this permutation representation explains through (5.3.2) the origin of the famous idempotents indexed by the $2A$-involutions of the Monster.

5.4 Elementary induced modules

In this section we discuss certain generalizations of the 3-dimensional representations of S_4 from Section 5.2. Throughout the section, H is a finite group, V is a finite dimensional H-module, and

$$\psi : H \to GL(V)$$

is the corresponding representation. The H-module V and the representation ψ are said to be *monomial* if H stabilizes a direct sum decomposition

$$V = \bigoplus_{i \in I} D_i,$$

where each summand is 1-dimensional. Taking a non-zero vector from every direct summand D_i we obtain a *monomial basis* of V in which $\psi(H)$ is realized by *monomial matrices* which means having one non-zero entry in every column and in every row.

Lemma 5.4.1 *In each of the following cases the action of H on V is monomial:*

(i) *V is a permutation module of H;*
(ii) *ψ is induced from a linear representation of a subgroup of H;*
(iii) *H contains an abelian normal subgroup E such that every E-eigenspace in V is 1-dimensional;*
(iv) *V is a direct sum of monomial modules.*

Proof. The assertion (ii) is a standard property of induced modules, while (i) is a special case of (ii) corresponding to the trivial linear representation. Since H preserves the set of eigenspaces of a normal subgroup, (iii) follows. Finally (iv) is also quite clear. □

If ψ is induced from a linear representation, then the 1-dimensional summands in the corresponding decomposition are transitively permuted by H. In fact this property characterizes such representations in the class of all monomial representations. We will deal with an even further specialized class of monomial induced representations.

Definition 5.4.2 *Let H be a finite group, let V be an H-module, and let $\psi : H \to GL(V)$ be the corresponding representation. Let E be a normal subgroup in H which is elementary abelian of exponent 2. Then V and ψ are said to be elementary induced or more specifically E-induced if*

(i) *there is a subgroup K of H and a linear representation ζ of K such that $|\zeta(K)| = 2$ and ψ is ζ induced from K to H;*
(ii) *E acts on V fixed-point freely and every E-eigenspace in V is 1-dimensional.*

The E-eigenspaces, their centralizers in E, and the E-eigenvectors in V are called weight spaces, *roots,* and weight vectors, *respectively.*

The 3-dimensional irreducible S_4-modules discussed in Section 5.2 are (elementary) $O_2(S_4)$-induced.

Lemma 5.4.3 *If V is E-induced, then V is irreducible self-dual.*

Proof. It follows from the elementary part of Clifford's theory that V possesses the direct sum decomposition

$$V = \bigoplus_{E_i \in \mathcal{E}} C_V(E_i),$$

where $\mathcal{E}$ is the set of roots. The direct summands afford pairwise distinct irreducible linear representations of E and H acting by conjugation permutes the roots transitively. This means that H permutes transitively the summands in the above decomposition of V, in particular V is irreducible. The subgroup K in (5.4.2 (i)) is the normalizer in H of one of the roots, say of the root E_0. Therefore, $E \leq K$ and $E \cap \ker(\zeta) = E_0$. Let d_0 be a non-zero weight vector centralized by E_0. Then the image of d_0 under an element from K is either d_0 or $-d_0$. This implies that the H-orbit of d_0 intersects every weigh space $C_V(E_i)$ in precisely two vectors; if d_i is one of the vectors in the intersection, then the other one is $-d_i$. Subject to this notation the union D of d_is taken for all the weight spaces is a monomial basis of V in which $\psi(H)$ is represented by ± 1-monomial matrices. By declaring D to be orthonormal, we define on V an H-invariant inner product.   □

If E is considered in the usual manner as a $GF(2)$-vector space, then the roots are precisely those hyperplanes in E whose centralizers in V are non-zero. The centralizers of roots are the weight spaces. Two roots can be added as elements of the dual space of E. If E_i and E_j are distinct roots, then $E_i + E_j$ is the third hyperplane (different from E_i and E_j) containing the codimension 2 subspace $E_i \cap E_j$ in E. This third hyperplane might or might not be a root, and if it is a root, then the triple $B = \{E_i, E_j, E_i + E_j\}$ will be called a *bunch* (of roots). The 3-dimensional subspace R spanned by the weight spaces corresponding to the roots in a bunch B will be called *reper*. The intersection F of (any two or all three) roots in a bunch will be called the *core* of the bunch. Notice that $R = C_V(F)$.

Definition 5.4.4 *Let V be an E-induced H-module, and let ψ be the corresponding representation. Let $B = \{E_i, E_j, E_k\}$ be a bunch of roots with core F and let $R = C_V(F)$ be the corresponding reper. Then B and R are called* symmetric *of type $\epsilon \in \{+, -\}$ if:*

(i) $N_H(E/F)/C_H(E/F) \cong GL(E/F) \cong S_3$;
(ii) $N_H(R)/C_H(R) \cong S_4$;
(iii) R *affords a representation of $N_H(R)/C_H(R)$ isomorphic to θ^ϵ.*

Since E induces on the reper R, the elementary abelian group E/F of order 4 whose non-zero elements are indexed by the roots in the bunch B and since $N_H(R)/C_H(R)$ splits over E/F by Frattini argument, (5.4.4 (i)) implies (5.4.4 (ii)) .

In the case of a 3-dimensional S_4-representation, the roots are the order 2 subgroups in $O_2(S_4)$ and all three of them constitute the unique bunch whose core is the identity subgroup; the whole of V_3 is the only reper.

Lemma 5.4.5 *Let V be an E-induced H-module, and let*

$$\star : (u, v) \mapsto u \star v$$

be an H-invariant algebra multiplication on V. Let $d_i \in C_V(E_i)$ and $d_j \in C_V(E_j)$ be non-zero weight vectors, and put $d = d_i \star d_j$. Then:

(i) *whenever $E_i = E_j$ we have $d = 0$, in particular all the weight vectors are nilpotent;*

(ii) *if $E_i \neq E_j$, then $d \in C_V(E_i + E_j)$, therefore $d = 0$ unless*

$$\{E_i, E_j, E_i + E_j\}$$

is a bunch of roots.

Proof. In hypothesis (i), the vector d is centralized by E_i, which implies the inclusion $d \in C_V(E_i)$. On the other hand, $e \in E \setminus E_i$ negates both d_i and d_j, therefore (because of the bilinearity) it centralizes d. This forces d to be the zero vector. In the hypothesis of (ii) for $e \in E$, let d^e be the image of d under e. Then again because of the bilinearity we have

$$d^e = \varepsilon_i \varepsilon_j d,$$

where ε_i is plus or minus 1 depending on whether or not e is contained in E_i and similarly for ε_j. This shows that d is an E-eigenvector, whose centralizer in E is $E_i + E_j$ and (ii) follows. □

By (5.4.5), a non-zero H-invariant algebra multiplication on an E-induced H-module can only exist if there are repers, and every reper is closed under such a multiplication.

Proposition 5.4.6 *Let V be an E-induced H-module and suppose that every reper in V is symmetric. Then:*

(i) *the space of H-invariant bilinear multiplications on V has dimension equal to the number of H-orbits on the set of repers; if all the repers are of the same type $\epsilon \in \{+, -\}$, then every invariant algebra is symmetric or alternating depending on whether ϵ is plus or minus;*

(ii) *the assertion (i) holds with trilinear forms in place of bilinear multiplication.*

Proof. Let $\star$ be an H-invariant algebra multiplication, and let R be a reper. Then by (5.4.5 (ii)), R is $\star$-closed, while by (5.2.1) and since R is symmetric, $\star|_R$ must be $\star_\epsilon$ properly rescaled (where ϵ is the type of R). In particular, on every reper R there is a non-zero $N_H(R)$-invariant algebra multiplication. On the other hand, as soon as we specify the subalgebra $(R, \star|_R)$, the product of any two vectors contained in the image of R under an element of H is specified because of the H-invariance. Since any two repers are either disjoint or share a single weight space and all the weight vectors are nilpotent in both the reper-subalgebras, the subalgebras induced on repers from different orbits can be rescaled independently. Finally, the product of any two weight vectors not contained in a common reper is always zero by (5.4.5) and the result follows in view of the fact that whenever the products of all the pairs of weight vectors are specified, the multiplication expands uniquely on the whole of V via the distributivity law. This completes the proof of (i). The proof of (ii) almost literally repeats the above argument (compare the proof of (iv) in (5.2.1)). $\square$

As an immediate consequence of (5.4.3) and (5.4.6), we obtain the following:

Lemma 5.4.7 *Let V be an E-induced H-module. Suppose that the set of repers is non-empty forming a single H-orbit and suppose further that every reper is symmetric of plus type. Then (H, V) is a Norton pair.* $\square$

5.5 $(\Omega_{10}^{+}(2), \Pi_{155}^{10})$ is a Norton pair

We follow notation introduced in Sections 4.3 and 4.5. Put $\bar{G}_{10} = G_{10}/Q_{10} \cong \Omega_{10}^{+}(2)$ and adopt the bar convention for the images in $\bar{G}_{10}$ of subgroups of G_{10}. For $1 \leq i \leq 5$, put

$$G_{10,i} = G_{10} \cap G_i$$

and (when it matters) expand the superscripts (t) and (l) from G_5 to $G_{10,5}$. By (4.3.7) and its proof

$$\{\bar{G}_{10,1}, \bar{G}_{10,2}, \bar{G}_{10,3}, \bar{G}_{10,5}^{(t)}, \bar{G}_{10,5}^{(l)}\}$$

is the amalgam of maximal parabolics corresponding to the action of $\bar{G}_{10}$ on the D_5-geometry associated with the orthogonal space (Z_{10}, κ), and we have the following

$$\bar{G}_{10,1} \cong 2^8 : \Omega_8^{+}(2), \quad \bar{G}_{10,2} \cong 2_{+}^{1+12} : (A_8 \times S_3),$$
$$\bar{G}_{10,3} \cong 2^{3+12} : (L_3(2) \times \Omega_4^{+}(2)), \quad \bar{G}_{10,5} \cong 2^{10} : L_5(2).$$

The above isomorphisms and further information on the structure of the parabolics $\bar{G}_{10,i}$ to be used in this section are rather standard and can be found for instance in Section 2.1 of [Iv04].

Lemma 5.5.1 *Under the action of $\bar{G}_{10,5}$ the module Π^{10}_{155} is elementary $O_2(\bar{G}_{10,5})$-induced. The weight spaces are indexed by the 3-dimensional subspaces of Z_5. There is a single orbit on the set of repers and every reper is symmetric of plus type.*

Proof. Let $E = O_2(\bar{G}_{10,5})$ and $L \cong L_5(2)$ be a Levi complement to E in $\bar{G}_{10,5}$. Then E is the exterior square of Z_5 so that the dual E^* of E is the exterior square of the dual Z_5^* of Z_5. Therefore, the shortest orbit of $\bar{G}_{10,5}$ on the set of hyperplanes in E has length 155 and it is indexed by the 3-dimensional subspaces in Z_5. Since 155 is precisely the dimension of Π^{10}_{155} (and since the action of $\bar{G}_{10,5}$ on Π^{10}_{155} is faithful), the module in question must be E-induced. The sum of two hyperplanes from the 155-orbit is again a hyperplane from the same orbit if and only if the corresponding 3-subspaces intersect in a 2-space (and/or are contained in a 4-space). Hence, there is a single reper-orbit. In order to show that the repers are symmetric of plus type, it is sufficient to produce an S_3-subgroup which stabilizes a reper and induces on it the natural permutation representation. Let $L(Z_4) \cong 2^4 : L_4(2)$ be the stabilizer of Z_4 in L and let $K \cong L_4(2)$ be a Levi complement in $L(Z_4)$. The subspace Ξ in Π^{10}_{155} spanned by the weight spaces indexed by the 3-subspaces contained in Z_4 is 15-dimensional and the stabilizer in K of a weight space from Ξ is isomorphic to $2^3 : L_3(2)$. Since the latter group does not contain index 2 subgroups, we conclude that the action of K on Ξ is permutation. If R is a reper contained in Ξ (so that R corresponds to a 2-space contained in Z_4), then the stabilizer of R in K clearly induces S_3 on the triple of weight spaces contained in R and its action on R is permutation since the whole action of K on Ξ is permutation. This completes the proof of the assertion. $\square$

It might be worth mentioning that in terms of the above proof the action of $L \cong L_5(2)$ on Π^{10}_{155} is strictly monomial (not permutation). This can be seen by comparing the 155-dimensional characters of $\Omega^+_{10}(2)$ and $L_5(2)$ in [CCNPW].

Lemma 5.5.2 $(\bar{G}_{10}, \Pi^{10}_{155})$ *is a Norton pair.*

Proof. By (5.5.1) and (5.4.7), $(\bar{G}_{10,5}, \Pi^{10}_{155})$ is a Norton pair, while by (5.3.3), we know that Π^{10}_{155} carries a $\bar{G}_{10}$-invariant inner product β and an invariant multiplication $\star$. Therefore, Π^{10}_{155} also carries a $\bar{G}_{10}$-invariant trilinear form

$$(u, v, w) \mapsto \beta(u \star v, w).$$

Thus the unique (up to rescaling) $\bar{G}_{10,5}$-invariant inner product, multiplication, and trilinear form are $\bar{G}_{10}$-invariant and the assertion is established. $\square$

In order to accomplish our proof that (G, Π) is a Norton pair, we need some information on the action on Π_{155}^{10} of $\bar{G}_{10,1}$ and of $\bar{G}_{10,2}$. For the latter case, the information is already available in (4.5.1) and we start by restating it in the terms convenient for the future usage.

Lemma 5.5.3 *For* $\bar{G}_{10,2} \cong 2_+^{1+12} : (A_8 \times S_3)$, *let* $\bar{S}$ *denote the largest solvable normal subgroup in* $\bar{G}_{10,2}$ *so that* $\bar{G}_{10,2}/\bar{S} \cong A_8$. *Then* Π_{155}^{10} *possesses the following decomposition into irreducible* $\bar{G}_{10,2}$-*submodules*

$$\Pi_{155}^{10} = \Pi_7 \oplus \Pi_{3 \cdot 28} \oplus \Pi_{2^6},$$

where:

(i) *the kernel of* Π_7 *is* $\bar{S}$, Π_7 *is the faithful irreducible constituent of the natural permutation module of* $\bar{G}_{10,2}/\bar{S} \cong A_8$ *and* $\Pi_7 = \Pi_{155}^{10} \cap \Pi_{23}^2$;
(ii) *the kernel of* $\Pi_{3 \cdot 28}$ *is* $Z(\bar{G}_{10,2})$ *(of order 2)*, $\Pi_{3 \cdot 28}$ *is elementary induced for* $O_2(\bar{G}_{10,2})/Z(\bar{G}_{10,2}) \cong 2^{12}$ *and* $\Pi_{3 \cdot 28} = \Pi_{155}^{10} \cap \Pi_{3 \cdot 276}^2$;
(iii) Π_{2^6} *is faithful and it is the intersection of* Π_{155}^{10} *with* $\Pi_{2^6 \cdot 759}^2$. $\square$

The next lemma is easy to deduce combining (5.5.3), the fusion patterns of the G_{12}-submodules in Π into G_1- and G_2-submodules given in the paragraph after (3.1.21) and the structure of $\bar{G}_{10,1}$.

Lemma 5.5.4 *The module* Π_{155}^{10} *possesses the following decomposition into irreducible submodules of* $\bar{G}_{10,1} \cong 2^8 : \Omega_8^+(2)$

$$\Pi_{155}^{10} = \Pi_{35} \oplus \Pi_{120},$$

where:

(i) *the kernel of* Π_{35} *is* $O_2(\bar{G}_{10,1})$ *and* $\Pi_{35} = \Pi_{155}^{10} \cap \Pi_{299}^1$;
(ii) Π_{120} *is faithful, elementary* $O_2(\bar{G}_{10,1})$-*induced and it is the intersection of* Π_{155}^{10} *with* $\Pi_{|\bar{\Lambda}_2|}^1$. $\square$

5.6 Allowances for subalgebras

According to the plan set up the beginning of the chapter, our next objection is to produce a one-parametric family $A^{(z)}$ of G_1-invariant algebras on $C_\Pi(Z_1)$ whose restrictions to $C_\Pi(Z_2)$ are G_2-invariant. Towards this end we calculate with the spaces A_1 and A_2 of G_1- and G_2-invariant algebras on $C_\Pi(Z_1)$ and $C_\Pi(Z_2)$, respectively. Constructing $A^{(z)}$ amounts to showing that the space A_{12} formed by the restrictions to $C_\Pi(Z_2)$ of the algebras from A_1 intersects A_2 in a 1-dimensional subspace. This does not seem to be achievable by a mere dimension comparison. On the other hand, since in (5.5.2) $(\bar{G}_{10}, \Pi^{10}_{155})$ was proved to be a Norton pair, the restrictions of A_1 and A_2 to $C_\Pi(Q_{10})$ projected onto Π^{10}_{155} share the 1-dimensional space of G_{10}-invariant algebras on Π^{10}_{155}. So we recover $A^{(z)}$ through a careful analysis of the way the algebras from A_1 and A_2 restrict to $C_\Pi(Q_{10})$ and how the restrictions project onto Π^{10}_{155}.

Let H be a finite group, let V, W, and U be finite dimensional H-modules. Let $A_H(V, W; U)$ denote the space of H-invariant (bilinear) algebra multiplications

$$V \times W \to U.$$

For the sake of simplicity, we write $A_H(V; U)$ instead of $A_H(V, V; U)$ and $A_H(V)$ instead of $A_H(V, V; V)$. We start by stating a standard result.

Lemma 5.6.1 *In the above notation suppose that U is irreducible. Then the dimension of $A_H(V, W; U)$ is equal to the multiplicity of U in the decomposition into H-irreducibles of the tensor product $V \otimes W$.* $\square$

Lemma 5.6.2 *Let U, V and W be irreducible self-dual H-modules and let σ be a permutation of $\{U, V, W\}$. Then the dimension of*

$$A_H(\sigma(U), \sigma(V); \sigma(W))$$

does not depend on σ.

Proof. By (5.6.1), since the modules are self-dual, the dimension of $A_H(U, V; W)$ is precisely the multiplicity of the trivial character of H in the tensor product $T = U \otimes V \otimes W$. Hence, $\dim(A_H(U, V; W)) = \dim(C_T(H))$. Since the isomorphism of T onto $\sigma(T) = \sigma(U) \otimes \sigma(V) \otimes \sigma(W)$ induced by σ maps $C_T(H)$ onto $C_{\sigma(T)}(H)$, the result follows. $\square$

If V is an H-module which is a direct sum

$$V = \bigoplus_{i \in I} V_i$$

of non-isomorphic self-dual irreducible submodules, then $A_H(V)$ is the direct sum of the subspaces $A_H(V_i, V_j; V_k)$ (called *ingredients* of $A_H(V)$) taken over all the ordered triples (i, j, k) of elements from I. If we put

$$A^{(s)}_H(V_i, V_j, V_k) = \sum_{\sigma \in S_3} A_H(\sigma(V_i), \sigma(V_j); \sigma(V_k)),$$

then in view of (5.6.2) the sum can be rearranged into $A^{(s)}_H(V_i, V_j; V_k)$s.

Lemma 5.6.3 *Let V be an E-induced module for H and let U be an irreducible self-dual H-module which is centralized by E and which appears once in the permutation module of H on the set of weight spaces in V. Then*

$$A_H(U; V) = 0, \ while \ \dim(A_H(V; U)) = 1.$$

Proof. Since E centralizes the tensor product $U \otimes U$ while acting faithfully on V, the first equality is obvious. Let K be the stabilizer in H of a weight space D from V, let $d \in D$ be a non-zero weight vector, and let $\star : U \times V \to V$ be an H-invariant multiplication. Then for every $u \in U$, the product $u \star d$ is centralized by $C_E(D)$ and hence it is contained in D. Therefore

$$u \star d = \lambda(u)d \text{ for some scalar } \lambda(u)$$

and $u \mapsto \lambda(u)$ is a K-invariant homomorphism of U. Applying the Frobenius reciprocity, we conclude from the hypothesis of the lemma that $C_U(K)$ is 1-dimensional and so is $A_H(V, U; V)$. Because of (5.6.2), the second equality follows. $\square$

Lemma 5.6.4 *Let (H, U) be one of the following group-representation pairs*

$$(G_1/Q_1 \cong Co_1, \Pi^1_{299}), \ \ (G_2/C_{G_2}(\bar{R}_2) \cong M_{24}, \Pi^2_{23}),$$
$$(\bar{G}_{10,1}/O_2(\bar{G}_{10,1}) \cong \Omega_8^+(2), \Pi_{35}), \ \ (\bar{G}_{10,2}/\bar{S} \cong A_8, \Pi_7).$$

Then (H, U) is a Norton pair, in particular $\dim(A_H(U)) = 1$.

Proof. In view of (5.1.2) the result can be achieved by calculating with the relevant irreducible characters of Co_1, M_{24}, $\Omega_8^+(2)$, and A_8 found in [CCNPW] (the modules Π_{35} and Π_7 are defined in (5.5.4)). $\square$

In terms of the above lemma if Σ is one of Π_{35}, Π^2_{23}, and Π_7, then Σ is an H-submodule in Π^1_{299} for a suitable subgroup H of G_1 and hence the multiplication with respect to an algebra from $A_H(\Sigma)$ is the restriction to Σ of a multiplication contained in $A_{G_1}(\Pi^1_{299})$.

5.7 G_1-invariant algebras on $C_\Pi(Z_1)$

In this section we calculate with the space $A_1 := A_{G_1}(C_\Pi(Z_1))$ of G_1-invariant algebras on $C_\Pi(Z_1) = \Pi^1_{299} \oplus \Pi^1_{|\bar{\Lambda}_2|}$.

Lemma 5.7.1 *The following assertions hold:*

(i) *the module $\Pi^1_{|\bar{\Lambda}_2|}$ is Q_1/Z_1-induced;*

(ii) *there is a single reper-orbit of G_1 in $\Pi^1_{|\bar{\Lambda}_2|}$ and every reper is symmetric of plus type;*

(iii) *$(G_1/Z_1, \Pi^1_{|\bar{\Lambda}_2|})$ is a Norton pair;*

(iv) *$A_{G_1}(\Pi^1_{|\bar{\Lambda}_2|}; \Pi^1_{299})$ is 1-dimensional.*

Proof. Directly from the definition of $\Pi^1_{|\bar{\Lambda}_2|}$ (cf. (3.1.21 (ii)) and (3.1.6)), we obtain (i). Identifying Q_1/Z_1 with $\bar{\Lambda}$ through the isomorphism induced by $\chi : Q_1 \to \bar{\Lambda}$, we see by (3.1.7) that the weight spaces and the roots are

$$\{\Pi(\bar{\mu}) \mid \bar{\mu} \in \bar{\Lambda}_2\} \text{ and } \{\bar{\mu}^\perp \mid \bar{\mu} \in \bar{\Lambda}_2\},$$

respectively. Since $\bar{\Lambda}$ is self-dual, a triple of distinct elements of $\bar{\Lambda}_2$ forms a bunch of roots if and only if their sum is zero. This means that the repers are indexed by the so-called 222-triangles in $\bar{\Lambda}$. By a standard result (cf. Lemma 4.10 in [Iv99]), there is a single Co_1-orbit on the set of such triangles with stabilizer isomorphic to $U_6(2) : S_3$. The fact that every reper is symmetric of plus type can be justified as follows. Let $T = \{p, q, r\}$ be a 3-element subset of $\mathcal{P}$. Then the Leech vectors

$$\mu = 4p - 4q, \quad \nu = 4r - 4p \text{ and } \mu + \nu = 4r - 4q$$

are contained in Λ_2, therefore

$$R = \langle \Pi(\bar{\mu}), \Pi(\bar{\nu}), \Pi(\bar{\mu} + \bar{\nu}) \rangle$$

is a reper. The stabilizer of T in the permutation complement $M \cong M_{24}$ contained the monomial subgroup of Co_1 is isomorphic to $P\Sigma L_3(4)$ and it induces S_3 on $\{\bar{\mu}, \bar{\nu}, \bar{\mu} + \bar{\nu}\}$. By (3.1.7), the subgroup $M(T)$ induces S_3 on R. Hence, (ii) follows by (5.2.1 (iii)). Now (iii) is a consequence of (i), (ii), and (5.4.7). The permutation module of Co_1 on $\bar{\Lambda}_2$ is known to involve a single copy of Π^1_{299} (cf. Lemma 2.22.1 (i) in [ILLSS]), therefore (iv) follows from (5.6.3). $\square$

Lemma 5.7.2 *Let $\pi_{1\to 10}$ be the restrictions map of the algebras from A_1 to $C_\Pi(Q_{10})$ projected onto Π^{10}_{155}. Then $\pi_{1\to 10}$ is an isomorphism of A_1 onto $A_{G_{10,1}}(\Pi^{10}_{155})$.*

Proof. Since every multiplication from A_1 is closed on $C_\Pi(Q_{10})$ and since Π^{10}_{155} is G_{10}-invariant, the map $\pi_{1\to 10}$ is well defined. By (5.5.4 (ii)), the module $\Pi_{120} = \Pi^{10}_{155} \cap \Pi^1_{|\bar{\Lambda}_2|}$ is $O_2(\bar{G}_{10,1})$-induced. Since $O_2(\bar{G}_{10,1})$ is a natural 8-dimensional $GF(2)$-module for

$$\bar{G}_{10,1}/O_2(\bar{G}_{10,1}) \cong \Omega^+_8(2),$$

it follows from the order consideration that the weight spaces correspond to non-isotropic vectors, while the repers correspond to the minus-lines. Since $Q_1 Q_{10} = O_2(G_{10,1})$, the repers in Π_{120} are also repers in $\Pi^1_{|\bar{\Lambda}_2|}$. Hence $\pi_{1 \to 10}$ maps $A_{G_1}(\Pi^1_{|\bar{\Lambda}_2|})$ bijectively onto $A_{G_{10,1}}(\Pi_{120})$. The permutation module of $\Omega_8^+(2)$ on the set of non-isotropic vectors involves a single copy of a 35-dimensional irreducible module (cf. [CCNPW]), therefore $\pi_{1 \to 10}$ establishes an isomorphism between the 1-dimensional subspaces $A_{G_1}(\Pi^1_{|\bar{\Lambda}_2|}; \Pi_{299})$ and $A_{G_{10,1}}(\Pi_{120}; \Pi_{35})$. Now in order to complete the proof it is sufficient to refer to (5.6.2) and (5.6.4). □

Although (5.7.2) is basically all we need to know about A_1, for the sake of completeness we include the following:

Lemma 5.7.3 $\dim(A_1) = 5$.

Proof. The number of irreducible constituents is 2, hence the number of ingredients is $2^3 = 8$. By (5.6.2) and (5.6.3), $A_{G_1}(\Pi^1_{299}; \Pi^1_{|\bar{\Lambda}_2|})$ and two its symmetric relatives are zero, while each of the remaining five ingredients is 1-dimensional by (5.7.1 (iii), (iv)) and (5.6.2). □

5.8 G_2-invariant algebras on $C_\Pi(Z_2)$

In this section we describe the G_2^s-invariant algebras on $C_\Pi(Z_2)$ and distinguish those of them which are G_{12}- and G_2-invariant. We start by reviewing some basic facts on the structure of G_2^s from Section 2.3 and on the action of G_2^s on $C_\Pi(Z_2)$ from Section 3.1.

The group G_2^s acts on

$$C_\Pi(Z_2) = \Pi_{23}^2 \oplus \Pi_{3 \cdot 276}^2 \oplus \Pi_{2^6 \cdot 759}^2$$

via $\bar{G}_2^s = G_2^s/Z_2$ and the latter quotient is a partial semidirect product of $\bar{Q}_2$ and $\bar{M}^\delta$ over $\bar{R}_2$. In their turn, $\bar{Q}_2$ is the product of three elementary abelian groups $\bar{Q}_2^{(a)}$ of order 2^{22} each intersecting in $\bar{R}_2$ and indexed by the elements $a \in Z_2^\#$, while $\bar{M}^\delta \cong C_{11}^* \cdot M_{24}$. In particular, $\bar{R}_2$ and $\bar{Q}_2^{(a)}/\bar{R}_2$ are isomorphic to C_{11}^* and C_{11}, respectively (as the modules for $\bar{M}^\delta/\bar{R}_2 \cong M_{24}$). Thus

$$\bar{G}_2^s/\bar{R}_2 \cong (\bar{Q}_2^{(a)} \times \bar{Q}_2^{(b)}) : (\bar{M}^\delta/\bar{R}_2) \cong (C_{11} \times C_{11}) : M_{24}.$$

The module Π_{23}^2 is the faithful irreducible constituent of the permutation module of $G_2^s/Q_2 \cong M_{24}$ on $\mathcal{P}$.

The module $\Pi^2_{3 \cdot 276}$ is the direct sum the irreducible 276-dimensional G_2^s-modules

$$\Pi^{(a)}_{276} := C_{\Pi^2_{3 \cdot 276}}(Q_2^{(a)})$$

taken for all three $a \in Z_2^\#$. Furthermore, $\Pi^{(a)}_{276}$, considered as a module for $G_2^s / Q_2^{(a)} \cong C_{11} : M_{24}$, is E-induced for $E = Q_2 / Q_2^{(a)} \cong Q_2^{(b)} / R_2 \cong C_{11}$, where $b \in Z_2^\# \setminus \{a\}$. The weight spaces $\Pi^{(a)}_{\{p,q\}}$ are indexed by the pairs from $\mathcal{P}$ and, if $x \in Z_2^\# \setminus \{z\}$, then

$$C_{Q_2^{(z)}}(\Pi^{(x)}_{\{p,q\}}) = \chi^{-1}((4p - 4q + 2\Lambda)^\perp \cap (\sum_{r \in \mathcal{P}} 2r + 2\Lambda)^\perp).$$

The module $\Pi^2_{2^6 \cdot 759}$ is $\bar{Q}_2^{(a)}$-induced for every $a \in Z_2^\#$ and, if $a = z$, then the corresponding weight spaces $\Pi(\mathcal{O}, \sigma)$ are indexed by the $\bar{\Lambda}_2$-elements

$$\bar{\lambda}(\mathcal{O}, \sigma) = \sum_{p \in s_1} 2p - \sum_{q \in s_2} 2q + 2\Lambda,$$

where $\mathcal{O}$ is an octad and $\sigma : \mathcal{O} = s_1 \cup s_2$ is a partition of $\mathcal{O}$ into two even subsets.

Lemma 5.8.1 *The following assertions hold:*

(i) $\dim(A_{G_2^s}(\Pi^2_{2^6 \cdot 759})) = 2$;
(ii) $A_{G_{12}}(\Pi^2_{2^6 \cdot 759})$ and $A_{G_2}(\Pi^2_{2^6 \cdot 759})$ are equal and 1-dimensional.

Proof. For each $a \in Z_2^\#$, the G_2^s-module $\Pi^2_{2^6 \cdot 759}$ is a $Q_2^{(a)} / Z_2$-induced, while considered as a G_{12}-module it is Q_1 / Z_2-induced. Hence we have to classify the relevant reper-orbits and apply (5.4.6). Without loss of generality, we assume that $a = z$. The weight spaces with respect to both Q_1 / Z_2- and $Q_2^{(z)} / Z_2$-induced structures are the subspaces

$$\Pi(\mathcal{O}, \sigma) = C_{\Pi^2_{2^6 \cdot 759}}(\chi^{-1}(\bar{\lambda}(\mathcal{O}, \sigma)^\perp),$$

where

$$\bar{\lambda}(\mathcal{O}, \sigma) = \sum_{p \in s_1} 2p - \sum_{q \in s_2} 2q + 2\Lambda,$$

with $\mathcal{O}$ being an octad and $\sigma : \mathcal{O} = s_1 \cup s_2$ being a partition of $\mathcal{O}$ into two even subsets.

A triple $\{\bar{\mu}_1, \bar{\mu}_2, \bar{\mu}_3\}$ of $\bar{\lambda}(\mathcal{O}, \sigma)$s corresponds to a frame with respect to $Q_2^{(z)} / Z_2$ if and only if

$$\bar{\mu}_1 + \bar{\mu}_2 + \bar{\mu}_3 = \varepsilon \bar{\lambda}_4,$$

where $\bar{\lambda}_4 = \sum_{p \in \mathcal{P}} 2p + 2\Lambda$ is the generator of $\chi(Z_2)$ and $\varepsilon \in \{0, 1\}$. The parameter ε will be called the *level* of the frame. The flames of level 0 correspond to the Q_1/Z_2-induced structure of $\Pi^2_{2^6 \cdot 759}$. In other terms, the frames of level 0 are the frames of $\Pi^1_{|\bar{\Lambda}_2|}$ with respect to Q_1/Z_1 which are contained in $\Pi^2_{2^6 \cdot 759}$.

It is a standard result from the Conway–Leech theory (which can be checked by elementary calculations) that G_2^s permutes transitively the frames in $\Pi^2_{2^6 \cdot 759}$ of a given level 0 or 1. The orbit on the frames of level 0 is represented by

$$\{\bar{\lambda}(\mathcal{O}_1, \sigma_1), \bar{\lambda}(\mathcal{O}_2, \sigma_2), \bar{\lambda}(\mathcal{O}_3, \sigma_3)\},$$

where

$$\mathcal{O}_3 = (\mathcal{O}_1 \cup \mathcal{O}_2) \setminus (\mathcal{O}_1 \cap \mathcal{O}_2)$$

so that $\mathcal{S} := \mathcal{O}_1 \cap \mathcal{O}_2$ is a 4-set; while σ_1 and σ_2 are trivial partitions (involving the empty set) and

$$\sigma_3 : \mathcal{O}_3 = (\mathcal{O}_1 \setminus \mathcal{S}) \cup (\mathcal{O}_2 \setminus \mathcal{S}).$$

The orbit on the frames of level 1 is represented by

$$\{\bar{\lambda}(\mathcal{O}_1, \sigma_1), \bar{\lambda}(\mathcal{O}_2, \sigma_2), \bar{\lambda}(\mathcal{O}_3, \sigma_3)\},$$

where $\{\mathcal{O}_1, \mathcal{O}_2, \mathcal{O}_3\}$ is a trio and all the partitions involved are trivial. Therefore, (i) follows from (5.4.6).

For $X = A_{G_2^s}(\Pi^2_{2^6 \cdot 759})$ and $a \in Z_2^{\#}$, let $X_0^{(a)}$ and $X_1^{(a)}$ be the subspaces of X corresponding to level 0 and 1 frames with respect to the $Q_2^{(a)}/Z_2$-induced structure of $\Pi^2_{2^6 \cdot 759}$. This means that a non-zero G_2^s-invariant multiplication $\star_\varepsilon$ is contained in $X_\varepsilon^{(a)}$ if the following holds: given two non-zero weight vectors u and v, the product $u \star_\varepsilon u$ is non-zero if and only if u and v are contained in a common $Q_2^{(a)}/Z_2$-frame of level ε. By (5.4.6) and (i), we have

$$X = X_0^{(a)} \oplus X_1^{(a)}.$$

We claim that the above decomposition is G_2-stable, which means that $X_\varepsilon^{(a)}$ is a-independent for $\varepsilon = 0$ and for $\varepsilon = 1$. First of all, since G_2 normalizes both G_2^s and $\Pi^2_{2^6 \cdot 759}$, the action of G_2 on X is well defined. The Q_2-irreducible submodules inside $\Pi^2_{2^6 \cdot 759}$ are 2^6-dimensional indexed by the octads and G_2 induces on the set of these irreducible submodules the natural action of M_{24}. In terms of the $Q_2^{(z)}/Z_2$-induced structure of $\Pi^2_{2^6 \cdot 759}$, the Q_2-irreducible corresponding to $\mathcal{O}$ is the sum of the weight spaces $\Pi(\mathcal{O}, \sigma)$ taken for all the 2^6 even partitions of $\mathcal{O}$. By the proof of (i), the weight spaces of every $Q_2^{(a)}/Z_2$-frame are contained in pairwise different Q_2-irreducibles. If a and b are distinct elements of $Z_2^\#$, then an element of G_2 which conjugates a onto b maps every $Q^{(a)}/Z_2$-frame onto a $Q_2^{(b)}/Z_2$-frame. On the other hand, the frames of levels 0 and 1 correspond, respectively, to triple octads with pairwise intersections of sizes 4 and 0, respectively. The group G_2 inducing M_{24} on the set of octads preserves the intersection sizes. Hence the claim follows and we can drop the superscripts (a) and write simply X_0 and X_1.

Since the frames of level 1 are not frames with respect to the Q_1/Z_2-induced structure and those of level 0 are still frames, we conclude that $G_{12}/G_2^s \cong 2$ negates X_1 and centralizes X_0. Since $G_2/G_2^s \cong S_3$ and since X_εs are proved to be G_2-invariant, we conclude that the whole of G_2/G_2^s centralizes X_0 and induces on X_1 an action of order 2. □

Let us now turn to the remaining ingredients of $A_{G_2^s}(C_\Pi(Z_2))$ corresponding to the following decomposition of $C_\Pi(Z_2)$ into G_2^s-irreducibles

$$C_\Pi(Z_2) = \Pi_{23}^2 \bigoplus_{a \in Z_2^\#} \Pi_{276}^{(a)} \oplus \Pi_{2^6.759}^2.$$

Lemma 5.8.2 *Let U be either Π_{23}^2 or $\Pi_{276}^{(a)}$ for some $a \in Z_2^\#$. Then*

$$\dim(A_{G_2^s}(\Pi_{2^6.759}^2; U)) = 1.$$

Proof. Since $\Pi_{2^6.759}^2$ is $Q_2^{(a)}/Z_2$-induced while U is centralized by $Q_2^{(a)}$ and because of (5.6.3) all we have to show is that U appears once in the permutation module of G_2^s/Q_2 on the set of weight spaces in $\Pi_{2^6.759}^2$. Without loss of generality we assume that $a = z$ and identify $G_2^s/Q_2^{(z)} \cong G_{12}/Q_1$ with the monomial subgroup $F = EM \cong C_{11} : M_{24}$ of Co_1. Subject to this identification, the stabilizer K of a weight space is a subgroup of index 2^6 in the octad stabilizer

$$F(\mathcal{O}) = EM(\mathcal{O}) \cong 2^{11} : 2^4 : A_8.$$

This condition specifies K up conjugation in F, particularly $K \cong 2_+^{1+8} : A_8$. Since $\Pi_{276}^{(z)}$ is induced from the linear representation of F with kernel $E_{\{p,q\}}M(\{p, q\})$ (cf. (3.1.18 (vi)) and (3.1.19)), it is straightforward to invoke the Frobenius reciprocity and to prove the assertion by showing that $C_U(K)$ is 1-dimensional. □

As a direct consequence of (5.8.2) and its proof we obtain the following:

Lemma 5.8.3 *Every algebra from $A_{G_2^s}(\Pi_{2^6.759}^2; \Pi_{23}^2)$ is G_2-invariant.* □

Lemma 5.8.4 *Each of $A_{G_2^s}(\Pi_{276}^{(a)}; \Pi_{23})$ and $A_{G_2}(\Pi_{3\cdot276}^2, \Pi_{23})$ has dimensional 1.*

Proof. Since the stabilizer in M_{24} of a pair from $\mathcal{P}$ has 1-dimensional centralizer in Π_{23}, the first assertion follows from (5.6.3). Thus the space $A_{G_2^s}(\Pi_{3\cdot276}^2)$ is 3-dimensional and $G_2/G_2^s \cong S_3$ induces on this space the natural permutation action. Since the centralizer of the action is 1-dimensional, the second assertion follows. □

Lemma 5.8.5 *The following assertions hold:*

(i) *the $\bar{G}_2$-module $\Pi_{3\cdot276}^2$ is $\bar{Q}_2$-induced;*
(ii) *there are precisely two $\bar{G}_2$-orbits on the set of repers in $\Pi_{3\cdot276}^2$ with representatives $R^{(i)}$ and $R^{(o)}$ such that $R^{(i)} \le \Pi_{276}^{(a)}$ for some $a \in Z_2^{\#}$ and $R^{(o)} \cap \Pi_{276}^{(a)} \ne 0$ for all $a \in Z_2^{\#}$;*
(iii) *both $R^{(i)}$ and $R^{(o)}$ are symmetric of plus type.*

Proof. Notice that $\bar{Q}_2$, as a module for $\bar{G}_2/Q_2 \cong M_{24} \times S_3$, is the tensor product of $\mathcal{C}_{11}$ and of the natural 2-dimensional module for $S_3 \cong GL_2(2)$. By the definition of $\Pi_{3\cdot276}^2$ in (3.1.21 (vi)), a root $\rho(\{p, q\}, a)$ is determined by $\{p, q\} \subseteq \mathcal{P}$ and by $a \in Z_2^{\#}$. The following are the only two cases where the sum of two roots is again a root

$$\rho(\{p, q\}, a) + \rho(\{p, r\}, a) = \rho(\{q, r\}, a),$$
$$\rho(\{p, q\}, a) + \rho(\{p, q\}, b) = \rho(\{p, q\}, c).$$

These cases correspond to the repers $R^{(i)}$ and $R^{(o)}$, respectively. This proves (i) and (ii). The assertion (iii) is easy to prove making use of the fact that $\bar{G}_2$ splits over $\bar{Q}_2$. □

Lemma 5.8.6 *The spaces $A_{G_2}(C_\Pi(Z_2))$ and $A_{G_{12}}(C_\Pi(Z_2))$ have dimensions 13 and 22, respectively.*

Proof. The dimensions in question can be calculated as sums of dimensions of the relevant ingredients and the data and organized in the table below. A row in the table corresponds to pair (V_i, V_j) of G_2-irreducibles in $C_\Pi(Z_2)$ and a column (except for the last one) corresponds to a further such irreducible V_k so that the corresponding entry is the pair

$$(\dim(A_{G_2}(V_i, V_j; V_k)), \dim(A_{G_{12}}(V_i, V_j; V_k))).$$

The last column indicates the total contribution of the ingredients corresponding to the triples (V_i, V_j, V_k) and (V_j, V_i, V_k) of G_2-irreducibles (here V_i and V_j are determined by the row, while V_k runs through the three irreducibles).

The pair at the bottom right gives the resulting dimensions and it is the componentwise sum of the pairs above it.

By (5.6.2), the triples (V_i, V_j, V_k) and (V_k, V_i, V_j) make equal contributions to the dimensions, therefore some information in the table is redundant.

Let us explain how the dimensions in the table come about. Since

$$\Pi_{23}^2 = C_\Pi(Q_2) \text{ and } \Pi_{23}^2 \oplus \Pi_{3\cdot276}^2 = C_\Pi(R_2),$$

every G_2^s-invariant multiplication is closed on Π_{23}^2 and on $\Pi_{23}^2 \oplus \Pi_{3\cdot276}^2$, which explains all the zeros in the table. The non-zero entries in the rows one to six follow from (5.8.1 (iii)), (5.8.2), (5.8.4), (5.8.5), (5.8.4), and (5.6.4), respectively. The second entry in the fourth row deserves a few further lines of explanations. Recall that $\Pi_{2^6\cdot759}^2$ and Π_{23}^2 stay irreducible when restricted to G_{12}, while $\Pi_{3\cdot276}^2$ splits into $\Pi_{276}^{12} = \Pi_{276}^{(z)}$ and $\Pi_{2\cdot276}^{12} = \Pi_{276}^{(x)} \oplus \Pi_{276}^{(y)}$. Under the action of G_{12}, the G_2-orbit of the reper $R^{(i)}$ splits into two orbits: the repers contained in Π_{276}^{12} and those contained in $\Pi_{2\cdot276}^{12}$; the reper $R^{(i)}$ stays symmetric. Therefore, by (5.4.6) the G_2-orbit of $R^{(i)}$ contributes 2 to the dimension of $A_{G_{12}}(C_\Pi(Z_2))$. On the other hand, the action of G_{12} on the G_2-orbit of the reper $R^{(o)}$ is transitive but the reper itself is no longer symmetric: its stabilizer on G_{12} induces only the dihedral group D_8. It is easy to deduce from the discussions in Section 5.2 that the space of D_8-invariant multiplications on $R^{(o)}$ is 3-dimensional. Hence the corresponding entry is in fact (2,5).

		$\Pi_{2^6\cdot759}^2$	$\Pi_{3\cdot276}^2$	Π_{23}^2	$(d_{G_2}, d_{G_{12}})$
$\Pi_{2^6\cdot759}^2$,	$\Pi_{2^6\cdot759}^2$	(1,1)	(1,2)	(1,1)	(3,4)
$\Pi_{2^6\cdot759}^2$,	$\Pi_{3\cdot276}^2$	(1,2)	(0,0)	(0,0)	(2,4)
$\Pi_{2^6\cdot759}^2$,	Π_{23}^2	(1,1)	(0,0)	(0,0)	(2,2)
$\Pi_{3\cdot276}^2$,	$\Pi_{3\cdot276}^2$	(0,0)	(2,5)	(1,2)	(3,7)
$\Pi_{3\cdot276}^2$,	Π_{23}^2	(0,0)	(1,2)	(0,0)	(2,4)
Π_{23}^2,	Π_{23}^2	(0,0)	(0,0)	(1,1)	(1,1)
					(13,22)

□

For H being G_{12} or G_2 let $B_H(C_\Pi(Z_2))$ be the sum of all the ingredients of $A_H(C_\Pi(Z_2))$ except for $A_H(\Pi_{2^6\cdot759}^2)$ and for those involved in

$$A_H^{(s)}(\Pi_{2^6\cdot759}^2, \Pi_{2^6\cdot759}^2, \Pi_{23}^2) = \sum_{\sigma \in S_3} A_H(\sigma(\Pi_{2^6\cdot759}^2), \sigma(\Pi_{2^6\cdot759}^2); \sigma(\Pi_{23}^2)).$$

Lemma 5.8.7 *For H being G_{12} or G_2, let $\pi_{2\to 10}$ be the restriction map of the algebras from $A_H(C_\Pi(Z_2))$ to Π_{155}^{10}. Then:*

(i) $\ker(\pi_{2\to 10}) = A_H(\Pi_{2^6 \cdot 759}^2) \oplus A_H^{(s)}(\Pi_{2^6 \cdot 759}^2, \Pi_{2^6 \cdot 759}^2; \Pi_{23}^2);$

(ii) *the restriction of $\pi_{2\to 10}$ to $B_H(C_\Pi(Z_2))$ is an isomorphism onto* $A_{G_{10}\cap H}(\Pi_{155}^{10})$.

Proof. Recall that Π_{2^6} is the Q_2-orbit on $\Pi_{2^6 \cdot 759}^2$ which corresponds to an octad $\mathcal{O}$. On the other hand, a Q_2-reper in $\Pi_{2^6 \cdot 759}^2$ intersects three such orbits (compare the proof of (5.8.1)). Therefore, $A_H(\Pi_{2^6 \cdot 759}^2)$ is in fact contained in the kernel of the projection map $\pi_{2\to 10}$. To prove (ii) we explore an analogy between the pairs $(G_2/Z_2, C_\Pi(Z_2))$ and $(\bar{G}_{10,2}, \Pi_{155}^{10})$.

		Π_{2^6}	$\Pi_{3 \cdot 28}$	Π_7	$(d_{G_{10,2}}, d_{G_{10,1}\cap G_{10,2}})$
Π_{2^6},	Π_{2^6}	(0,0)	(1,2)	(0,0)	(1,2)
Π_{2^6},	$\Pi_{3 \cdot 28}$	(1,2)	(0,0)	(0,0)	(2,4)
Π_{2^6},	Π_7	(0,0)	(0,0)	(0,0)	(0,0)
$\Pi_{3 \cdot 28}$,	$\Pi_{3 \cdot 28}$	(0,0)	(2,5)	(1,2)	(3,7)
$\Pi_{3 \cdot 28}$,	Π_7	(0,0)	(1,2)	(0,0)	(2,4)
Π_7,	Π_7	(0,0)	(0,0)	(1,1)	(1,1)
					(9,18)

Let $\bar{G}_{10,2}^s$ and $\bar{P}$ be the images in $\bar{G}_{10,2}$ of $G_2^s \cap G_{10}$ and Q_2, respectively. Then $\bar{P} \cong 2_+^{1+12}$ with $Z(\bar{P})$ being the image of R_2 and if $\bar{P}^{(a)}$ is the image of $Q_2^{(a)}$ for $a \in Z_2^\#$, then $\bar{P}^{(a)}$ is elementary abelian of order 2^7, $\bar{P}^{(a)}$ is normal in $\bar{G}_{10,2}^s$, and $\bar{G}_{10,2}^s/\bar{P}^{(a)} \cong 2^6 : A_8$. The module Π_{2^6} affords a faithful irreducible representation of the extra-special group $\bar{P}$. It can be deduced from the basic properties of this representation that the $\bar{G}_{10,2}^s$-module Π_{2^6} is $\bar{P}^{(a)}$-induced for every $a \in Z_2^\#$. The roots are all the hyperplanes disjoint from $Z(\bar{P})$, in particular there are no repers. The stabilizer of a weight space is an A_8-complement to $\bar{P}$ in $\bar{G}_{10,2}^s$. The permutation module of $\bar{G}_{10,2}^s/\bar{P}^{(a)} \cong 2^6 : A_8$ on the cosets of an A_8-complement has three irreducible constituents of degree 1, 28, and 35. In particular, it does not involve Π_7. Therefore, $A_{G_{10}\cap H}(\Pi_{2^6}, \Pi_7) = 0$, and hence the kernel of $\pi_{2\to 10}$ is as stated.

If we put $\Pi_{28}^{(a)} = \Pi_{276}^{(a)} \cap \Pi_{155}^{10}$ for $a \in Z_2^\#$, then the $\bar{G}_{10,2}^s$-module $\Pi_{28}^{(a)}$ is $\bar{P}^{(b)}/Z(\bar{P})$-induced for $b \in Z_2^\# \setminus \{a\}$ with weight spaces indexed by the 2-element subsets of the octad $\mathcal{O}$. It is easy to deduce from this description that there is a single reper-orbit of $\bar{G}_{10,2}^s$ inside $\Pi_{28}^{(a)}$. The $\bar{G}_{10,2}$-module $\Pi_{3 \cdot 28}$ is $\bar{P}/Z(\bar{P})$-induced with two reper-orbits as in (5.8.5). With this information in hand, it is rather straightforward to make use of (5.5.3) and (5.6.4) to calculate the dimensions of the ingredients of $A_{\bar{G}_{10,2}}(\Pi_{155}^{10})$ and $A_{\bar{G}_{10,1} \cap \bar{G}_{10,2}}(\Pi_{155}^{10})$. The dimensions are organized into the above table. □

5.9 Producing $A^{(z)}$

In this section we show that the space $A^{(z)}$ of G_1-invariant algebras on $C_{\Pi}(Z_1)$ whose restrictions to $C_{\Pi}(Z_2)$ are G_2-invariant has dimension 1. The most important tool is the following consequence of the results established in the previous section.

Lemma 5.9.1 *Let* $\star$ *be a* G_{12}-*invariant product on* $C_{\Pi}(Z_2)$ *whose restriction to* $C_{\Pi}(Q_{10})$ *projected onto* Π_{155}^{10} *is* $G_{10,2}$-*invariant. Then* $\star$ *is* G_2-*invariant.*

Proof. Let us represent $\star$ as the sum of a multiplication $\star_A$ from $\ker(\pi_{2 \to 10})$ and a multiplication $\star_B$ from the sum $B_{G_{12}}(C_{\Pi}(Z_2))$ of the ingredients not involved in $\ker(\pi_{2 \to 10})$. This means that the product $u \star_A v$ is always contained in $\Pi_{2^6 \cdot 759}^2$ and it is non-zero only if $u, v \in \Pi_{2^6 \cdot 759}^2$, while the product $u \star_B v$ is contained in $\Pi_{3 \cdot 2765}^2 \oplus \Pi_{23}^2$ whenever $u, v \in \Pi_{2^6 \cdot 759}^2$. By (5.8.1 (ii)) and (5.8.3), the multiplication $\star_A$ is G_2-invariant (since it is G_{12}-invariant). By (5.8.7), the mapping $\pi_{2 \to 10}$ establishes an isomorphism of $B_{G_{12}}(C_{\Pi}(Z_2))$ onto $A_{\bar{G}_{10,1} \cap \bar{G}_{10,2}}(\Pi_{155}^{10})$, and $\pi_{2 \to 10}$ maps the subspace $B_{G_2}(C_{\Pi}(Z_2))$ onto $A_{\bar{G}_{10,2}}(\Pi_{155}^{10})$ isomorphically. Therefore, $\star_B$ is G_2-invariant (since it is G_{12}-invariant by the hypothesis of the lemma). This implies that $\star$ itself is G_2-invariant and the result follows. □

By (5.5.2), $A_{G_{10}}(\Pi_{155}^{10})$ is 1-dimensional, and by (5.7.2) the mapping $\pi_{1 \to 10}$ is an isomorphism of A_1 onto $A_{G_{10,1}}(\Pi_{155}^{10})$. Since $A_{G_{10}}(\Pi_{155}^{10})$ is a subspace of $A_{G_{10,1}}(\Pi_{155}^{10})$, we put

$$A^{(z)} = \pi_{1 \to 10}^{-1}(A_{G_{10}}(\Pi_{155}^{10})).$$

Lemma 5.9.2 *If* $\star$ *is a multiplication from* $A^{(z)}$, *then the restriction of* $\star$ *to* $C_{\Pi}(Z_2)$ *is* G_2-*invariant.*

Proof. Since $A^{(z)}$ is G_1-invariant, the restriction $\star_2$ of $\star$ to $C_{\Pi}(Z_2)$ is clearly G_{12}-invariant. On the other hand, the equality

$$\pi_{1 \to 10}(\star) = \pi_{2 \to 10}(\star_2)$$

holds. Therefore, the restriction of $\star_2$ to $C_{\Pi}(Q_{10})$ projected onto Π_{155}^{10} is $\bar{G}_{10,2}$-invariant, since $\pi_{1 \to 10}(\star) \in A_{G_{10}}(\Pi_{155}^{10})$ by the definition of $A^{(z)}$. Now the result is immediate from (5.9.1). □

The main result of the construction accomplished so far is the following:

Proposition 5.9.3 *The set of G_1-invariant multiplications on $C_\Pi(Z_1)$ whose restrictions to $C_\Pi(Z_2)$ are G_2-invariant coincides with $A^{(z)}$.*   □

5.10 Expanding $A^{(z)}$

By the result of the previous section, we have a 1-dimensional family $A^{(z)}$ of G_1-invariant algebras on $C_\Pi(Z_1)$ whose restrictions to $C_\Pi(Z_2)$ are G_2-invariant.

Let $\star_z$ be a multiplication from $A^{(z)}$, let a be a $2B$-involution in G, and let g be an element of G which conjugates z onto a, so that $C_\Pi(a) = \{g(u) \mid u \in C_\Pi(z)\}$. Define a multiplication $\star_a^g$ on $C_\Pi(a)$ by the following rule

$$u \star_a^g v = g(g^{-1}(u) \star_z g^{-1}(v)),$$

where $u, v \in C_\Pi(a)$.

Lemma 5.10.1 *In the above terms, $\star_a^g$ is (gG_1g^{-1})-invariant multiplication on $C_\Pi(a)$ and $\star_a^g$ does not depend of the choice of g within the same left G_1-coset.*

Proof. Let h be another element in G which conjugates z onto a, and suppose that $h^{-1}g \in G_1$. Applying h^{-1} to both sides of the defining equality of $\star_a^g$, we obtain

$$h^{-1}(u \star_a^g v) = h^{-1}g(g^{-1}(u) \star_z g^{-1}(v)).$$

Since $h^{-1}g$ is contained in G_1 and $\star_z$ is G_1-invariant, the right-hand side of the above equality is equal to $h^{-1}(u) \star_z h^{-1}(v)$, which is $h^{-1}(u \star_a^h v)$ by the definition of $\star_a^h$. This proves both the independence and the invariance of $\star_a^g$. □

Lemma 5.10.2 *In the hypothesis of (5.10.1), suppose that $a \in Z_2^\#$ and that $g \in G_2$. Then $\star_a^g$ does not depend of the choice of such g and the restrictions of $\star_z$ and $\star_a^g$ to $C_\Pi(Z_2) = C_\Pi(z) \cap C_\Pi(a)$ coincide.*

Proof. Since $C_{G_2}(z) = G_1 \cap G_2$, the left coset of G_1 in G containing g is uniquely determined by the choice of a, which proves the g-independence by (5.10.1). Since the restriction of $\star_z$ to $C_\Pi(Z_2)$ is G_2-invariant, the second assertion also follows.   □

Lemma 5.10.3 *In the hypothesis of (5.10.1), suppose that $a \in Z_4^\#$ and that $g \in G_3 \cup G_4$. Then $\star_a^g$ does not depend on the choice of such g. Furthermore, the restrictions of $\star_z$ and $\star_a^g$ to $C_\Pi(z) \cap C_\Pi(a)$ coincide and are (hG_2h^{-1})-independent, where h is an element from G_{14} which conjugates x onto a.*

Proof. Let $m \in \{3, 4\}$. Since $C_{G_m}(z) = G_1 \cap G_m$, the left coset of G_1 in G containing g is uniquely determined by the choice of a and m. If $a \in Z_3$ and both $g \in G_3$ and $f \in G_4$ conjugate z onto a, then there exists $k \in G_{34}$ which does the same. By the above and (5.10.1) we have

$$\star_a^g = \star_a^k = \star_a^f$$

and the g-independence follows. The element h as in the hypothesis of the lemma conjugates the pair (z, x) onto the pair (z, a). Hence h^{-1} maps the pair $(\star_z, \star_a^g)$ of multiplications onto the pair $(\star_z, \star_x^{gh^{-1}})$ and the second assertion follows from (5.10.2). □

For a fixed multiplication $\star_z$ on $C_\Pi(z)$ contained in $A^{(z)}$, define a multiplication $\star$ on Π by the following rule: whenever $u, v \in C_\Pi(a)$ for some $a \in Z_4^{\#}$ we put

$$u \star v = u \star_a^g v = g(g^{-1}(u) \star_z g^{-1}(v))$$

for an element $g \in G_3 \cup G_4$ which conjugates z onto a, and expand by distributivity.

Lemma 5.10.4 *The multiplication* $\star$ *is well defined on the whole of* Π *and it is uniquely determined by* $\star_z$.

Proof. By (5.10.3), $\star$ does not depend on the choices of gs. The space Π contains a basis such that any pair of basis vectors is contained in $C_\Pi(a)$ for some $a \in Z_4^{\#}$. This follows from (4.5.2) and the obvious fact that in a $GF(2)$-vector space of dimension at least 3, any two 1-subspaces are contained in a common hyperplane (notice that the assertion is not be true for Z_2 in place of Z_4). Finally, if $a, b \in Z_4^{\#}$, then there is an element $g \in G_4$ which conjugated (z, x) onto (a, b). Then g maps $(\star_z, \star_x)$ onto $(\star_a^g, \star_b^g)$ and therefore $\star_a$ and $\star_b$ coincide on $C_\Pi(a) \cap C_\Pi(b)$ by (5.10.2). □

Lemma 5.10.5 *The product* $\star$ *is preserved by every* $h \in G_3 \cup G_4$.

Proof. Let $u, v \in C_\Pi(a)$ for some $a \in Z_4^{\#}$. Then

$$h(u \star v) = hg(g^{-1}(u) \star_z g^{-1}(v)).$$

Let a be the g-conjugate of z, b be the h-conjugate of a, and let k be an element from G_4 which conjugates z onto b. Then $k^{-1}hg \in G_1$ and by the G_1-invariance of $\star_z$, we have

$$h(u \star v) = k(k^{-1}hg)(g^{-1}(u) \star_z g^{-1}(v)) =$$
$$k(k^{-1}(h(u)) \star_z k^{-1}(h(v)))) = h(u) \star_b^k h(v) = h(u) \star h(v)$$

and the claim follows. $\qquad\qquad\square$

Proposition 5.10.6 *The G-invariant algebras on Π form a 1-space and the restrictions of these algebras to $C_\Pi(z)$ constitute $A^{(z)}$.*

Proof. By (5.10.4) for every algebra $\star_z$ from $A^{(z)}$, its expansion $\star$ is $\langle G_3, G_4 \rangle$-invariant. Since $G_i = \langle G_{i3}, G_{i4} \rangle$ for $i = 1$ and 2, the multiplication $\star$ is G-invariant. $\qquad\qquad\square$

Next we formulate the principal outcome of the chapter.

Proposition 5.10.7 (G, Π) *is a Norton pair.*

Proof. The appearance of Π in $\Pi \otimes \Pi$ with multiplicity one follows from (5.10.6). Since the multiplications in A are commutative, Π is contained in $S^2(\Pi)$. Since Π is the unique minimal module which affords a representation of a faithful completion of $\mathcal{M}$, it is clearly self-dual. Since the bilinear form which arranges the self-duality is symmetric, so is its restriction to Π_{299}, say. $\square$

The pair $(\Pi, \star)$, where $\star$ is a non-zero G-invariant algebra product on Π as in (5.10.6), is called the *Griess algebra*. By (5.1.2), the trilinear form

$$(u, v, w) \mapsto (u \star v, w)_\Pi$$

is symmetric and therefore the inner product $(\,,\,)_\Pi$ is associative with respect to the Griess algebra in the sense that

$$(u \star v, w)_\Pi = (u, v \star w)_\Pi$$

for all $u, v, w \in \Pi$ (compare p. 2 in [MN93]).

6

Automorphisms of Griess algebra

In this chapter we analyse the automorphism group $\mathbb{A}$ of the Griess algebra $(\Pi, \star)$. We prove the crucial feature of $\mathbb{A}$ which is the equality

$$C_{\mathbb{A}}(\varphi(Z_1)) = \varphi(G_1) \cong 2_+^{1+24}.Co_1.$$

As a consequence of this equality, we deduce that $\varphi(G)$ is a non-abelian finite simple group. Thus, according to traditional definition, $\varphi(G)$ is the Monster group.

6.1 Trace form

Let H be a (finite) group and let U be an H-module. In some cases we can recover an H-invariant inner product from an H-invariant algebra structure on U. Let

$$\star : (u, v) \mapsto u \star v$$

be an H-invariant (bilinear) algebra multiplication on U. For a vector $v \in U$, define the *adjoint transformation* ad(v) associated with v by the following rule

$$\text{ad}(v) : u \mapsto u \star v.$$

Since $\star$ is bilinear, ad(v) is a linear operator on U.

Definition 6.1.1 *Let $\star$ be an H-invariant bilinear algebra multiplication on an H-module U. The* trace form *on U associated with $\star$ is defined by*

$$(u, v)_\star = \text{tr}(\text{ad}(u) \cdot \text{ad}(v)),$$

where tr *denotes the trace function on the set of linear operators of U and $\cdot$ stands for the usual composition of linear operators.*

Lemma 6.1.2 *The form* $(\ ,\ )_\star$ *is bilinear symmetric and H-invariant.*

Proof. The result follows from the basic properties of the trace function together with the H-invariance hypothesis. $\qquad\square$

Lemma 6.1.3 *Let* $\star$ *be a Griess algebra. Then the trace form is non-zero and therefore up to rescaling it is the unique G-invariant one.*

Proof. Since there are non-zero eigenvectors of adjoint transformations, there are non-isotropic vectors and hence the form is indeed non-zero. $\qquad\square$

Lemma 6.1.4 *Let V be an elementary induced H-module with every reper being symmetric of plus type, let* $v_1, \ldots, v_n$ *be a V basis of weight vectors, and let* $\star$ *be an H-invariant multiplication on V such that* $v_i \star v_j = \pm v_k$ *whenever* v_i, v_j, *and* v_k *span a reper. Then* $(v_i, v_j)_\star = \delta_{ij} 2m$, *where m is the number of repers containing a given weight space.*

Proof. The result is immediate by trace calculation in the given basis. $\qquad\square$

6.2 Some automorphisms

By (5.2.1 (iii) (a)), S_4 is the full automorphism group of $(V_3^+, \star_+)$. This is a special case of the following result whose brilliant proof can be found in the last section of [DG98].

Lemma 6.2.1 *Let* $\mathcal{A}_{n-1}$ *be the* $(n-1)$-*dimensional Krein algebra associated with the natural permutation representation of the symmetric group* S_n. *Then* S_n *is the full automorphism group of* $\mathcal{A}_{n-1}$. $\qquad\square$

Let E be an elementary abelian group of order 2^n for $n \geq 3$. Let H be the semidirect product of E and $L \cong GL(E) \cong L_n(2)$ with respect to the natural action, and let V be an E-induced H-module. If E_0 is an index 2 subgroup in E, then $L(E_0) \cong 2^{n-1} : L_{n-1}(2)$ possesses an index 2 subgroup only when $n = 3$ and in that case H contains two classes of $L_3(2)$-complements fused in the automorphism group of H. Therefore, for every $n \geq 3$ the pair (H, V) is determined uniquely up to isomorphism.

In the above described situation, every hyperplane in E is a root and the centralizer in V of every codimension 2 subspace from E is a symmetric reper. By (5.4.7), (H, V) is a Norton pair and therefore up to rescaling V carries a unique H-invariant algebra structure $\mathcal{L}_n$. Notice that $\mathcal{L}_n$ is commutative, $\mathcal{L}_2 = (V_3^+, \star_+)$, while $\mathcal{L}_3$ possesses the following elegant description in terms of the

Fano plane. With every point p of the plane we associate a vector v_p and define the multiplication $\star$ by the following rule

$$v_p \star v_p = 0 \text{ while } v_p \star v_q = v_r \text{ whenever } \{p, q, r\} \text{ is a line.}$$

Grigory Kobatyanski pointed out the following:

Lemma 6.2.2 $\mathrm{Aut}(\mathcal{L}_n) \cong S_{2^n}$.

Proof. The group $H \cong AGL_n(2)$ acts doubly transitively 2^n-element point-set of the n-dimensional affine $GF(2)$-space. By the uniqueness assertion, $\mathcal{L}_n$ is precisely the Krein algebra on the faithful irreducible constituent of the per-mutational module, which coincides with the Krein algebra of the symmetric group of the point-set. Now (6.2.1) applies. □

Consider an arbitrary elementary E-induced H-module V. Let P be the set of roots and L be the set of bunches of roots. Then (P, L) is a point-line incidence system with three points per a line and if $\{v_p \mid p \in P\}$ is a basis of weight vectors, then an H-invariant multiplication on V satisfies

$$v_p \star v_q = \varepsilon_{pqr} v_r,$$

where ε_{pqr} is non-zero only if $\{p, q, r\}$ is a line. Recall that the universal abelian representation group $\mathcal{R}$ of (P, L) has the following definition in terms of generators and relations

$$\mathcal{R} = \langle e_p, \ p \in P \mid e_p^2 = [e_p, e_q] = e_p e_q e_r = 1, p, q \in P, \{p, q, r\} \in L \rangle$$

and notice that the dual group of E is a factor group of R.

Lemma 6.2.3 *Let V be an E-induced H-module such that every reper is symmetric of plus type, and let $\star$ be an H-invariant algebra multiplication on V. Let $\mathbb{A}^{(E)}$ be the subgroup in $\mathrm{Aut}(V, \star)$ which stabilizes the decomposition of V into E-eigenspaces and let $\mathbb{A}_0^{(E)}$ be the subgroup in $\mathbb{A}^{(E)}$ which stabilizes every term in this decomposition. Then*

(i) *the dual of $\mathbb{A}_0^{(E)}$ is a subgroup of $\mathcal{R}$;*
(ii) $\mathbb{A}^{(E)}/\mathbb{A}_0^{(E)} \leq \mathrm{Aut}(P, L)$.

Proof. A non-zero H-invariant bilinear form $(\, , \,)_\star$ on V can be recovered as the trace form associated with $\star$. With respect to this form, E-eigenspaces are non-singular pairwise perpendicular. The subgroup $\mathbb{A}_0^{(E)}$ preserving every E-eigenspaces and the restriction of $(\, , \,)_\star$ to it induces on the eigenspace an ac-tion of order at most 2. Therefore, $\mathbb{A}_0^{(E)}$ is an elementary abelian 2-group and its dual is generated by elements indexed by the point-set P. The restriction of $\star$ to a reper R turns R into a $(V_3^+, \star_+)$-algebra, whose automorphism group is S_4 by (5.2.1 (iii) (a)). In particular, $\mathbb{A}_0^{(E)}$ induces on R an action of order at most 4 and hence the line relations of $\mathcal{R}$ hold in the dual of $\mathbb{A}_0^{(E)}$, proving (i). Since the product of two weight vectors can be non-zero only if they correspond to distinct collinear roots, we also have (ii).   □

6.3 Involution centralizer

Let $\star$ be the Griess algebra on Π and let $\mathbb{A}$ be the automorphism group of $(\Pi, \star)$, which is the group of invertible linear transformations a of Π such that

$$a(u) \star a(v) = a(u \star v)$$

for all $u, v \in \Pi$. By (6.1.3), $\mathbb{A}$ also preserves the associated trace form $(\ ,\ )_\star$ on Π. Since $\star$ is G-invariant, $\varphi(G) \leq \mathbb{A}$. Let $\mathbb{A}_z$ be the centralizer of $\varphi(z)$ in $\mathbb{A}$, where as above z is the non-dentity element of $Z_1 = Z(G_1)$. Clearly, $\mathbb{A}_z$ contains $\varphi(G_1) \cong 2_+^{1+24}.Co_1$. The central result of the chapter is the following:

Proposition 6.3.1 $\mathbb{A}_z = \varphi(G_1)$.

The proof of the proposition will be achieved in a series of lemmas spread over the remainder of the chapter.

Lemma 6.3.2 $\mathbb{A}_z$ *preserves the decomposition*

$$\Pi = \Pi_{299}^1 \oplus \Pi_{|\bar{\Lambda}_2|}^1 \oplus \Pi_{24 \cdot 2^{12}}^1$$

of Π into G_1-irreducibles.

Of course, $\mathbb{A}_z$ preserves the centralizer-commutator decomposition of π with respect to the action of z, which means that $\mathbb{A}_z$ stabilizes both

$$C_\Pi(Z_1) = \Pi_{299}^1 \oplus \Pi_{|\bar{\Lambda}_2|}^1 \text{ and } [\Pi, Z_1] = \Pi_{24 \cdot 2^{12}}^1.$$

The separation of Π_{299}^1 and $\Pi_{|\bar{\Lambda}_2|}^1$ is achieved via the following result:

Lemma 6.3.3 *Let $\star_z$ and $(\ ,\ )_z$ be the restrictions to $C_\Pi(Z_1)$ of $\star$ and $(\ ,\ )_\star$, respectively and let $(\ ,\ )_{\star_z}$ be the trace form associated with $\star_z$. Then $(\ ,\ )_z$ and $(\ ,\ )_{\star_z}$ are not proportional.*   □

It appears problematic to prove (6.3.3) without choosing a particular Griess algebra among the scalar multiples and this is one of the reasons for introducing a more explicit version of $\star$ in the next section. Having such a version it is very tempting to start using it and we do so already in this section, although (6.3.3) will be proven only in Subsection 6.4.5.

Since both Π^1_{299} and $\Pi^1_{|\bar{\Lambda}_2|}$ are irreducible under $\varphi(G_1) \leq \mathbb{A}_z$, unless $\mathbb{A}_z$ preserves the decomposition

$$C_\Pi(Z_1) = \Pi^1_{299} \oplus \Pi^1_{|\bar{\Lambda}_2|}$$

it acts irreducibly on $C_\Pi(Z_1)$ and preserves on it, up to rescaling, at most one non-zero bilinear form. By (6.3.3), this is not the case and hence (6.3.2) follows from (6.3.3) and the paragraph preceding it.

Lemma 6.3.4 *The following assertions hold:*

(i) $\mathbb{A}_z$ *preserves the decomposition of* $\Pi^1_{|\bar{\Lambda}_2|}$ *into the weight spaces of* Q_1;
(ii) $C_{\mathbb{A}_z}(\Pi^1_{|\bar{\Lambda}_2|}) = C_{\mathbb{A}_z}(C_\Pi(Z_1))$;
(iii) *the action of* $\mathbb{A}_z$ *on* $C_\Pi(Z_1)$ *is isomorphic to* $\varphi(G_1)/\varphi(Z_1) \cong 2^{24} \cdot Co_1$.

Proof. Whenever $u \in \Pi^1_{299}$ and $v \in \Pi^1_{|\bar{\Lambda}_2|}$, the product $u \star v$ is always in $\Pi^1_{|\bar{\Lambda}_2|}$. Therefore, there is a $\mathbb{A}_z$-homomorphism $\psi : \Pi^1_{299} \rightarrow End(\Pi^1_{|\bar{\Lambda}_2|})$ defined via

$$\psi(u) : v \mapsto u \star v.$$

By (5.6.3) (compare (6.4.7 (iii)) below), the weight spaces of Q_1 in $\Pi^1_{|\bar{\Lambda}_2|}$ are precisely the weight spaces of $Im(\psi)$ and hence (i) follows from (6.3.2). By (6.4.7 (ii)), Π^1_{299} is spanned by the squares of the weight vectors in $\Pi^1_{|\bar{\Lambda}_2|}$, which gives (ii). So $\mathbb{A}_z$ acts on $C_\Pi(Z_1)$ as a subgroup of

$$Aut\left(\Pi^1_{|\bar{\Lambda}_2|}, \star_{|\bar{\Lambda}_2|}\right)$$

which stabilizes the set of Q_1-weight spaces. Let Θ be the point-line incidence system of 222-triangles in $\bar{\Lambda}_2$ and suppose that the incidence is via inclusion. Then by (6.1.4), in order to deduce (iii) from (i) and (ii) it is sufficient to show that $\bar{\Lambda}$ and Co_1 are the universal abelian representation and the automorphism groups of Θ, respectively. These two assertions are proved in Section 6.5 as (6.5.2) and (6.5.3), respectively. $\square$

The action of $\mathbb{A}_z$ on $C_\Pi(Z_1)$ is now identified with $\varphi(G_1)/\varphi(Z_1)$. Therefore, in order to complete the proof of (6.3.1) it only remains to identify $\varphi(Z_1)$ with the kernel of the action with.

Lemma 6.3.5 *The group* $\mathbb{A}_z$ *acts on* $C_\Pi(Z_1)$ *with kernel* $\varphi(Z_1)$.

Proof. By (3.1.21), $[\Pi, Z_2]$ is the only faithful irreducible constituent of G_2 in Π (of dimension $3 \cdot 24 \cdot 2^{11}$) and restricted to G_2^s it splits into the direct sum of three irreducibles $C_{[\Pi,Z_2]}(a)$, $a \in Z_2^\#$ of equal dimensions, so that

$$[\Pi, Z_1] = C_{[\Pi, Z_2]}(x) \oplus C_{[\Pi, Z_2]}(y) \text{ and } C_{[\Pi, Z_2]}(z) = \bigoplus_{\bar{\mu} \in \bar{\Lambda}_2^3} \Pi(\bar{\mu}).$$

Let K be the centralizer of $C_\Pi(Z_2)$ in $\mathbb{A}_z$. In order to prove the lemma, it is sufficient to show that K induces on $[\Pi, Z_2]$ an action of order four (thus coinciding with $\varphi(Z_2)$). First we analyse the possibilities for the action of $k \in K$ on $C_{[\Pi, Z_2]}(z)$. Since k centralizes Π_{299}^1, it stabilizes every weight subspace of Q_1 in $C_\Pi(Z_1)$. Since $\bar{\Lambda}_2^3$ is the complement to the hyperplane formed by the elements of $\bar{\Lambda}_2$ perpendicular to $\chi(x) = \sum_{r \in \mathcal{P}} 2r + 2\Lambda$, by (6.2.3 (ii)) and (6.5.2) k acts on $C_{[\Pi, Z_2]}(z)$ as a ± 1-scalar operator. By the obvious symmetry on each $C_{[\Pi, Z_2]}(a)$ for $a \in Z_2^{\#}$, the action of K has order 2 and it only remains to show that the action on the union of any two of the centralizers is faithful. In order to achieve this it is sufficient to find a non-zero vector in $C_{[\Pi, Z_2]}(z)$ and a non-zero vector in $C_{[\Pi, Z_2]}(x)$ product to a non-zero vector in $C_{[\Pi, Z_2]}(y)$. Since G_3 induces on Z_3 the full linear group $GL(Z_3)$, we can and will shift the calculations into $C_\Pi(Z_1)$ and show that for Y being a complement to Z_1 in Z_3 with $Y^{\#} = \{x, u, xu\}$ there is a 222-triangle $\{\bar{\mu}, \bar{v}, \bar{\mu} + \bar{v}\}$ in $\bar{\Lambda}_2$ such that

$$(\chi(x), \bar{\mu}) = (\chi(u), \bar{v}) = 1, \quad (\chi(u), \bar{\mu}) = (\chi(x), \bar{v}) = 0.$$

We take

$$\chi(x) = \sum_{r \in \mathcal{P}} 2r + 2\Lambda, \quad \chi(u) = \sum_{r \in T} 4r + 2\Lambda,$$

where T is a tetrad defining Z_3 and put

$$\bar{\mu} = 3p - \sum_{r \in \mathcal{P} \setminus \{p\}} r + 2\Lambda, \quad \bar{v} = \sum_{r \in \mathcal{O}} 2r + 2\Lambda,$$

where $\mathcal{O}$ is an octad disjoint from p and intersecting T in a single element. It is easy to check that all the imposed conditions hold. $\square$

6.4 Explicit version of $A^{(z)}$

In this section (which comprises a few subsection) we produce a more explicit description the module

$$C_\Pi(Z_1) = \Pi_{299}^1 \oplus \Pi_{|\bar{\Lambda}_2|}^1$$

along with certain G_1-invariant inner $(\ ,\ )_z$ and algebra $\star_z$ products whose restrictions to $C_\Pi(Z_2)$ are G_2-invariant.

6.4.1 Jordan algebra on $S^2(\Lambda)$

By (3.1.21 (iii)), if $S^2(\Lambda)$ denotes the symmetric square of $\Lambda \otimes \mathbb{R}$, then $S^2(\Lambda) = \Pi_1 \oplus \Pi_{299}^1$ for a trivial 1-dimensional Co_1-module Π_1.

As in Section 1.8 we treat $\mathcal{P}$ as a basis of $\Lambda \otimes \mathbb{R}$ which is orthonormal with respect to the form $(\ ,\ )_\Lambda$. Then $S^2(\Lambda)$ has a basis consisting of the vectors

$$p^2 := p \otimes p$$

taken for all $p \in \mathcal{P}$ together with the vectors

$$(pq)_z := p \otimes q + q \otimes p$$

taken for all 2-element subsets $\{p, q\}$ in $\mathcal{P}$. This basis will be denoted by $S^2(\mathcal{P})$ and it is orthogonal with p^2 and $(pq)_z$ being of the (squared) norm 1 and 2 with respect to the form $(\ ,\ )_{S^2(\Lambda)}$ which is the tensor square of the form $(\ ,\ )_\Lambda$ restricted to symmetric part.

The space $S^2(\Lambda)$ can be considered as a G_1-module with Q_1 being the kernel and the action of $G_1/Q_1 \cong Co_1$ being the natural one (Co_0 acts on $\Lambda \otimes \mathbb{R}$ and $Z(Co_0)$ consisting of the (± 1)-scalar transformations act trivially on the tensor square of $\Lambda \otimes \mathbb{R}$). When restricted to the monomial subgroup $G_{12}/Q_1 = E : M \cong \mathcal{C}_{11} : M_{24}$, the basis $S^2(\mathcal{P})$ becomes monomial with

$$\Pi_1 \oplus \Pi_{23}^{12} = \langle p^2 \mid p \in \mathcal{P} \rangle$$

being the 24-dimensional permutation module of M_{24} and

$$\Pi_{276}^{12} = \Pi_{276}^{(z)} = \langle (pq)_z \mid p, q \in \mathcal{P}, p \neq q \rangle,$$

coming with a monomial basis with respect to the E-induced structure.

The form $(\ ,\ )_\Lambda$ establishes an isomorphism between $\Lambda \otimes \mathbb{R}$ and the dual space $(\Lambda \otimes \mathbb{R})^\star$, while $(\Lambda \otimes \mathbb{R}) \otimes (\Lambda \otimes \mathbb{R})^\star$ can be treated as the space of linear operators on $\Lambda \otimes \mathbb{R}$. Therefore, we can identify $S^2(\Lambda)$ with the space of linear operators on $\Lambda \otimes \mathbb{R}$ which are self-adjoint with respect to $(\ ,\ )_\Lambda$. If $L(s)$ is the linear operator attached to $s \in S^2(\Lambda)$ by the above identification, then the action in the basis $\mathcal{P}$ is the following, where $p, q, r \in \mathcal{P}$ and $p \neq q$

$$L(p^2) : r \mapsto \delta_{p,r} r,$$
$$L((pq)_z) : r \mapsto \delta_{p,r} q + \delta_{q,r} p.$$

In these terms the 1-dimensional G_1-invariant subspace Π_1 consists of scalar transformations so that

$$\Pi_1 = \left\{ \alpha \sum_{p \in \mathcal{P}} p^2 \mid \alpha \in \mathbb{R} \right\}.$$

The subspace Π_{299}^1 is a G_1-invariant complement to Π_1, formed by the self-adjoint traceless operators on $\Lambda \otimes \mathbb{R}$. In particular, Π_{299}^1 is spanned by the elements $(pq)_z$ and $p^2 - q^2$ taken for all distinct $p, q \in \mathcal{P}$. In these terms the tensor square form can be redefined via

$$(s_1, s_2)_{S^2(\Lambda)} = \text{tr}(L(s_1) \cdot L(s_2)),$$

in particular the decomposition $S^2(\Lambda) = \Pi_1 \oplus \Pi_{299}^1$ is orthogonal.

Define the *Jordan multiplication* $\star = \star_{S^2(\Lambda)}$ on $S^2(\Lambda)$ via

$$L(s_1 \star s_2) = 2(L(s_1) \cdot L(s_2) + L(s_2) \cdot L(s_1)),$$

where $s_1, s_2 \in S^2(\Lambda)$ and $\cdot$ stands for the usual composition of linear operators.

Lemma 6.4.1 *The following assertions hold:*

(i) *if* $\lambda_1, \lambda_2, \mu_1, \mu_2 \in \Lambda \otimes \mathbb{R}$, *then*

$$(\lambda_1 \otimes \lambda_2 + \lambda_2 \otimes \lambda_1) \star (\mu_1 \otimes \mu_2 + \mu_2 \otimes \mu_1)$$
$$= 2 \sum_{i=1,2} \sum_{j=1,2} (\lambda_{3-i}, \mu_{3-j})_\Lambda (\lambda_i \otimes \mu_j + \mu_j \otimes \lambda_i);$$

(ii) *the following equalities hold:*

$$p^2 \star q^2 = \delta_{p,q} \, 4 \, p^2; \quad p^2 \star (pq)_z = 2 \, (pq)_z \text{ and}$$
$$p^2 \star (qr)_z = 0 \text{ if } p \notin \{q, r\};$$
$$(pq)_z \star (pq)_z = 4 \, (p^2 + q^2); \quad (pq)_z \star (qr)_z = 2 \, (pr)_z \text{ if } q \neq r \text{ and}$$
$$(pq)_z \star (rs)_z = 0 \text{ if } \{p, q\} \cap \{r, s\} = \emptyset;$$

Proof. Since the right-hand side in (i) is linear in each of $\lambda_1, \lambda_2, \mu_1$, and μ_2, it is sufficient to multiply the pairs of basis elements as is done in (ii). $\quad\square$

6.4.2 Structure of $\Pi_{|\bar{\Lambda}_2|}^1$

In this subsection we discuss the G_1-module $\Pi_{|\bar{\Lambda}_2|}^1$ together with G_1-invariant inner and algebra products on it. The module being elementary induced for $Q_1/Z_1 \cong \bar{\Lambda} \cong 2^{24}$ possesses a direct sum decomposition $\Pi_{|\bar{\Lambda}_2|}^1 = \bigoplus_{\bar{\mu} \in \bar{\Lambda}_2} \Pi(\bar{\mu})$. The quotient $G_1/Q_1 \cong Co_1$ permutes the summands as it does the elements of $\bar{\Lambda}_2$, while $r \in Q_1$ stabilizes every summand $\Pi(\bar{\mu})$ centralizing or inverting it depending on whether $(\chi(r), \bar{\mu})_{\bar{\Lambda}}$ is zero or not (3.1.7). This means that for $\vartheta \in Q_1$ with $\chi(\vartheta) = \bar{\mu} \in \bar{\Lambda}_2$, an element from G_1 which centralizes ϑ centralizes $\Pi(\bar{\mu})$, while an element which swaps ϑ and ϑz inverts $\Pi(\bar{\mu})$. Therefore, ϑ and ϑz can be assigned to a vector from $\Pi(\bar{\mu})$ and to its negative. In order to make this assignment consistent with the group action, we consider $\Pi^1_{|\bar{\Lambda}_2|}$ as a quotient of the permutation module V of G_1 on $\chi^{-1}(\bar{\Lambda}_2)$. If we consider $\chi^{-1}(\bar{\Lambda}_2)$ as a permutation basis of V, then $\Pi^1_{|\bar{\Lambda}_2|}$ is the quotient of V over the subspace spanned by the sums $\vartheta + \vartheta z$ taken for all $\vartheta \in \chi^{-1}(\bar{\Lambda}_2)$. If $\pi(\vartheta)$ denotes the image of ϑ in $\Pi^1_{|\bar{\Lambda}_2|}$ under the natural homomorphism $V \to \Pi^1_{|\bar{\Lambda}_2|}$, then

$$\pi(\vartheta) = -\pi(\vartheta z)$$

as desired.

Let us now discuss the G_1-invariant inner and algebra products on $\Pi^1_{|\bar{\Lambda}_2|}$. The inner product can be defined simply by

$$(\pi(\vartheta_1), \pi(\vartheta_2))_{\Pi^1_{|\bar{\Lambda}_2|}} = \begin{cases} 1 & \text{if } \vartheta_1 = \vartheta_2; \\ 0 & \text{if } \langle \vartheta_1, z \rangle \neq \langle \vartheta_2, z \rangle. \end{cases}$$

The following lemma is a reformulation of (5.7.1) and its proof.

Lemma 6.4.2 *Let* $\star$ *be a* G_1-*invariant algebra multiplication on* $\Pi^1_{|\bar{\Lambda}_2|}$, *let* $\vartheta_1, \vartheta_2 \in \chi^{-1}(\bar{\Lambda}_2)$ *and suppose that the product* $\pi(\vartheta_1) \star \pi(\vartheta_2)$ *is non-zero. Then the following assertions hold, where* $\bar{\mu}_1 = \chi(\vartheta_1)$, $\bar{\mu}_2 = \chi(\vartheta_2)$, *and* $\bar{\mu}_3 = \bar{\mu}_1 + \bar{\mu}_2$:

(i) $\bar{\mu}_3$ *is contained in* $\bar{\Lambda}_2$, *so that* $\{\bar{\mu}_1, \bar{\mu}_2, \bar{\mu}_3\}$ *is a 222-triangle;*
(ii) *if* ϑ_3 *is the product of* ϑ_1 *and* ϑ_2 *in* Q_1, *then* $\chi(\vartheta_3) = \bar{\mu}_3$ *and there is a non-zero* γ *such that* $\pi(\vartheta_1) \star \pi(\vartheta_2) = \gamma\pi(\vartheta_3)$;
(iii) *the multiplication* $\star$ *is uniquely determined by the choice of* γ. $\square$

Let $\star_{\Pi^1_{|\bar{\Lambda}_2|}}$ denote the multiplication in (6.4.2) corresponding to the choice $\gamma = 1$, so that for $\vartheta_1, \vartheta_2 \in \chi^{-1}(\bar{\Lambda}_2)$ we have

$$\pi(\vartheta_1) \star_{\Pi^1_{|\bar{\Lambda}_2|}} \pi(\vartheta_2) = \begin{cases} \pi(\vartheta_1\vartheta_2) & \text{whenever } \vartheta_1\vartheta_2 \in \chi^{-1}(\bar{\Lambda}_2); \\ 0 & \text{otherwise}. \end{cases}$$

Since $G_1/Z_1 \cong 2^{24} \cdot Co_1$ is non-split, it is difficult to find a nice basis in $\Pi^1_{|\bar{\Lambda}_2|}$ consisting of weight vectors $\pi(\vartheta)$. The situation improves when we restrict to G_{12} acting on $\chi^{-1}(\bar{\Lambda}_2^4)$, since in this case the acting group

$$G_{12}/R_2 \cong (\mathcal{C}_{11} \oplus \mathcal{C}_{11}) : (M_{24} \times 2)$$

splits. The set $\chi^{-1}(\bar\Lambda_2^4)$ is contained in R_2 and the latter is elementary abelian of order 2^{13}. When R_2 is considered as a module for $M^\delta/R_2 \cong M_{24}$, it is isomorphic to the direct sum $X \oplus Z_2$, where $X \cong R_2/Z_2 \cong C_{11}^*$ (cf. (2.3.3)). Therefore, the intersection $\chi^{-1}(\bar\Lambda_2^4) \cap X$ is precisely the 276-orbit of M_{24} in X indexed by the 2-element subsets of $\mathcal{P}$.

Lemma 6.4.3 *Let* $p, q \in \mathcal{P}$ *with* $p \neq q$ *and let* $\vartheta(pq)$ *denote the corresponding element in* $\chi^{-1}(\bar\Lambda_2^4) \cap X$. *Then:*

(i) *if* $\{p, q, r\}$ *is a 3-subset of* $\mathcal{P}$, *then* $\vartheta(pq)\vartheta(qr) = \vartheta(pr)$;
(ii) $\{\vartheta(pq), \vartheta(pq)z\} = \chi^{-1}(4p - 4q + 2\Lambda)$ *and*
$\{\vartheta(pq)x, \vartheta(pq)y\} = \chi^{-1}(4p + 4q + 2\Lambda)$, *where* $\{z, x, y\} = Z_2^\#$.

Proof. Since X is a subgroup in Q_1, (i) follows. Since $(4p - 4q) + (4q - 4r) = 4p - 4r$, the image of $\chi^{-1}(\bar\Lambda_2^4) \cap X$ under χ is

$$\{4p - 4q + 2\Lambda \mid p, q \in \mathcal{P}, p \neq q\},$$

which gives (ii). □

Lemma 6.4.4 *For* $a \in Z_2 \setminus Z_1$, *put*

$$(pq)_a = \pi(\vartheta(pq)) + \pi(\vartheta(pq)a).$$

Then $\{(pq)_a \mid p, q \in \mathcal{P}, p \neq q\}$ *is a basis of* $\Pi_{276}^{(a)} = C_{\Pi_{2\cdot276}^{12}}(Q_2^{(a)})$ *consisting of weight vectors with respect to the* Q_2-*induced structure.*

Proof. The vector $(pq)_a$ is centralized by $Q_2^{(a)}$ and every element from $Q_2^{(z)}$ either centralizes or negates it. Hence the result follows from the dimension consideration. □

Lemma 6.4.5 *If* $\{p, q, r\}$ *is a 3-subset in* $\mathcal{P}$ *and* $a \in Z_2 \setminus Z_1$, *then*

$$(pq)_a \star_{\Pi_{|\bar\Lambda_2|}^1} (qr)_a = 2\,(pr)_a.$$

Proof. The result follows from (6.4.3 (i)) and the fact that a is an involution. □

6.4.3 $\Pi_{|\bar\Lambda_2|}^1 \times \Pi_{|\bar\Lambda_2|}^1 \to S^2(\Lambda)$

Define a bilinear multiplication $\star_\alpha$ on $\Pi_{|\bar\Lambda_2|}^1$ with products contained in $S^2(\Lambda)$ by the following conditions, where $\vartheta, \vartheta' \in \chi^{-1}(\bar\Lambda_2)$ and α is a parameter:

(a) $\pi(\vartheta) \star_\alpha \pi(\vartheta') = 0$ unless $\vartheta' = \vartheta$ or $\vartheta' = \vartheta z$;
(b) $\pi(\vartheta) \star_\alpha \pi(\vartheta) = \alpha(\mu \otimes \mu)$, where $\mu \in \Lambda_2$ and $\chi(\vartheta) = \mu + 2\Lambda$.

Lemma 6.4.6 *The following assertions hold:*

(i) *the operation $\star_\alpha$ is well defined and G_1-invariant;*

(ii) *every G_1-invariant multiplication on $\Pi^1_{|\bar{\Lambda}_2|}$ with products contained in Π^1_{299} is the projection of $\star_\alpha$ for some α.*

Proof. There are exactly two elements in Λ_2 equal to $\chi(\vartheta)$ modulo 2Λ and they are negatives of each other. Hence the right-hand side in (b) is independent of the choice of μ satisfying the conditions and $\star_\alpha$ is indeed well defined. Since χ is G_1-invariant, so is $\star_\alpha$. By (5.7.1 (iv)), the assertion (ii) follows. $\square$

6.4.4 $(,)_z$ and $\star_z$

Now we are well prepared to prove the following main result of the section:

Proposition 6.4.7 *Let $(,)_z$ and $\star_z$ be G_1-invariant inner and algebra products on $C_\Pi(Z_1) = \Pi^1_{299} \oplus \Pi^1_{|\bar{\Lambda}_2|}$ such that:*

(a) *the restrictions to $C_\Pi(Z_2)$ are G_2-invariant;*

(b) *the restriction of $(,)_z$ to Π^1_{299} coincides with the restriction of the tensor square form $(,)_{S^2(\Lambda)}$;*

(c) *the projected restriction of $\star_z$ to Π^1_{299} coincides with that of the Jordan algebra $\star_{S^2(\Lambda)}$.*

Then

(i) *the irreducible constituents of G_1 are perpendicular with respect to $(,)_z$ and the restriction of $(,)_z$ to $\Pi^1_{|\bar{\Lambda}_2|}$ coincides with $(,)_{\Pi^1_{|\bar{\Lambda}_1|}}$: the weight vectors $\pi(\vartheta)$ have norm one and are perpendicular unless proportional;*

(ii) *if $\vartheta_1, \vartheta_2 \in \chi^{-1}(\bar{\Lambda}_2)$, then*

$$
\pi(\vartheta_1) \star_z \pi(\vartheta_2) = \begin{cases} \pi(\vartheta_1 \vartheta_2) & \text{if } \vartheta_1 \vartheta_2 \in \chi^{-1}(\bar{\Lambda}_2); \\ \frac{\varepsilon}{8}(\mu \otimes \mu)_{299} & \text{if } \pi(\vartheta_1) = \varepsilon\pi(\vartheta_2) \text{ and} \\ & \quad \chi(\vartheta_1) = \mu + 2\Lambda \text{ for } \mu \in \Lambda_2; \\ 0 & \text{otherwise} \end{cases}
$$

(where $(\mu \otimes \mu)_{299}$ denotes the projection of the tensor square of μ to Π^1_{299} and $\varepsilon = \pm 1$);

(iii) *if $\vartheta \in \chi^{-1}(\bar{\Lambda}_2)$ with $\mu = \chi(\vartheta)$ mod 2Λ and $s \in \Pi^1_{299}$, then*

$$
s \star_z \pi(\vartheta) = \frac{(s, \mu \otimes \mu)_{S^2(\Lambda)}}{8} \pi(\vartheta).
$$

The products $(\ ,\ )_z$ and $\star_z$ are the restrictions to $C_\Pi(Z_1)$ of a non-zero G-invariant inner product $(\ ,\ )_\Pi$ and of a Griess algebra $\star$.

Proof. By (5.10.6), the space of Griess algebras projects isomorphically onto $A^{(z)}$, while by (5.9.3) and (5.6.4), the latter projects isomorphically onto $A_{G_1}(\Pi^1_{299})$. Therefore, $\star_z$ exists and is unique (similarly for $(\ ,\ )_z$).

Let $T = \{p, q, r\}$ be a 3-subset of $\mathcal{P}$. Let S be the stabilizer of T in M^δ, so that $R_2 < S$ and $S/R_2 \cong P\Sigma L_3(4)$ is the triple stabilizer in $M^\delta/R_2 \cong M_{24}$. Since $G_2/R_2 \cong (\mathcal{C}_{11} \oplus \mathcal{C}_{11}) : (M_{24} \times S_3)$, we conclude that $N_{G_2}(S)$ induces on $Z_2^\#$ the symmetric group S_3. Hence there is an element σ of order 3 in $N_{G_2}(S)$ which commutes with S and induces the permutation (z, x, y). Then

$$\sigma : (pq)_z \mapsto (pq)_x, \quad \sigma : (qr)_z \mapsto (qr)_x, \quad \sigma : (pr)_z \mapsto (pr)_x.$$

Since $(pq)_z$ has norm 2 with respect to $(\ ,\ )_{S^2(\Lambda)}$, $(pq)_x$ must be of norm 2 with respect to $(\ ,\ )_z$. By (6.4.4)

$$(pq)_x = \pi(\vartheta(pq)) + \pi(\vartheta(pq)x)$$

and (i) follows. Comparing the product $(pq)_z \star_{S^2(\Lambda)} (qr)_z$ given in (6.4.1 (ii)) and the value of $(pq)_x \star_{\Pi^1_{|\bar{\Lambda}_2|}} (qr)_x$ in (6.4.5), we conclude that for $u, v \in \Pi^1_{|\bar{\Lambda}_2|}$ the $\Pi^1_{|\bar{\Lambda}_2|}$th projection of $u \star_z v$ is precisely $u \star_{\Pi^1_{|\bar{\Lambda}_2|}} v$, while by (6.4.6) the Π^1_{299}-th projection coincides with that of $u \star_\alpha v$ for some α. Hence it only remains to determine α. This can be achieved by calculating $(pq)_x \star_z (pq)_x$ and equalizing the result with $(pq)_z \star_{S^2(\Lambda))} (pq)_z = 4(p^2 + q^2)$. By (6.4.3 (ii))

$$\chi(\vartheta(pq)) = 4p - 4q + 2\Lambda, \ \text{and} \ \chi(\vartheta(pq)x) = 4p + 4q + 2\Lambda.$$

Since $\vartheta(pq)\vartheta(pq)x = x \notin \chi^{-1}(\bar{\Lambda}_2)$, we obtain

$$(pq)_x \star_z (pq)_x = \pi(\vartheta(pq)) \star_\alpha \pi(\vartheta(pq)) + \pi(\vartheta(pq)x) \star_\alpha \pi(\vartheta(pq)x)$$

$$= \alpha((4p - 4q) \otimes (4p - 4q) + (4p + 4q) \otimes (4p + 4q)) = \alpha(32p^2 + 32q^2).$$

Hence $\alpha = \frac{1}{8}$ as claimed. The associativity of $(\ ,\ )_z$ with respect to $\star_z$ implies (iii) (notice that the decomposition $S^2(\Lambda) = \Pi_1 \oplus \Pi^1_{299}$ is $(\ ,\ )_{S^2(\Lambda)}$-orthogonal). $\qquad\square$

Lemma 6.4.8 *Let $\{a, b, c\} = Z_2^\#$ and let $p, q \in \mathcal{P}$ with $p \neq q$. Then*

$$(pq)_a \star_z (pq)_b = -4(pq)_c.$$

Proof. Since $\star_z$ restricted to $C_\Pi(Z_2)$ is G_2-invariant, in the considered situation we have the full symmetry between the elements of $Z_2^\#$. Therefore, we can put $a = x, b = y$, in which case

$$(pq)_x \star_z (pq)_y = (\pi(\vartheta(pq)) + \pi(\vartheta(pq)x)) \star_{\frac{1}{8}} (\pi(\vartheta(pq)) + \pi(\vartheta(pq)y))$$

$$= \pi(\vartheta(pq)) + \pi(\vartheta(pq)x)) \star_{\frac{1}{8}} (\pi(\vartheta(pq)) - \pi(\vartheta(pq)x))$$

$$= \pi(\vartheta(pq)) \star_{\frac{1}{8}} \pi(\vartheta(pq)) - \pi(\vartheta(pq)x) \star_{\frac{1}{8}} \pi(\vartheta(pq)x))$$

$$= \frac{1}{8}((4p - 4q) \otimes (4p - 4q) - (4p + 4q) \otimes (4p + 4q))$$

$$= -4(p \otimes q + q \otimes p) = -4(pq)_z.$$

Since $(pq)_z$ is already in Π_{299}, there is no need to apply the projection mapping. □

6.4.5 Calculating traces

Here we apply the explicit version of $(\ ,\ )_z$ and $\star_z$ built up in the previous section to prove (6.3.3). We achieve this by showing that $(pq)_z$ and $(pq)_x$, both of norm 2 with respect to $(\ ,\ )_z$, have different norms with respect to the trace form $(\ ,\ )_{\star_z}$.

Consider the centralizer-commutator decomposition of $C_\Pi(Z_1)$ with respect to x

$$C_\Pi(Z_1) = C_\Pi(Z_2) \oplus \Pi^{(z)}_{24 \cdot 2^{11}},$$

where $\Pi^{(z)}_{24 \cdot 2^{11}} = \sum_{\bar{\mu} \in \bar{\Lambda}_2^3} \Pi(\bar{\mu})$ (compare (3.1.8)).

Lemma 6.4.9 *The operators* ad$(pq)_z$ *and* ad$(pq)_x$ *stabilize the commutator-centralizer decomposition of* $C_\Pi(Z_1)$ *with respect to* x.

Proof. Since $C_\Pi(Z_2)$ is closed under $\star_z$ and contains $(pq)_z$ and $(pq)_x$, it is clearly preserved by the adjoint operators. Since $(pq)_z \in \Pi^1_{299}$ in the case of ad$(pq)_z$, the assertion follows directly from (6.4.7 (iii)). Recall from (6.4.4) that

$$(pq)_x = \pi(\vartheta(pq)) + \pi(\vartheta(pq)x),$$

where $\chi(\vartheta(pq)) = 4p - 4q + 2\Lambda$ and $\chi(\vartheta(pq)x) = 4p + 4q + 2\Lambda$. In view of (6.4.7 (ii)) in order to prove the assertion it is sufficient to show that whenever $\{\bar{\lambda}, \bar{\mu}, \bar{\nu}\}$ is a 222-triangle in $\bar{\Lambda}$ such that $\bar{\lambda} = 4p \pm 4q + 2\Lambda$ and $\bar{\mu} \in \bar{\Lambda}_2^3$ the vector $\bar{\nu}$ is also in $\bar{\Lambda}_2^3$. But this is true, since $\bar{\Lambda}_2^3$ is the intersection with $\bar{\Lambda}_2$ of the complement to the hyperplane in $\bar{\Lambda}$ formed by the vectors perpendicular to

$$\chi(x) = \sum_{p \in \mathcal{P}} 2p + 2\Lambda$$

with respect to $(\ ,\ )_{\bar{\Lambda}}$, since $\bar{\lambda}$ is contained in that hyperplane, and since $\{\bar{\lambda}, \bar{\mu}, \bar{\nu}\}$ is a line.   □

We have seen in the proof of (6.4.7) that G_2 contains an element σ which maps $(pq)_z$ onto $(pq)_x$. Since σ stabilizes $C_\Pi(Z_2)$, we have the following:

Lemma 6.4.10 *The squares of* $\mathrm{ad}(pq)_z$ *and* $\mathrm{ad}(pq)_x$ *acting on* $C_\Pi(Z_2)$ *have equal traces.*   □

Next we calculate the traces on $\Pi^{(z)}_{24 \cdot 2^{11}}$. Let Λ_2^3 be the set of Leech vectors which are the images under the monomial subgroup $F_0 = C_{12} : M_{24}$ of the vector

$$\lambda_2 = 3p - \sum_{r \in \mathcal{P} \setminus \{p\}} r.$$

The subgroup $E_0 \cong C_{12}$ of sign changes acts regularly on Λ_2^3 having 24 orbits $\Lambda^{(r)}$ indexed by the elements of $\mathcal{P}$, so that $\mu \in \Lambda^{(r)}$ precisely when the rth coordinate of μ in the basis $\mathcal{P}$ is the one with the absolute value 3.

Lemma 6.4.11 *Let* $\mu \in \Lambda^{(r)}$, $\bar{\mu} = \mu + 2\Lambda$ *and*

$$\bar{\nu}_\varepsilon = \bar{\mu} + (4p + \varepsilon 4q + 2\Lambda)$$

for $\varepsilon = \pm 1$. *Then* $\bar{\nu}_\varepsilon \in \bar{\Lambda}_2^3$ *if and only if* $r \in \{p, q\}$, *in which case* ε *is uniquely determined by* $\bar{\mu}$. *Furthermore,* $\bar{\nu}_\varepsilon \in \bar{\Lambda}^{(s)}$, *where* $\{r, s\} = \{p, q\}$.

Proof. The result follows from the fact that for $\lambda, \mu \in \Lambda_2$, the inclusion $\bar{\lambda} + \bar{\mu} \in \bar{\Lambda}_2$ holds if and only if $(\lambda, \mu)_\Lambda = \pm 16$.   □

Lemma 6.4.12 *For* $a \in \{z, x\}$, *let* t_a *be the trace of the square of* $\mathrm{ad}(pq)_a$ *acting on* $\Pi^{(z)}_{24 \cdot 2^{11}}$. *Then:*

(i) $t_z = 5 \cdot 2^{10}$ *and*
(ii) $t_x = 2^{12}$.

In particular, $(pq)_z$ *and* $(pq)_x$ *have different norms with respect to the trace form of* $\star_z$.

Proof. Let Θ be a subset of Q_1 such that $\mathcal{B} = \{\pi(\vartheta) \mid \vartheta \in \Theta\}$ is a basis of $\Pi^{(z)}_{24 \cdot 2^{11}}$. By (6.4.7 (iii)), the action of $\mathrm{ad}(pq)_z$ in the basis $\mathcal{B}$ is diagonal with $\pi(\vartheta)$ being multiplied by $\frac{3}{4}$ or $\frac{1}{4}$ depending on whether or not $\chi(\vartheta) \in \bar{\Lambda}^{(p)} \cup \bar{\Lambda}^{(q)}$ (remember that the norm of $(pq)_z$ with respect to $(\ ,\ )_{S(\Lambda)}$ is 2). Hence

$$t_z = \frac{9}{16} \cdot 2 \cdot 2^{11} + \frac{1}{16} \cdot 22 \cdot 2^{11} = \frac{18 + 22}{16} \cdot 2^{11} = 5 \cdot 2^{10}.$$

By (6.4.11), the square of $\text{ad}(pq)_x$ stabilizes $\pi(\vartheta)$ if $\chi(\vartheta) \in \bar{\Lambda}^{(p)} \cup \bar{\Lambda}^{(q)}$ and annihilates $\pi(\vartheta)$ otherwise, which gives (ii). $\qquad\qquad\square$

6.5 222-triangle geometry

In this section Θ denotes the point line incidence system whose point-set is $\bar{\Lambda}_2$ and the line set is

$$\{\{\bar{\lambda}, \bar{\mu}, \bar{\nu}\} \mid \bar{\lambda}, \bar{\mu}, \bar{\nu} \in \bar{\Lambda}_2, \bar{\lambda} + \bar{\mu} + \bar{\nu} = 0\}$$

(which is the set of 222-triangles in $\bar{\Lambda}$). Our first objection is to identify the universal abelian representation group of Θ with $\bar{\Lambda} = \Lambda/2\Lambda$. A machinery for calculating representation groups was developed in [ISh02].

We start with the following preliminary result:

Lemma 6.5.1 *Let $(\mathcal{B}, \mathcal{L})$ be a point-line incidence system, having octads as points with a triple of them forming a line if and only if their sum in the Golay code $\mathcal{C}_{12}$ is contained in $\{\emptyset, \mathcal{P}\}$. Then $\mathcal{C}_{11}$ is the universal abelian representation group of that system.*

Proof. Let U be the universal abelian representation group. It is immediate from the definition that $\mathcal{C}_{11}$ is a quotient of U. So in order to prove the assertion it is sufficient to bound the order of U. For a subset $\mathcal{S}$ of $\mathcal{B}$ let $U(\mathcal{S})$ denote the subgroup of U generated by the elements of $\mathcal{S}$ (identified in the obvious manner with the standard generators of U). Let $\mathcal{O} \in \mathcal{B}$ be an octad and let $\mathcal{B}_i$ denote the set of octads intersecting $\mathcal{O}$ in i elements. Then by the basic property of the $S(5, 8, 24)$-Steiner system we have

$$\mathcal{B}_8 = \{\mathcal{O}\}, \ |\mathcal{B}_0| = 30, \ |\mathcal{B}_4| = 280, \ |\mathcal{B}_2| = 448.$$

Clearly, $U(\mathcal{B}_8)$ has order 2. Let $\mathcal{Q}$ be the set 15 octads refining a sextet. Applying the line relations and the well-known structure of the representation group of the generalized quadrangle of order $(2, 2)$ (compare Section 3.4 in [ISh02]), we conclude that $U(\mathcal{Q})$ has order 2^4. This, together with (3.1.3) in [ISh02] shows that $U(\mathcal{B}_0)$ has order 2^5, while $U(\mathcal{B}_4)/U(\mathcal{B}_0)$ has order 2^6. In order to achieve the required bound we show that $U(\mathcal{B}_2)$ is contained in $U(\mathcal{B}_4)$. Consider a sextet $\mathcal{Q}$, 5 of whose refining octads are in $\mathcal{B}_4$ and 10 are in $\mathcal{B}_2$. The existence of such a sextet can be seen, for instance, from the diagram on p. 125 in [Iv99]. Since $|U(\mathcal{Q})| = 2^4$, the former 5 octads generate the whole of $U(\mathcal{Q})$. $\qquad\qquad\square$

Lemma 6.5.2 $\bar{\Lambda}$ *is the universal abelian representation group of Θ.*

Proof. Let R be the universal abelian representation group of Θ. Since $\bar{\Lambda}$ is clearly a factor group of R, it is sufficient to show that the order of R is at most 2^{24}. For a subset $\bar{\Xi}$ od $\bar{\Lambda}_2$ by $R(\bar{\Xi})$, we denote the subgroup in R generated by the elements-generators contained in $\bar{\Xi}$. Put

$$\bar{\Lambda}_2^{4-} = \{4p - 4q + 2\Lambda \mid p, q \in \mathcal{P}\}, \quad \bar{\Lambda}_2^4 = \{4p \pm 4q + 2\Lambda \mid p, q \in \mathcal{P}\},$$

$$\bar{\Lambda}_2^2 = \left\{ \sum_{p \in \mathcal{O}} 2p + 2\Lambda \mid \mathcal{O} \in \mathcal{B} \right\}.$$

The required bound will be achieved in four steps:

Step 1. $R(\bar{\Lambda}_2^{4-}) \cong \mathcal{C}_{11}^*$. The mapping

$$\{p, q\} \mapsto 4p - 4q + 2\Lambda$$

is easily seen to be a homomorphism of the even half of the $GF(2)$-permutation module V_{24} of the Mathieu group M_{24} (which is the setwise stabilizer of the basis $\mathcal{P}$ of Λ). Furthermore, the sum over an octad maps onto the zero element of $\bar{\Lambda}$. Hence the stated isomorphism follows directly from the submodule structure of V_{24}.

Step 2. $[R(\bar{\Lambda}_2^4) : R(\bar{\Lambda}_2^{4-})] = 2$. Let Δ be a graph on

$$\bar{\Lambda}_2^{4+} = \bar{\Lambda}_2^4 \setminus \bar{\Lambda}_2^{4-} = \{4p + 4q + 2\Lambda \mid p, q \in \mathcal{P}\},$$

in which two vertices are adjacent if there is a line of Θ containing them, whose third point is in $\bar{\Lambda}_2^{4-}$. It is easy to check that Δ is connected which gives the required equality.

Step 3. $R(\bar{\Lambda}_2^2)/R(\bar{\Lambda}_2^4) \cong \mathcal{C}_{11}$. We consider the quotient point-line incidence system Σ whose points are the images in $R(\bar{\Lambda}_2^2)/R(\bar{\Lambda}_2^4)$ of the point-generators contained in $\bar{\Lambda}_2^2$ and the lines are the images of those contained in that set. By elementary calculations in the Leech lattice we show that Σ is precisely the system in (6.5.1) and obtain the required isomorphism.

Step 4. $[R : R(\bar{\Lambda}_2^2)] = 2$. The equality follows from the connectivity of the graph on

$$\bar{\Lambda}_2^3 = \bar{\Lambda}_2^2 \setminus (\bar{\Lambda}_2^2 \cup \bar{\Lambda}_2^4),$$

where two elements are adjacent if they are contained in a 222-triangle whose third element is in $\bar{\Lambda}_2^2 \cup \bar{\Lambda}_2^4$. $\square$

Lemma 6.5.3 *The automorphism group of Θ is isomorphic to Co_1.*

Proof. The automorphism group in question is that of the *short vector graph* Δ (cf. Section 4.11 in [Iv99]). The fact that $\mathrm{Aut}(\Theta) \cong Co_1$ is well known and easy to prove. The local graph of Δ when the quotient over the equivalence relation with classes of size 2, induced by the lines containing the distinguished vertex, is the famous strongly regular graph on 2300 vertices having Co_2 as the full automorphism group (cf. Table 10A1 in [Gri98]). $\square$

6.6 Finiteness and simplicity of $\varphi(G)$

Let (H, ψ) be a faithful generating completion of the Monster amalgam $\mathcal{M}$, so that H is a group and

$$\psi : \mathcal{M} \to H$$

is a bijection, the restriction of ψ to G_i is an isomorphism for every $1 \leq i \leq 3$, and H is generated by $\mathrm{Im}(\psi)$. We start with the following:

Lemma 6.6.1 *Let* (H, ψ) *be a faithful generating completion of* $\mathcal{M}$. *Then* $\psi(G_i) \cap \psi(G_j) = \psi(G_{ij})$ *for all* $1 \leq i < j \leq 3$.

Proof. As $\psi(G_i) \cong G_i$ and $\psi(G_i) \cap \psi(G_j) \geq \psi(G_{ij})$ since the completion is faithful, the only thing to be shown is that $\psi(G_i) \cap \psi(G_j)$ cannot contain $\psi(G_{ij})$ as a proper subgroup. But this is rather clear, since G_{ij} is maximal in G_j (with index 3 or 7) and since G_i does not contain subgroups isomorphic to G_j. $\square$

Lemma 6.6.2 *Let* (H, ψ) *be a faithful generating completion of* $\mathcal{M}$, *and let* N *be a proper normal subgroup in* H. *Then:*

(i) $N \cap \psi(G_i) = 1$ *for every* $1 \leq i \leq 3$;
(ii) *if* $\psi_N : a \mapsto \psi(a)N$ *for* $a \in \mathcal{M}$, *then* $(H/N, \psi_N)$ *is a faithful generating completion of* $\mathcal{M}$.

Proof. Making use of the following chief series of G_is

$$1 < Z_1 < Q_1 < G_1;$$
$$1 < Z_2 < R_2 < Q_2 < G_2^s < G_2;$$
$$1 < Z_3 < R_3 < T_3 < Q_3 < G_3^s < G_3;$$

it is rather easy to prove the following assertion for every $1 \leq i < j \leq 3$: let M_i be a non-identity normal subgroup in G_i, let M_j be the normal closure in G_j of $M_i \cap G_j$, and let $M_i^{(1)}$ be the normal closure in G_i of $M_j \cap G_i$; then

$M_i^{(1)}$ contains M_i is a proper subgroup. This assertion can be used to prove that whenever N contains Z_1, it contains the whole of H. After that in order to establish (i), it is sufficient to notice that Z_1 is contained in every non-trivial subgroup normal in G_2 or G_3. The assertion (ii) follows from (i) and (6.6.1). $\square$

Lemma 6.6.3 *Let (H, ψ) be a faithful generating completion of $\mathcal{M}$ such that*

$$\psi(G_1) = C_H(\psi(Z_1)).$$

Then:

(i) *H contains no non-trivial abelian normal subgroups;*
(ii) *H contains no non-trivial solvable normal subgroup.*

Proof. By considering the last term in the derived series we deduce (ii) from (i). Let N be a non-trivial abelian normal subgroup in H. By (6.6.2), $N \cap \psi(G_1) = 1$, and hence by the hypothesis the element $\psi(z)$ (where z is the generator of Z_1) negates every element of N. Since $\psi(Z_2)$ contains three H-conjugates of $\psi(z)$ whose product is the identity element, this is impossible. $\square$

Lemma 6.6.4 *Let (H, ψ) be a faithful generating completion of $\mathcal{M}$ such that:*

(a) *$\psi(G_1) = C_H(\psi(Z_1))$ and*
(b) *H is an algebraic group.*

Then H is a finite group.

Proof. By (6.6.3 (ii)) and (b), the group H is a reductive algebraic. Therefore, $\psi(Z_1)$ being an finite abelian subgroup in a reductive group is contained in a torus T. Since T is abelian, it centralizes $\psi(Z_1)$ and by (a) the torus T must be finite. By a fundamental property of reductive groups, this implies that the whole H is finite. $\square$

Lemma 6.6.5 *Let (H, ψ) be a faithful completion of $\mathcal{M}$ such that:*

(a) *$\psi(G_1) = C_H(\psi(Z_1))$ and*
(b) *H is finite.*

Then H is simple.

Proof. Let N be a non-trivial normal subgroup in H. If the order of N is odd, then N is solvable by the Feit–Thompson theorem and the assertion follows from (6.6.3 (ii)). If $|N|$ is even, then $\psi(Z_1)$ normalizes a Sylow 2-subgroup S in N (since the number of Ss is odd) and $\psi(Z_1)$ centralizes a non-identity element s in S (since the number of such elements is odd). Since

$N \cap \psi(G_1) = 1$ by (6.6.1 (i)), $\langle s, \psi(G_1) \rangle$ properly contains $\psi(G_1)$ and centralizes $\psi(Z_1)$ contrary to (a).   □

Now we arrive with the principal result of the section.

Proposition 6.6.6 *The group* $\varphi(G)$ *is finite and non-abelian simple.*

Proof. The automorphism group of the Griess algebra is clearly an algebraic group. Since $\varphi(G_1) = C_{\varphi(G)}(\varphi(Z_1))$ by (6.3.1), we deduce the finiteness of $\varphi(G)$ from (6.6.4) and its simplicity from (6.6.5).   □

Another finiteness proof due to Conway will be given in Section 8.6 as (8.6.5).

7

Important subgroups

We continue with revealing the subgroup structure of the universal completion G of the Monster amalgam $\mathcal{M} = \{G_1, G_2, G_3\}$. It was shown in Chapter 4 that G is also the universal completion of $\mathcal{M}$ enriched by $G_4, G_5^{(t)}, G_5^{(l)}, G_{10}$, and $^2G_2$. The enriched amalgam can be described as follows: when X runs through

$$G_1, G_2, G_3, G_4, G_5^{(t)}, G_5^{(l)}, G_{10}, \text{ and } ^2G_2,$$

$Z(O_2(X))$ runs through

$$Z_1, Z_2, Z_3, Z_4, Z_5^{(t)}, Z_5^{(l)}, Z_{10}, \text{ and } Y_2$$

with $|Z(O_2(X))| = 2^i$, where i is the subscript in the name of X. The subgroups $Z(O_2(X))$ are contained in Z_{10} and, except for Y_2, they are totally isotropic with respect to the G_{10}-invariant quadratic form κ on Z_{10}, and

$$Z_1 < Z_2 < Z_3 < Z_4 = Z_5^{(t)} \cap Z_5^{(l)},$$

while Y_2 is a minus line in (Z_{10}, κ). Furthermore, $G_4 = N_{G_5^{(t)}}(Z_4)$ and $G_5^{(l)} = N_{G_{10}}(Z_5^{(l)})$, while the remaining members of the enriched amalgam have the following shapes

$$G_1 \sim 2_+^{1+24}.Co_1, \quad G_2 \sim 2^{2+11+22}.(M_{24} \times S_3),$$

$$G_3 \sim 2^{3+6+12+18}.(3 \cdot S_6 \times L_3(2)), \quad G_5^{(t)} \sim 2^{5+10+20}.(S_3 \times L_5(2)),$$

$$G_{10} \sim 2^{10+16}.\Omega_{10}^+(2), \quad ^2G_2 \sim 2^2.(^2E_6(2)) : S_3.$$

In this chapter we enrich $\mathcal{M}$ further by adjoining

$$F_{3+} \cong 3 \cdot Fi_{24} \text{ and } F_{2+} \cong 2 \cdot BM$$

where Fi_{24} is the largest exceptional Fischer's 3-transposition group, whose index 2 commutator subgroup is a sporadic simple group and BM is the Baby Monster sporadic simple group.

7.1 Trident groups

This appears to be the perfect place to put the construction of the group $\bar{G}_2$ accomplished in Section 2.4 into a general setting to include the three bases subgroup from Section 1.11. The setting generalizes Section 4.5 of [Iv04].

Let H be a finite group, let U and V be a pair of faithful $GF(2)$-modules for H, and let

$$\pi : U \times U \to V$$

be a symmetric H-invariant mapping which satisfies the cocycle condition

$$\pi(u_1 + u_2, u_3) + \pi(u_1, u_2) = \pi(u_1, u_2 + u_3) + \pi(u_2, u_3)$$

for all $u_1, u_2, u_3 \in U$ (this is a weakened version of the bilinearity condition). The following result is standard (cf. Section 15.7 in [H59]).

Lemma 7.1.1 *Let $\mathcal{W}(U, V, \pi)$ be the direct sum of U and V with addition defined by*

$$(u_1, v_1) + (u_2, v_2) = (u_1 + u_2, v_1 + v_2 + \pi(u_1, u_2)).$$

Then $\mathcal{W}(U, V, \pi)$ is a $GF(2)$-module for H which is an extension of V by U. The extension splits if and only if π is a coboundary. □

Put $\mathcal{S}_U = \{(u, 0) \mid u \in U\}$ and $\mathcal{S}_V = \{(0, v) \mid v \in V\}$. Then $\mathcal{S}_V$ is an H-submodule in $\mathcal{W}(U, V, \pi)$ isomorphic to V, while $\mathcal{S}_U$ is an H-set isomorphic to U which maps bijectively onto the quotient over $\mathcal{S}_V$. In what follows when referring to an element $(0, v) \in \mathcal{S}_V$ we simply write v. The addition in $\mathcal{W}(U, V, \pi)$ is not closed on $\mathcal{S}_U$ whenever π is non-zero, although if π is a coboundary, $\mathcal{S}_U$ can be always 'shifted' onto an H-invariant complement to $\mathcal{S}_V$.

Let us now proceed with the main definition of the section. Let $Q^{(a)}, Q^{(b)}$ and R be elementary abelian 2-groups, and let

$$\psi^{(x)} : \mathcal{W}(U, V, \pi) \to Q^{(x)}, \ x \in \{a, b\} \text{ and } \psi : \mathcal{S}_V \to R$$

be group isomorphisms. Let Q be a group satisfying the following:

(T1) $Q = Q^{(a)} Q^{(b)}$ with $Q^{(a)} \cap Q^{(b)} = R$;

(T2) the restrictions of $\psi^{(a)}$ and $\psi^{(b)}$ to S_V coincide with ψ;
(T3) $[\psi^{(a)}(u_1, v_1), \psi^{(b)}(u_2, v_2)] = \psi(\pi(u_1, u_2))$ for all $u_1, u_2 \in U$ and $v_1, v_2 \in V$.

Lemma 7.1.2 *The group Q satisfying $(T1)$ to $(T3)$ exists and unique up to isomorphism. Furthermore:*

(i) *Q is a 2-group of order $|V| \cdot |U|^2$;*
(ii) *$R \leq Z(Q)$, $R \leq [Q, Q]$ and $Q/R \cong U \oplus U$.*

Proof. Since π is a cocycle, the result follows from the standard part of the group extension theory.   □

The subgroups $Q^{(a)}$ and $Q^{(b)}$ of Q are known as *dents*. The third dent $Q^{(c)}$ can be seen in the following way:

(1) observe that the product $\psi^{(a)}(u, v)\psi^{(b)}(u, v)$ does not depend on the choice of v;
(2) put $\psi^{(c)}(u, v) = \psi^{(a)}(u, 0)\psi^{(b)}(u, 0)\psi(0, v)$;
(3) define $Q^{(c)} = \{\psi^{(c)}(u, v) \mid u \in U, v \in V\}$.

Lemma 7.1.3 *The mapping which sends (u, v) onto $\psi^{(c)}(u, v)$ for all $u \in U$ and $v \in V$ is an isomorphism of $\mathcal{W}(U, V, \pi)$ onto $Q^{(c)}$.*

Proof. It is straightforward to check using (T3) and the addition rule in (7.1.1) that the composition in $Q^{(c)}$ is controlled by the same cocycle π (compare the paragraph after the proof of (2.4.2)).   □

Next we discuss certain automorphisms of Q.

Lemma 7.1.4 *For $h \in H$ and $s \in Sym\{a, b, c\}$, define*

$$\rho_h : \psi^{(x)}(u, v) \mapsto \psi^{(x)}(h(u), h(v)),$$

$$\rho_s : \psi^{(x)}(u, v) \mapsto \psi^{(s(x))}(u, v)$$

for all $x \in \{a, b, c\}$, $u \in U$ and $v \in V$. Then ρ_h and ρ_s are commuting automorphisms of Q, so that

$$\varphi : (h, s) \mapsto (\rho_h, \rho_s)$$

is a monomorphism of the direct product $H \times Sym\{a, b, c\}$ into $\mathrm{Aut}(Q)$.

Proof. The result follows from the defining properties (T1) to (T3) of Q together with (7.1.3). Notice that $\ker(\varphi)$ is trivial since U and V are faithful H-modules.   □

A *trident group* is a group D containing Q as a constrained normal subgroup (which means $C_D(Q) \leq Q$) such that $D/Q \cong H \times Sym\{a, b, c\}$. The easiest trident group is the semidirect product defined via the monomorphism φ as in (7.1.4) we denote it by

$$\mathcal{D}^{(0,0)}(H, U, V, \pi).$$

The superscript is reserved for various 'twisted' trident groups we are about to introduce.

Since R is in the centre of Q, the subgroups H and RH acting via conjugation have the same image in the automorphism group of Q. This observation suggests the first type of twisting. Let H^α be an extension of V by H specified by an element $\alpha \in H^2(H, V)$ (so that $\alpha = 0$ corresponds to the split extension) and let

$$\mathcal{D}^{(\alpha,0)}(H, U, V, \pi)$$

be the partial semidirect product of $H^\alpha \times Sym\{a, b, c\}$ and Q over $V \leq H^\alpha$ identified with $R \leq Q$ via an H-isomorphism. This is again a trident group.

In order to perform the second type of twisting, observe that RH acts on Q via H, and hence every automorphism of RH which centralizes both R and RH/R extends uniquely to an automorphism of $\mathcal{D}^{(0,0)}(H, U, V, \pi)$ centralizing Q. Let V^β be an H-module, which is an extension of V by the trivial 1-dimensional $GF(2)$-module, specified by an element $\beta \in H^1(H, V)$. Let H^β be the semidirect product of V^β and H with respect to the natural action. An element $\gamma \in V^\beta \setminus V$ acting by conjugation induces an automorphism which centralizes both V and VH/V (γ is not the identity whenever $\beta \neq 0$). The partial semidirect product X of Q and H^β over R (identified with V) is an extension of Q by

$$H \times \langle \gamma \rangle \times Sym\{a, b, c\} \cong H \times 2 \times S_3$$

and we define

$$\mathcal{D}^{(0,\beta)}(H, U, V, \pi)$$

to be the subgroup of index 2 in X which contains $Q.(H \times Alt\{a, b, c\})$ and distinct both from $Q.(H^\beta \times Alt\{a, b, c\})$ and from $\mathcal{D}^{(0,0)}(H, U, V, \pi)$.

Now we combine the above twistings to obtain

$$\mathcal{D}^{(\alpha,\beta)}(H, U, V, \pi).$$

This is achieved by taking the 'diagional' index 2 subgroup in the partial semidirect product over R of Q and $H^{(\alpha,\beta)}$, where the latter is an extension of V^β by H which contains H^α as an index 2 subgroup, such that

$$H^{(\alpha,\beta)}/(V^\beta \cap H^\alpha) \cong H \times 2.$$

Now (2.4.2) and (1.11.2) can be restated as follows:

Lemma 7.1.5 *Let* $\bar{G}_2 = G_2/Z_2$, *and let* T *be the three bases subgroup of* Co_1 *as in Section 1.11. Then:*

(i) $\bar{G}_2 \cong \mathcal{D}^{(1,1)}(M_{24}, \mathcal{C}_{11}, \mathcal{C}_{11}^*, \pi)$, *where* $\pi : (u, v) \mapsto (u \cap v)^*$;

(ii) $T \cong \mathcal{D}^{(0,1)}(3 \cdot S_6, \mathcal{H}, U_4, \theta)$, *where* $\theta : (u, v) \mapsto \vartheta(u + v)$ *and* ϑ *is the unique surjective mapping of* $\mathcal{H}$ *onto* U_4 *which commutes with the action of the hexacode group.* □

7.2 Tri-extraspecial groups

In this section we review some results from [ISh05] on construction and characterization of the tri-extraspecial groups. Let $L \cong L_3(2) \cong SL_3(2)$, let V_3 be the natural 3-dimensional $GF(2)$-module for L, and let $V_3^* = V_3 \wedge V_3$ be the dual of V_3. For $i \in \{0, 1\} = GF(2)$ let

$$\pi_i : V_3 \times V_3 \to V_3^*$$

be defined by $\pi_i : (u, v) \mapsto i(u \wedge v)$, and let

$$W_6^i = \mathcal{W}(V_3, V_3^*, \pi_i).$$

Since π_0 is the zero map, W_6^0 is the direct sum of V_3 and V_3^*. We will make use of the following cohomological data.

Lemma 7.2.1 *The following assertions hold:*

(i) *the group* $H^\alpha(L, X)$ *has order 2 for* $\alpha = 1$ *or 2 and for* $X = V_3$ *or* V_3^*;

(ii) W_6^1 *is the unique indecomposable extension of* V_3^* *by* V_3;

(iii) $H^1(L, W_6^1)$ *has order 2.*

Proof. (i) and (ii) are well known with the standard referencies being [Be78] and [JP76]. The assertion (iii) is less famous but still very elementary (cf. Lemma 4.4.3 in [Iv04]). □

For $\alpha \in H^2(L, V_3^*)$ let L^α be the extension of V_3^* by L corresponding to α. Then L^0 is the semidirect product with respect to the natural action, while $L^1 \cong 2^3 \cdot L_3(2)$ is the unique non-split extension.

Lemma 7.2.2 *Let* X^α *be the partial semidirect product over* V_3^* *of* W_6^1 *and* L^α *with respect to the natural action of* $L^\alpha/V_3^* \cong L$. *Then the isomorphism type of* X^α *is independent of* α.

Proof. The group X^0 is the semidirect product of W_6^1 and L with respect to the natural action. By (7.2.1 (i)) the factor group

$$X^0/V_3^* \cong L^0 \cong 2^3 : L_3(2)$$

contains two classes of $L_3(2)$-complements, while by (7.2.2) there is only two such class in the whole of X^0 and both of them intersect V_3^*L (compare (7.2.1 (i))). Therefore, X^0 contains the non-split extension $L^1 \cong 2^3 \cdot L_3(2)$ which proves the isomorphism $X^0 \cong X^1$. □

Let U_{2n} be a $2n$-dimensional $GF(2)$-space, let q be a non-singular quadratic form on U_{2n} of type $\varepsilon = \varepsilon(q)$ (which is plus or minus), and let β be the associated bilinear form. Define $T^\varepsilon(2n)$ as a semidirect product of L and a group Q of order 2^{3+6n}, where Q is the product of groups $Q^{(u)}$, one for every $u \in U_{2n}^\#$ satisfying the following:

(TE1) every $Q^{(u)}$ is elementary abelian of order 2^6 normalized by L and possessing an L-module isomorphism

$$\psi^{(u)} : W_6^{q(u)} \to Q^{(u)},$$

where $q(u) \in \{0, 1\}$ is the value of q on u;
(TE2) $R = Q^{(u_1)} \cap Q^{(u_2)}$ does not depend on the choice of the non-zero vectors u_1 and u_2 from U_{2n}, as long as they are distinct;
(TE3) there is an L-module isomorphism

$$\psi : V_3^* \to R$$

which coincides with the restiction of $\psi^{(u)}$ to V_3^* for every $u \in U_{2n}^\#$;
(TE4) if $\{u_1, u_2, u_3\}$ is a line of the projective space associated with U_{2n}, then

$$\psi^{(u_1)}(v_1, v_2^*)\psi^{(u_2)}(v_1, v_2^*)\psi^{(u_3)}(v_1, v_2^*) = 1$$

for every $v_1 \in V_3, v_2^* \in V_3^*$;
(TE5) $[\psi^{(u_1)}(v_1, v_3^*), \psi^{(u_2)}(v_2, v_4^*)] = \psi(\beta(u_1, u_2)(v_1 \wedge v_2))$ for all $u_1, u_2 \in U_{2n}^\#, v_1, v_2 \in V_3$ and $v_3^*, v_4^* \in V_3^*$.

Since

$$q(u_1 + u_2) = q(u_1) + q(u_2) + \beta(u_1, u_2),$$

the commutator relation in (TE5) shows that

$$\psi^{(u_1+u_2)} : (v_1, v_2^*) \mapsto \psi^{(u_1)}(v_1, 0)\psi^{(u_2)}(v_1, 0)\psi(v_2^*).$$

Therefore, if g is a linear transformation of U_{2n} which preserves the quadratic form q, then the mapping

$$\rho_g : \psi^{(u)}(v_1, v_2^*) \mapsto \psi^{(g(u))}(v_1, v_2^*)$$

is an automorphism of Q which commutes with the action of L. Hence there is an isomorphism of the orthogonal group $O(U_{2n}, q) \cong O_{2n}^{\varepsilon}(2)$ into Aut(Q), whose image is contained in $C_{T^{\varepsilon}(2n)}(L)$.

The isomorphism type of $T^{\varepsilon}(2n)$ depends on the type of q rather than on q itselt. In fact, let $u \in U_{2n}^{\#}$ be isotropic with respect to q, so that

$$Q^{(u)} \cong W_6^0 \cong V_3 \oplus V_3^*$$

(as a module for L). By (7.2.1 (i)), $L(\psi^{(u)}(V_3)) \cong 2^3 : L_3(2)$ contains an $L_3(2)$-subgroup $L^{(u)}$ which is not in the class of L. Performing some elementary calculations in the trident group $\mathcal{D}^{(0,0)}(V_3, V_3^*, \pi_1)$, we observe that, if $w \in U_{2n}^{\#}$, then $Q^{(w)}$, when considered as a module for $L^{(u)}$, is isomorphic to $W_6^{\gamma(w)}$, where

$$\gamma(w) = q(w) + \beta(u, w).$$

Then γ is another quadratic form on U_{2n} having β as the associated bilinear from. Since u is isotropic, γ and q have the same type. Finally, taking various choices for u we obtain in this way all the quadratic forms whose type is that of q and having β as the associated bilinear form. This shows the q-independence of $T^{\varepsilon}(2n)$ and also the fact that Aut$(T^{\varepsilon}(2n))$ induces the full automorphism group $Sp_{2n}(2)$ of the symplectic space (U_{2n}, β). There is also an automorphism of $T^{\varepsilon}(2n)$ which centralizes Q and permutes the two classes of $L_3(2)$-complements in $RL \cong 2^3 : L_3(2)$. This gives the full automorphism group of $T^{\varepsilon}(2n)$ as calculated in [ISh05].

Lemma 7.2.3 *The following assertions hold:*

(i) Out$(T^{\varepsilon}(2n)) \cong Sp_{2n}(2) \times 2$;

(ii) Out$(T^{\varepsilon}(2n))$ *permutes transitively the conjugacy classes of subgroups in* $T^{\varepsilon}(2n)$ *isomorphic to* $L_3(2)$; *the stabilizer of one of these classes is isomorphic to* $O_{2n}^{\varepsilon}(2)$. $\qquad\qquad\square$

The following characterization is the main result of [ISh05].

Proposition 7.2.4 *Let T be a group satisfying the following conditions, where $Q = O_2(T)$ and $R = Z(Q)$:*

(i) *T is a split extension of Q by $L_3(2)$;*

(ii) *Q/R is elementary abelian and each of the chief factors of T inside Q/R is isomorphic to the natural module of $T/Q \cong L_3(2)$;*

(iii) *$R \cong 2^3$ is isomorphic to the dual natural module of $T/Q \cong L_3(2)$.*

Then $T \cong T^\varepsilon(2n)$ for some $n \geq 1$ and $\varepsilon \in \{+, -\}$. □

The following result served as one of the main motivations for introducing the tri-extraspecial groups at the first place.

Lemma 7.2.5 *Let $X = O_3(Y)$, where $Y \cong 3 \cdot S_6$ is the hexacode group complementing G_3^s in G_3, and let $S = N_{G_3}(X)/X$. Then:*

(i) *$S/O_2(S) \cong G_3/O_{2,3}(G_3) \cong S_6 \times L_3(2)$;*

(ii) *if T is the preimage of the $L_3(2)$-direct factor of $S/O_2(S)$, then $T \cong T^-(4)$;*

(iii) *S is an index 2 subgroup in the automorphism group of T.*

Proof. Since X is a Sylow 3-subgroup in $O_{2,3}(G_3)$, (i) can be deduced from the shape of G_3 by a Frattini argument. The hypothesis of (7.2.4) holds because of (2.8.8) and (2.13.1 (iv)), therefore $T \cong T^\varepsilon(4)$. By (7.2.4 (ii)), the centralizer of an $L_3(2)$-subgroup in the automorphism group of T is either $O_4^-(2) \cong S_5$ or $O_4^+(2) \cong S_3 \wr S_2$ depending on whether the type of T is minus or plus. By (2.13.1), $L_3(2)$ can be found in the centralizer of an element of order 5 in S. Hence T is of minus type. Finally, (iii) follows from (i) and (7.2.4 (i)). □

7.3 Parabolics in $2^{11} \cdot M_{24}$

By (2.14.1 (II)), the subgroup G_2^s is a partial semidirect product of $Q_2 = O_2(G_2^s)$ (whose order is 2^{35}) and $M^\delta \cong Z_2 \times \bar{M}^\delta$ over the subgroup $R_2 = Z(Q_2)$ of order 2^{13} containing Z_2, where $\bar{M}^\delta \cong 2^{11} \cdot M_{24}$ is the unique non-split extension by M_{24} of the irreducible Todd module $\mathcal{C}_{11}^*$. The whole group G_2 is the semidirect product of G_2^s and an S_3-subgroup which (i) acts faithfully on $Z_2^{\#}$; (ii) centralizes a complement to Z_2 in R_2; (iii) acts fixed-point freely on $Q_2/R_2 \cong \mathcal{C}_{11} \oplus \mathcal{C}_{11}$; (iv) normalizes M^δ inducing on it an outer automorphism group of order 2.

In what follows, the symbol $\bar{M}^\delta$ denotes the quotient M^δ/Z_2 as well as a complement to T in $C_{G_2}(T)$, where T is a Sylow 3-subgroup in $O_{2,3}(G_2)$ which normalizes M^δ ($\bar{R}_2 = O_2(\bar{M}^\delta)$ in both cases). The group $\bar{M}^\delta$ acts naturally on the Steiner system $(\mathcal{P}, \mathcal{B})$ with $\bar{R}_2$ being the kernel.

We are going to identify the isomorphism types of some important subgroup in $\bar{M}^\delta$ including the sextet, trio, octad, pair, and triple stabilizers, whose images of $\bar{M}^\delta / \bar{R}_2 \cong M_{24}$ are isomorphic to

$$2^6 : 3 \cdot S_6, \quad 2^6 : (S_3 \times L_3(2)), \quad 2^4 : A_8, \quad M_{22} : 2 \text{ and } P\Sigma L_3(4),$$

respectively. It follows from the action of G on the maximal parabolic geometry mentioned at the end of Section 4.3 that

$$G_{23} = G_2(\mathcal{S}), \quad G_2 \cap G_5^{(t)} = G_2(\mathcal{T}), \quad G_{10,2} = G(\mathcal{O}),$$

where $\mathcal{S}$, $\mathcal{T}$, and $\mathcal{O}$ are sextet, trio, and octad forming a flag in the M_{24}-geometry.

Lemma 7.3.1 *Let $\mathcal{S}$ the sextet such that $G_{23} = G_2(\mathcal{S})$. Then $\bar{M}^\delta(\mathcal{S}) \cong 2_+^{1+12}.(2^4 : 3 \cdot S_6)$. Furthermore:*

(i) *$\bar{M}^\delta(\mathcal{S})$ is a semidirect product of $O_2(\bar{M}^\delta(\mathcal{S}))$ and the hexacode group $Y \cong 3 \cdot S_6$;*

(ii) *if $X = O_3(Y)$, then $[O_2(\bar{M}^\delta(\mathcal{S})), X]$ is an extraspecial group of order 2^{13}, whose Frattini factor is isomorphic to $[W_{24}, X]$ as a Y-module;*

(iii) *$C_Q(X) \cong V_{1\backslash 4} \cong C_{C_{11}^*}(X)$.*

Proof. The shape $\bar{M}^\delta(\mathcal{S}) \sim 2_+^{1+12}.2^4.3 \cdot S_6$, the isomorphism types of the chief factors of $\bar{M}^\delta(\mathcal{S})$ and assertion (iii) follow from the isomorphism type of $\bar{M}^\delta(\mathcal{S})/\bar{R}_2$ and from the way $M(\mathcal{S})$ acts on $\mathcal{C}_{11}$ (cf. (1.10.6)). Notice that in the case of split extension $\mathcal{C}_{11} : M_{24}$, the Frattiti factor of the extraspecial group is the direct sum of $\mathcal{H}^\circ$ and $\mathcal{H}$. By the choice of $\mathcal{S}$ we have $\bar{M}^\delta(\mathcal{S})T = C_{G_{23}}(T)$. Since T acts on Z_2 fixed-point freely, $Z_2 = [T, Z_3]$, which gives the equality

$$\bar{M}^\delta(\mathcal{S})T = C_{G_3}(T).$$

Now in order to prove the lemma it only remains to identify T inside G_3 and to make use of the information on the structure of G_3 in (2.8.8). It follows directly from the shape of $\bar{M}^\delta(\mathcal{S})$ that

$$\bar{M}^\delta(\mathcal{S})C_{G_3}(R_3/Z_3) = G_3.$$

Since R_3/Z_3 is the only chief factor of G_3 isomorphic to the dual hexacode module $\mathcal{H}^\circ$ and since this factor appears inside $\bar{M}^\delta(\mathcal{S})$, we conclude that $T \leq C_{G_3}(R_3/Z_3)$. Since $C_{G_3}(R_3/Z_3)/Q_3 \cong L_3(2)$, whose 3-part is of order 3, the subgroup T is uniquely specified in G_3 up to conjugation. Now the result is easy to deduce from (2.8.8). The following observation simplifies the task even further. If $\langle u \rangle = C_{Z_3}(T)$, then $C_{G_3}(T)$ is contained in $C_{G_3}(u)$ and the latter centralizer is a conjugate of G_{13} in G_3. Therefore, (ii) follows from (1.10.8). $\square$

Lemma 7.3.2 *Let $\mathcal{T}$ be the trio such that $G_2 \cap G_5^{(t)} = G_2(\mathcal{T})$. Then $\bar{M}^\delta(\mathcal{T}) \cong 2^{3+12}.(L_3(2) \times S_4)$ and the preimage of the $L_3(2)$-direct factor is the tri-extraspecial group of minus type and rank 4.*

Proof. By the choice of $\mathcal{T}$, since $Z_2 = [Z_5^{(t)}, \mathcal{T}]$ and $G_2 \cap G_5^{(t)} = N_{G_5^{(t)}}(Z_2)$, we have $\bar{M}^\delta(\mathcal{T})T = C_{G_5^{(t)}}(T)$. The centralizer $C_{Z_5^{(t)}}(T)$ is a complement to Z_2 in $Z_5^{(t)}$ and since $G_5^{(t)}/C_{G_5^{(t)}}(Z_5^{(t)}) \cong GL(Z_5^{(t)})$, the subgroups $C_{Z_5^{(t)}}(T)$ and Z_3 are $G_5^{(t)}$-conjugate. Hence $\bar{M}^\delta(\mathcal{T})T$ is a $G_5^{(t)}$-conjugate of $C_{G_3}(T') \cap G_5^{(t)}$ for a subgroup T' of order 3 in G_3. Since T centralizes a complement to Z_2 in R_2, it is easy to check that T' is a Sylow 3-subgroup of $O_{2,3}(G_3)$ and the assertion follows from (7.2.5). $\qquad\square$

Lemma 7.3.3 *Let $\mathcal{O}$ be the octad such that $G_{10,2} = G_2(\mathcal{O})$. Then $\bar{M}^\delta(\mathcal{O})$ is an index three subgroup in*

$$C_{G_{10}}(T)/T \cong \mathcal{D}^{(0,1)}(K_{\mathcal{O}}, Q_{\mathcal{O}}, Q_{\mathcal{O}} \wedge Q_{\mathcal{O}}, \xi) \cong 2^{6+8}.(A_8 \times S_3),$$

where $\xi : (u, v) \mapsto u \wedge v$.

Proof. Because of the choice of $\mathcal{O}$ we have $\bar{M}^\delta(\mathcal{O})T \leq C_{G_{10}}(T)$ and our nearest goal is to locate T inside $G_{10} \cong 2^{10+16}.\Omega_{10}^+(2)$. Since T acts fixed-point freely on Q_2/R_2, we have $C_{Z_{10}}(T) = C_{R_2}(T) \cap Z_{10}$. We know that $C_{R_2}(T)$ is a complement to Z_2 in R_2 and, therefore, $C_{Z_{10}}(T)$ is a 6-dimensional natural orthogonal module for $K_{\mathcal{O}} \cong \Omega_6^+(2)$. Hence T stabilizes a decomposition of Z_{10} into

$$Z_6^+ := C_{Z_{10}}(T) \text{ and } Z_4^- := [Z_{10}, T],$$

where $(Z_m^\varepsilon, \kappa)$ is a non-singular m-dimensional orthogonal space of type ε and the summands are orthogonal with respect to the bilinear form associated with κ. Thus the image of T in $\bar{G}_{10} = G_{10}/Q_{10} \cong \Omega_{10}^+(2)$ is specified. The index of $\bar{M}^\delta(\mathcal{O})$ in $C_{G_{10}}(T)/T$ is three, since Z_2 is one of the three minus lines inside Z_4^-. $\qquad\square$

Lemma 7.3.4 *For $i = 1, 2, 3$ let $\bar{M}_i^\delta$ be the stabilizer in $\bar{M}^\delta$ of an i-subset in the element set $\mathcal{P}$ of the Steiner system, so that $\bar{M}_i^\delta$ is an extension of $\bar{R}_2$ by*

$$M_{23}, \quad M_{22} : 2 \text{ and } P\Sigma L_3(4),$$

respectively. Then for $i = 1$ the extension does not split and it splits in the remaining two cases.

Proof. We assume that the three i-sets are contained in the octad $\mathcal{O}$ as in (7.3.3). Since the number of octads is odd while $\bar{R}_2$ is a 2-group, by Gaschütz theorem $\bar{M}_i^\delta$ splits if and only if so does $\bar{M}_i^\delta(\mathcal{O})$. Let $K_{\mathcal{O},i}$ be the stabilizer of the i-set in the complement $K_{\mathcal{O}}$ to $Q_{\mathcal{O}}$ in $M(\mathcal{O})$. Then $K_{\mathcal{O},i}$ is isomorphic to

$$A_7, \; S_6 \text{ and } (3 \times A_5) : 2$$

for $i = 1, 2,$ and 3, respectively. By (7.3.3) every dent in

$$[\bar{M}^\delta(\mathcal{O}), \bar{M}^\delta(\mathcal{O})] \cong 2^{6+8} : A_8$$

when considered as a module for $K_{\mathcal{O}}$ is isomorphic to

$$\mathcal{W}(Q_{\mathcal{O}}, Q_{\mathcal{O}} \wedge Q_{\mathcal{O}}, \xi) \cong [Q_{\mathcal{O}}, C_{11}^*].$$

Therefore, the splitness question is reduced to decomposing the latter when considered as a module for $K_{\mathcal{O},i}$. This amounts to some pleasant elementary calculations. Notice that $\mathcal{W}(Q_{\mathcal{O}}, Q_{\mathcal{O}} \wedge Q_{\mathcal{O}}, \xi)$ is the $GF(2)$-permutation module of $K_{\mathcal{O}}$ acting on the set of non-zero vectors in $Q_{\mathcal{O}}$ factorized over the subspace spanned by the characteristic functions of the hyperplanes. Finally, since a Sylow 2-subgroup of $\bar{M}_2^\delta$ contains a Sylow 2-subgroup of $\bar{M}_3^\delta$, another application of Gaschutz theorem saves the calculations for $i = 3$.

The original non-splitness proof for $\bar{M}^\delta$ in (1.6.9) implies the following:

Lemma 7.3.5 *The dodecad stabilizer in $\bar{M}^\delta$ does not split over $\bar{R}_2$.*

7.4 $3 \cdot Fi_{24}$-subgroup

Let X be a Sylow 3-subgroup of $O_{2,3}(G_3)$ and T be a Sylow 3-subgroup of $O_{2,3}(G_2)$ (recall that $O_{2,3}(H)$ is the preimage in H of $O_3(H/O_2(H))$). Since

$$G_2 \sim 2^{2+11+22}.(M_{24} \times S_3) \text{ and } G_3 \sim 2^{3+6+12+18}.(3 \cdot S_6 \times L_3(2)),$$

both X and T have order 3. The following result could be inspired by the proof of (7.3.2).

Lemma 7.4.1 *The subgroups X and T are conjugate in $G_5^{(t)}$.*

Proof. It follows from the definition of $Z_5^{(t)}$ in the paragraph preceding (4.1.4) that

$$Z_2 < Z_3 < Z_5^{(t)} < R_3 < R_2.$$

Furthermore, $Z_2 = [Z_5^{(t)}, T] = [R_2, T]$ by (2.3.1) and $Z_3 = C_{Z_5^{(t)}}(T) = C_{R_3}(T)$ by (2.8.8). Since

$$\dim(Z_5^{(t)}) = \dim(Z_2) + \dim(Z_3),$$

the centralizer-commutator decompositions of $Z_5^{(t)}$ with respect to X and T have summands of equal dimensions. Since $G_5^{(t)}$ induces on $Z_5^{(t)}$ the full linear group $GL(Z_5^{(t)})$, the result follows.   □

Let $g = g_{T,X}$ be an element in $G_5^{(t)}$ which conjugates T onto X and let G_2^g denote the image of G_2 under g acting via conjugation, so that X is a Sylow 3-subgroup of $O_{2,3}(G_2^g)$.

The following result was established in Sections 5.6 to 5.8 in [Iv99] culminated in Lemma 5.8.8.

Proposition 7.4.2 *Let F_{3+} be the subgroup in G generated by $N_{G_1}(X)$ and $N_{G_2}(X)$, where X is a Sylow 3-subgroup of $O_{2,3}(G_3)$. Then:*

(i) *F_{3+}/X is the largest exceptional Fischer 3-transposition group Fi_{24} of order*

$$2^{22} \cdot 3^{16} \cdot 5^2 \cdot 7^3 \cdot 11 \cdot 13 \cdot 17 \cdot 23 \cdot 29;$$

(ii) *F_{3+} does not split over X.*   □

In this section we review the proof of (7.4.2) given in [Iv99] adopting it to our current notations.

Lemma 7.4.3 $N_{G_1}(X)/X \cong 2_+^{1+12} : 3 \cdot U_4(3).2^2.$

Proof. The result is achieved via (a) identifying the image $\widetilde{X}$ of X in $G_1/Q_1 \cong Co_1$; (b) calculating the normalizer of $\widetilde{X}$ in Co_1, and (c) analysing the centralizer-commutator decomposition of $Q_1/Z_1 \cong \Lambda/2\Lambda$ with respect to the action of $\widetilde{X}$ (cf. (4.14.9), (4.14.11), and (5.6.1) in [Iv99]).   □

Lemma 7.4.4 *Let $N = N_{G_2^g}(X)$. Then $N \cong 2^{11}.(S_3 \times M_{24})$, $N/X \cong 2^{12} \cdot M_{24}$ is the unique non-split extension of C_{12}^* by M_{24} and N splits over X.*

Proof. By (7.4.1), $N \cong N_{G_2}(T)$, so that the result is immediate from the structure of G_2 exposed in (2.14.1 (II)) (compare (5.6.2) in [Iv99]).   □

Lemma 7.4.5 *$N_{G_3}(X)/X \cong 2^{3+12}.(S_6 \times L_3(2))$, the preimage of the $L_3(2)$-direct factor is the tri-extraspecial group $T^-(4)$ of order 2^{15} and minus type, and $N_{G_3}(X)$ does not split over X.*

Proof. The assertion is an extraction from (7.2.5) together with the observation that $N_{G_3}(X)$ contains the hexacode group $3 \cdot S_6$ which does not split over X.   □

Lemma 7.4.6 *$N_{G_{10}}(X)/X \cong 2^{7+8}.(A_8 \times S_3)$ and $N_{G_{10}}(X)$ splits over X.*

Proof. Since

$$N_{G_{10}}(Z_5^{(t)})/C_{G_{10}}(Z_5^{(t)}) \cong N_{G_5^{(t)}}(Z_5^{(t)})/C_{G_5^{(t)}}(Z_5^{(t)}) \cong GL(Z_5^{(t)}),$$

the element $g_{T,x}$ introduced after the proof of (7.4.1) can be chosen inside G_{10}. Therefore, the result follows from (7.3.3) and its proof together with the observation that here we deal with the normalizer of X rather than with its centralizer. The splitness follows from the triviality of the 3-parts of the Schur multipliers of A_8 and S_3. □

Define F_{3+} to be the subgroup in G generated by the normalizers $N_A(X)$ taken for A equal to G_1, G_2^g, G_3, and G_{10}.

Lemma 7.4.7 *Let $\mathcal{F}_{3+}$ be the coset geometry of F_{3+}/X with respect to the subgroups $N_A(X)/X$, where A runs through G_1, G_2, G_{10}, and G_2^2. Then $\mathcal{F}_{3+}$ is described by the following diagram:*

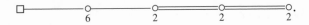

The nodes from the left to the right, with the leftmost fake node ignored, correspond to the subgroups A in the above order.

Proof. The diagram of $\mathcal{F}_{3+}$ was identified on p. 229 in [Iv99] by considering a connected component of subgeometry induced by the elements fixed by X in the tilde geometry $\mathcal{G}(G)$ defined in Section 4.2. □

Starting with $\mathcal{F}_{3+}$ another geometry, called $\mathcal{T}$, was recovered which falls under T. Meixner's characterization [Mei91] of extended polar spaces. Below is an outline of how $\mathcal{T}$ was recovered.

Call the elements of $\mathcal{F}_{3+}$ corresponding to the rightmost and the second right nodes of the diagram *points* and *lines*, respectively, and denote by Φ the collinearity graph of $\mathcal{F}_{3+}$. Then the vertices of Φ are the (right) cosets of $N_{G^g}(X)/X$ in F_{3+}/X and two such cosets are adjacent whenever they intersect a coset of $N_{G_{10}}(X)/X$. The group F_{3+} acts naturally on Φ by translations as a group of automorphisms and if v is a vertex of Φ, then its stabilizer $F_{3+}(v)$ is a conjugate of $N_{G_2^g}(X)$, in particular X is the kernel of the action. An arbitrary element u in $\mathcal{F}_{3+}$ can be identified with the subgraph $\Phi[u]$ induced by the vertices incident to u. Subject to this identification the incidence is via symmetrized inclusion. The elements of the first, second, third, and fourth types from the right of the diagram are realized by vertices, triangles, 15-vertex collinearity subgraphs of the generalized quadrangle of order

(2,2), and by 567-vertex subgraphs isomorphic to the collinearity graph of the residual $U_4(3)$-geometry with the diagram

$$\overset{2}{\circ}\!=\!=\!=\!=\!\overset{2}{\circ}\!=\!=\!=\!=\!\overset{2}{\circ}.$$

The elements of the forthcoming geometry $\mathcal{T}$ are also certain subgraphs in Φ with the symmetrized inclusion as the incidence relation.

If u is a vertex of Φ such that $F_{3+}(u)/X = N_{G_2}(X)/X$, then $O_2(F_{3+}(u)/X)$ is the Todd module and the 24-element set $\mathcal{P}$ can be identified with the unique 24-orbit of $F_{3+}(u)/X$ on $O_2(F_{3+}(u)/X)$. Consider a chain

$$A_1 \subset A_2 \subset A_3 \subset A_4 \subset \mathcal{P},$$

of subsets, where $|A_i| = i$ and identify A_i with the subgroup of order 2^i in $O_2(F_{3+}(u)/X)$ generated by the elements of A_i (considered as involutions from the 24-orbit). Define Φ_i to be the connected component containing u of the subgraph in Φ containing all the vertices v such that

$$A_i \leq O_2(F_{3+}(v)/X)$$

(compare the paragraph before (5.7.7) in [Iv99].)

Lemma 7.4.8 Φ_4 *is the Schläfli graph which is the collinearity graph of the generalized quadrangle of order* $(4, 2)$ *associated with* $U_4(2)$.

Proof. The assertion is (5.7.8) in [Iv99] and its proof is based on the fact that Φ_4 is contained in $\Phi[v]$ for an element of type 4 in $\mathcal{F}_{3+}$ such that the unique non-identity element in

$$Z(O_2(F_{3+}(v)/X))$$

is the product of the four involutions in A_4.   □

The vertex u is contained in 759 triangles realizing the lines of $\mathcal{F}_{3+}$ incident to u. These triangles are indexed by the octads of the Steiner system $(\mathcal{P}, \mathcal{B})$ on $\mathcal{P}$ preserved by $F_{3+}(u)$ and they are permuted naturally by the M_{24}-factor group of F_{3+}; an element $x \in O_2(F_{3+}(u))$, considered as a coset $x + \mathcal{C}_{12}$ of the Golay code $\mathcal{C}_{12}$ in $2^{\mathcal{P}}$ fixes vertex-wise the triangles whose octads intersect x evenly and induces an action of order two on each of the remaining triangles (compare (5.7.2) in [Iv99]).

Lemma 7.4.9 *The valency of* Φ_i *is* $2 \cdot 253, 2 \cdot 77, 2 \cdot 21,$ *and* $2 \cdot 5$ *for* $i = 1, 2, 3,$ *and* 4, *respectively.*

Proof. This is (5.7.7 (iii)) in [Iv99] and the valency of Φ_4 follows already from (7.4.8).   □

Define $\mathcal{T}$ to be the geometry whose elements are the vertices and triangles of Φ which are the points and lines of $\mathcal{F}_{3+}$ together with the images under F_{3+} of the subgraphs Φ_i for $1 \leq i \leq 4$. The incidence is via the symmetrized inclusion. It follows from (7.4.8) and (7.4.9) that $\mathcal{T}$ belongs to the following diagram

and that the action of F_{3+}/X on $\mathcal{T}$ is faithful and flag-transitive. The flag-transitive geometries with the above diagram were identified by T. Meixner in [Mei91] with those related to the Fischer group Fi_{24} (also denoted by $M(24)$) and with its triple cover $3 \cdot F_{24}$. The Fischer group geometry can be defined as follows.

The Fischer group Fi_{24} being a 3-transposition group acts as a group of automorphisms on 3-transposition graph Σ whose vertices form a conjugacy class of certain involutions (called 3-transpositions) in Fi_{24} and two 3-transpositions are adjacent in Σ if and only if they commute. The 3-transposition property implies that the order of the product of non-adjacent vertices is always 3. The centralizer in Fi_{24} of a 3-transposition is $2 \times Fi_{23}$ where Fi_{23} is a smaller Fischer 3-transposition group.

It is a basic consequence of the general theory of 3-transpositions groups (cf. [F69], [F71], and [A97]) that Σ is strongly regular and that Fi_{24} induces on Σ a rank 3 action (acts transitively on the adjacent and non-adjacent pairs of vertices). The specific parameters of Σ are the following

$$n = 306,936, \quad k = 31,671, \quad \lambda = 3,510, \quad \mu = 3,240,$$

where $n = |\Sigma| = [Fi_{24} : 2 \times Fi_{23}]$ is the number of vertices, k is the valency of a vertex, while λ and μ are the numbers of common neighbours of a pair of adjacent and non-adjacent vertices, respectively.

Let Θ_{24} be a maximal clique in Σ, that is a maximal set of pairwise commuting transpositions. Then:

(i) $|\Theta_{24}| = 24$;

(ii) the setwise stabilizer N_{24} of Θ_{24} in Fi_{24} is the non-split extension of the Todd module $\mathcal{C}_{12}^*$ by M_{24}, the subgroup $O_2(N_{24})$ is the vertex-wise stabilizer of Θ_{24} in Fi_{24}, and $N_{24}/O_2(N_{24}) \cong M_{24}$ acts on Θ_{24} as on the element set of the $S(5, 8, 24)$-Steiner system;

(iii) if Θ_8 is a maximal subset of Θ_{24} with the property that Θ_{24} is not the unique maximal clique containing it, then Θ_8 consists of eight

3-transpositions forming an octad and there are precisely three maximal cliques in Σ containing Θ_8.

Let $\Theta_1 \subset \Theta_2 \subset \Theta_3 \subset \Theta_4 \subset \Theta_8$ be a chain of subsets such that $|\Theta_i| = i$ and let N_i be the stabilizer of Θ_i in Fi_{24}. Then

$$N_1 \cong 2 \times Fi_{23}, \quad N_2 \cong (2 \times 2 \cdot Fi_{22}).2, \quad N_3 \cong (2 \times 2^2.U_6(2)).S_3,$$

$$N_4 \cong 2^{1+12}_+.3 : (U_4(2).2 \times 2) \text{ and } N_8 \cong 2^{7+8}.(A_8 \times S_3).$$

Furthermore, $N_{24} \cong N_{G_2^8}(X)/X$ and $N_8 \cong N_{G_{10}}(X)/X$. The intersection of any two maximal cliques in Σ, if non-empty, is the image under an element from Fi_{24} of a clique Θ_i for some $i \in \{1, 2, 3, 4, 8, 24\}$.

Let $\mathcal{G}(Fi_{24})$ be a geometry whose elements are the images under Fi_{24} of the cliques Θ_i for $i = 1, 2, 3, 4, 8,$ and 24, let the type of an element be determined by its size, and let the incidence be the symmetrized inclusion. Then Fi_{24} acts on $\mathcal{G}(Fi_{24})$ flag-transitively and the diagram of $\mathcal{G}(Fi_{24})$ coincides with that of $\mathcal{T}$. The geometry $\mathcal{G}(Fi_{24})$ possesses a triple cover $\mathcal{G}(3 \cdot Fi_{24})$ associated with the non-central non-split extension by Fi_{24} of a group of order 3. The group $3 \cdot Fi_{24}$ acts on a triple cover of the 3-transposition graph Σ and the elements of $\mathcal{G}(3 \cdot Fi_{24})$ are cliques in the covering graph. Now we are in a position to state Meixner's characterization.

Proposition 7.4.10 *Let $\mathcal{A}$ be a geometry with the diagram*

and let A be a flag-transitive automorphism group of $\mathcal{A}$. Then one of the following holds:

(i) $\mathcal{A} \cong \mathcal{G}(Fi_{24})$ *and A is either the Fischer group Fi_{24} or its index 2 simple commutator subgroup Fi'_{24};*

(ii) $\mathcal{A} \cong \mathcal{G}(3 \cdot Fi_{24})$ *and A is either $3 \cdot Fi_{24}$ or $3 \cdot Fi'_{24}$.* $\qquad\qquad\square$

Some weaker characterizations were known before [Mei91] and some stronger ones have been achieved afterwards (the relevant referencies can be found in [Iv99]), but (7.4.10) fits perfectly our purpose.

Now it only remains to exclude three out of four options for $(\mathcal{A}, A)$ in (7.4.10). Since the stabilizer of a maximal clique of Σ (respectively of the covering graph of Σ) in Fi'_{24} (respectively in $3 \cdot Fi'_{24}$) is $C^*_{11} \cdot M_{24}$, rather than $C^*_{12} \cdot M_{24}$, the groups Fi'_{24} and $3 \cdot Fi'_{24}$ are immediately excluded.

Suppose that $F_{3+}/X \cong 3 \cdot Fi_{24}$, so that the geometyry $\mathcal{T}$ is simply connected. We reach a contradiction by constructing a triple cover of $\mathcal{T}$ on which

$\mathcal{F}_{3+}$ acts faithfully. By (7.4.4) and (7.4.6) both the point and line stabilizers associated with the action of F_{3+} on $\mathcal{F}_{3+}$ split over X. Let u be a vertex of Φ and let $t = \{u, v, w\}$ be a triangle in Φ realizing a line incident to u. Let $F_{3+}(u)^c$ and $F_{3+}(t)^c$ be complements to X in $F_{3+}(u)$ and $F_{3+}(t)$, respectively which contain a Sylow 2-subgroup of $F_{3+}(u) \cap F_{3+}(t)$. Define $3.\Phi$ to be a graph on the set of cosets of $F_{3+}(u)^c$ in F_{3+} where two cosets are adjacent if they intersect a coset of $F_{3+}(t)^c$. Then $3.\Phi$ is a covering of Φ and the action of F_{3+} on $3.\Phi$ is faithful.

Let $3.\mathcal{F}_{3+}$ be the geometry whose elements are the connected components of the preimages in $3.\Phi$ of the subgraphs realizing the elements of $\mathcal{F}_{3+}$ with the incidence defined via the symmetrized inclusion. Then $\mathcal{F}_{3+}$ belongs to the diagram

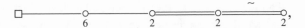

in particular the preimage in $3.\Phi$ of the 15-vertex subgraph realizing an element of the middle type is connected: it is the 45-vertex subgraph associated with the hexacode group $3 \cdot S_6$. Since the stabilizer in F_{3+} of a vertex from $3.\Phi$ is $C_{12}^* \cdot M_{24}$, we can apply the above procedure to obtain a geometry with the diagram as in (7.4.10) Since in this case the flag-transitive automorphism group F_{3+} contains a normal subgroup of order 3, (7.4.10 (ii)) with $A \cong 3 \cdot Fi_{24}$ is the only option for $(3.\mathcal{F}_{3+}, F_{3+})$. This completes the review of the proof of (7.4.2) given in [Iv99]. Some further information on geometries of the Fischer groups can be found in [Iv95].

7.5 $2 \cdot BM$-subgroup

In terms introduced in the heading to the chapter let Y_1 be a subgroup of order 2 in Y_2, so that Y_1 is a non-isotropic point in (Z_{10}, κ). The main result of this section is the following statement proved in [Iv99] as (5.10.22).

Proposition 7.5.1 *Let F_{2+} be the subgroup in G generated by $N_{G_1}(Y_1)$ and $N_{G_2}(Y_1)$. Then:*

(i) *F_{2+}/Y_1 is the Baby Monster sporadic simple group BM of order*

$$2^{41} \cdot 3^{13} \cdot 5^6 \cdot 7^2 \cdot 11 \cdot 13 \cdot 17 \cdot 19 \cdot 23 \cdot 31 \cdot 47;$$

(ii) *F_{2+} does not split over Y_1.*

Here we review the principal steps in the proof of the above proposition adapting them to the current notation. The idea of the proof is to identify F_{2+} with the unique faithful completion of the *Baby Monster amalgam*

$$\mathcal{BM} = \{C_{G_1}(Y_1), C_{G_2}(Y_1), C_{G_3}(Y_1)\}.$$

The members of $\mathcal{BM}$ are the centralizers of Y_1 in the members of the Monster amalgam $\mathcal{M}$. Since the latter has already been enriched, it is convenient to work with enriched Baby Monster amalgam formed by the centralizers of Y_1 in the members of the enriched Monster amalgam.

Lemma 7.5.2 *The following assertions hold:*

(i) $C_{G_1}(Y_1)/Y_1 \cong 2_+^{1+22}.Co_2;$

(ii) $C_{G_2}(Y_1)/Y_1 \cong 2^{2+10+20}.(\mathrm{Aut}(M_{22}) \times S_3);$

(iii) $C_{G_3}(Y_1)/Y_1 \cong 2^3.[2^{32}].(S_5 \times L_3(2));$

(iv) $C_{G_{10}}(Y_1)/Y_1 \cong 2^{9+16}.Sp_8(2);$

(v) $C_{2G_2}(Y_1)/Y_1 \cong 2 \cdot{}^2E_6(2) : 2;$

(vi) $C_{F_{3+}^g}(Y_1)/Y_1 \cong 2 \cdot Fi_{22}.2$ *for some* $g \in G_5^{(t)}$ *and* $C_{F_{3+}^{gh}}(Y_1)/Y_1 \cong Fi_{23}$
    for some $h \in G_1$.

Proof. Let $\mathcal{T}$ be the 4-subset of $\mathcal{P}$ which determines the sextet whose stabilizer in G_2 is G_{23} and let $\mathcal{Q}$ be the 3-subset in $\mathcal{T}$ which determines $\langle Y_2, Z_1 \rangle$. Then we may and will assume that

$$\chi(Y_1) = \langle 4p - 4q + 2\Lambda \rangle_{\bar{\Lambda}},$$

where $\{p, q\}$ is a 2-subset in $\mathcal{Q}$. Then $\chi(C_{Q_1}(Y_1))$ is the orthogonal complement of $4p - 4q + 2\Lambda$ with respect to the form $(\ ,\ )_{\bar{\Lambda}}$, while $C_{G_1}(Q_1)Q_1/Q_1$ is the stabilizer in Co_1 of $4p+4q+2\Lambda$. Since the latter element is contained in $\bar{\Lambda}_2$, its stabilizer is the second Conway group Co_2, which gives (i). In order to establish (ii) notice that Y_1 is contained in the M^δ-invariant complement to Z_2 in R_2, isomorphic to C_{11}^*. More specifically, Y_1 is the element in the 276-orbit inside the complement which corresponds to $\{p, q\}$. Turning to (iii) notice that Y_1 is contained in the $3 \cdot S_6$-invariant complement to Z_3 in R_3 and it is in the 18-orbit of the hexacode group on the element-set of this complement. Since Y_1 is a non-isotropic point in the orthogonal space (Z_{10}, κ), its centralizer in G_{10} contains Q_{10} and

$$C_{G_{10}}(Y_1)/Q_{10} \cong O_9(2) \cong Sp_8(2),$$

which is (iv). Since $Y_1 \leq O_2(^2G_2)$, (v) is immediate from the isomorphism $^2G_2 \cong 2^2 \cdot (^2E_6(2)) : S_3$.

The subgroup F_{3+} is generated by $N_{G_1}(X)$ and $N_{G_2}(X)$, where X is a Sylow 3-subgroup of $O_{2,3}(G_3)$, so that $X = O_3(F_{3+})$. By (7.4.1), there is $g \in G_5^{(t)}$, which conjugates X onto a Sylow 3-subgroup T of $O_{2,3}(G_2)$. Therefore, $T = O_3(F_{3+}^g)$ and

$$(F_{3+}^g \cap G_2)/T \cong C_{12}^* \cdot M_{24}.$$

As in the paragraph before(7.4.8) we identify $\mathcal{P}$ with the unique 24-orbit of M_{24} in $O_2(F_{3+}^g \cap G_2)/T \cong C_{12}^*$ and let $A_1 \subset A_2 \subset A_3 \subset A_4 \subset \mathcal{P}$ be a chain of subsets in $\mathcal{P}$, where $|A_i| = i$. Then, assuming that Y_1 is generated by the product of the elements in A_2, we deduce the first assertion in (iv) from the isomorphism type of N_2 given before (7.4.10). Without loss we can assume that the unique element in A_1 is

$$3p - \sum_{r \in \mathcal{P} \setminus \{p\}} r + 2\Lambda.$$

Since this is an element of $\bar{\Lambda}_2$, the subgroup A_1 is a G_1-conjugate of Y_1 and the second assertion in (vi) follows.   □

It is well-known and easy to check that $\bar{\Lambda}$ is an indecomposable Co_2-module, which shows that already $C_{G_1}(Y_1)$ does not split over Y_1, and therefore gives (7.5.1 (ii)).

Since G_{10} contains $G_5^{(l)}$ together with an index 3 subgroup of $G_5^{(t)}$, while the latter contains G_4, we can easily deduce from (7.5.2 (iv)) that

$$C_{G_4}(Y_1)/Y_1 \sim 2^4.[2^{30}].(L_4(2) \times 2), \quad C_{G_5^{(l)}}(Y_1)/Y_1 \sim 2^{5+10+10+5}.L_5(2).$$

The coset geometry of F_{2+} associated with

$$\{C_A(Y_1) \mid A = G_1, G_2, G_3, G_4, G_5^{(l)}\}$$

is a rank 5 Petersen geometry with the diagram

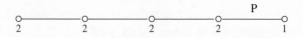

and essentially (7.5.1) is a consequence of the classification of the flag-transitive Petersen geometries accomplished in [Iv99] and [ISh02]. In the Baby Monster geometry case, the strategy is the following. We consider the action of F_{2+} on the cosets of $C_{2G_2}(Y_1) \cong 2^{2 \cdot 2} E_6(2) : 2$ and define a graph structure Ω on these cosets. It follows from (4.4.3) that

$$G_1 \cap C_{2G_2}(Y_1) = N_{G_1}(Y_2) \cap C_{G_1}(Y_1) \sim 2^2.2_+^{1+20}.U_6(2).2$$

is the preimage in $^2G_2$ of the stabilizer of an element of the leftmost type in the action of $^2G_2$ on the associated F_4-geometry with the diagram

(compare the proof of (4.4.4)). On the other hand, $C_{G_1 \cap {}^2G_2}(Y_1)Q_1/Q_1 \cong U_6(2) : 2$ is an index 3 subgroup in the stabilizer in Co_1 of a 222-triangle. The image $\chi(C_{G_1 \cap {}^2G_2}(Y_1) \cap Q_1)$ is the orthogonal complement of $\chi(Y_2) \cong 2^2$, while $\chi(C_{Q_1}(Y_1))$ is the orthogonal complement of $\chi(Y_1) \cong 2$. This gives the following:

Lemma 7.5.3 *The subgroup*

$$(G_1 \cap C_{{}^2G_2}(Y_1))C_{Q_1}(Y_1) \cong 2^2.2_+^{1+20}.(2 \times U_6(2).2)$$

contains $G_1 \cap C_{{}^2G_2}(Y_1)$ *as a normal subgroup of index 2.*   □

Let Ω be a graph on the set of cosets of $C_{{}^2G_2}(Y_1)$ in F_{2+} in which two cosets are adjacent if they intersect a coset of $(G_1 \cap C_{{}^2G_2}(Y_1))C_{Q_1}(Y_1)$. Let w denote $C_{{}^2G_2}(Y_1)$ considered as a vertex of Ω, and let $\Omega_1(w)$ be the set of vertices adjacent to w in Ω.

Lemma 7.5.4 *The following assertions hold:*

(i) F_{2+} *acts on* Ω *as an edge- and vertex-transitive automorphism group with kernel* Y_1;

(ii) *the valency of* Ω *is* $3, 968, 055$;

(iii) *the action of* $F_{2+}(w) \cong 2^2.{}^2E_6(2).2$ *on* $\Omega_1(w)$ *is similar to the action by conjugation on the set of 2-central involutions in*

$$F_{2+}(w))/Y_2 \cong {}^2E_6(2) : 2$$

with centralizer $2_+^{1+20} : U_6(2) : 2$;

(iv) *the* $C_{G_{10}}(Y_1)$*-orbit* Ξ *of* w *is doubly transitive comprising of* 120 *vertices, and the subgraph in* Ω *induced by* Ξ *is complete;*

(v) *the subgraph induced by* $\Omega_1(w)$ *is isomorphic to the commutation graph on the central involutions in* $^2E_6(2)$.

Proof. The assertions (i) to (iii) follow from the definition of Ω together with the well-known fact (cf. [CCNPW]) that the stabilizer in $^2E_6(2) : 2$ of an element of the leftmost type in the associated F_4-geometry is a maximal subgroup of index $3, 968, 055$ which is the centralizer of a 2-central involution, isomorphic to $2_+^{1+20} : U_6(2) : 2$. To prove (iv) notice that

$$C_{G_{10}}(Y_1) \cap F_{2+}(w) = C_{G_{10}}(Y_2) \cong 2^{10+16}.\Omega_8^-(2) : 2,$$

and that the action of $Sp_8(2) \cong C_{G_{10}}(Y_1)/Q_{10}$ on the cosets of $\Omega_8^-(2)$ is well-known to be doubly transitive of degree 120. Since $C_{G_{10}}(Y_1) \geq C_{Q_1}(Y_1)$, the orbit Ξ contains an edge. Therefore, the subgraph induced by Ξ is complete. The set $\Xi \setminus \{w\}$ is a 119-subset of $\Omega_1(w)$ formed by the elements of the left-most type in the F_4-geometry incident to a given element of the rightmost type. By the structure of the vertex-stabilizer in the action of F_{2+} on Ω we see that for every $u \in \Omega$ the subgroup $Z(F_{2+}(u)/Y_1)$ is of order 2 whose generator we denote by $\iota(u)$. By the structure of the edge stabilizer the involutions corresponding to adjacent vertices are distinct and always commute. By (iii), the involutions $\iota(u)$ taken for all $u \in \Omega_1(w)$ are mapped bijectively onto the set of 2-central involutions in

$$^2E_6(2) : 2 \cong F_{2+}(w)/Y_2.$$

The 2-central involutions in $^2E_6(2) : 2$ (or rather in $^2E_6(2)$) form a class of $\{3, 4\}$-transpositions and two of them commute if and only if the corresponding elements of the leftmost type in the associated F_4-geometry are incident to a common element of the rightmost type. By (iv) if u and v are vertices from $\Omega_1(w)$ such that $(\iota(u), \iota(v))$ map onto a pair of commuting involutions in $F_{2+}(w)/Y_2$, then u and v are adjacent in Ω. Thus (v) holds. $\square$

The assertion (7.5.4 (v)) can be restated by saying that Ω is locally the commutating graph of the 2-central involutions in $^2E_6(2)$. It was shown in [IPS01] that there exists a unique such (connected) graph. The characterization in the context of the Monster amalgam can be achieved along the following lines.

It is one of the basic properties of $^2E_6(2) : 2 \cong F_{2+}(w)/Y_2$ that $F_{2+}(w)$ acts transitively on the set of ordered pairs of 2-central involutions whose product has order α for $\alpha = 3$ and for $\alpha = 4$.

Lemma 7.5.5 *The group $F_{2+}(w)$ has precisely two orbits $\Omega_2^{(3)}(w)$ and $\Omega_2^{(4)}(w)$ on the set of vertices at distance 2 from w in Ω. Furthermore, for $\alpha = 3$ or 4 and $u \in \Omega_2^{(\alpha)}(w)$ the following hold:*

(i) *$F_{2+}(w) \cap F_{2+}(u)$ acts transitively on $\Omega_1(w) \cap \Omega_1(u)$;*
(ii) *$\langle \iota(w), \iota(u) \rangle \cong D_{2\alpha}$.*

Proof. By the paragraph before the lemma, we obtain (i) together with the isomorphism

$$\langle \iota(w), \iota(u), \iota(v) \rangle / \langle \iota(v) \rangle \cong D_{2\alpha}$$

for every common neighbour v of u and w in Ω. Since there are more than one common neighbour, (ii) follows. $\square$

Lemma 7.5.6 *Let $u \in \Omega_2^{(3)}(w)$. Then:*

(i) *the subgraph of Ω induces by $\Omega_1(u) \cap \Omega_1(w)$ is the transposition graph of the Fischer group Fi_{22};*

(ii) *$|\Omega_1(u) \cap \Omega_1(w)| = 3,510$ and $(F_{2+}(u) \cap F_{2+}(w))/Y_1 \cong Fi_{22}.2$;*

(iii) *every vertex from $\Omega_1(w)$ is at distance at most 2 from u in Ω.*

Proof. This is (5.10.10) in [Iv99]. The subgroup $F_{2+}(u) \cap F_{2+}(w)$ can be embedded into a conjugate of F_{3+} as follows. Let a and b be a pair of commuting transpositions in F_{3+}^g as in (7.5.2(vi)) so that $A_2 = \{a, b\}$, $Y_1 = \langle ab \rangle$ and

$$C_{F_{3+}^g}(\langle a, b \rangle) \cong 2^2 \cdot Fi_{22}.2.$$

Let c be the element b conjugated by an element of order 3 in $O_3(F_{3+}^g)$, so that the product ac is an order 6 element squared to a generator of $O_3(F_{3+}^g)$, cubed to the generator of Y_1, and

$$C_{F_{3+}^g}(\langle a, b \rangle)/Y_1 \cong Fi_{22}.2.$$

Now putting $\iota(w) = a$, $\iota(u) = c$ we achieve the required embedding. $\square$

Let $\Omega(Z_1)$ be the subgraph in Ω induced by the images of w under the subgroup $C_{G_1}(Y_1) \cong 2.2_+^{1+22}.Co_2$.

Lemma 7.5.7 *The following assertions hold:*

(i) *$\Omega(Z_1)$ contains 4,600 vertices and every orbit of $C_{Q_1}(Y_1)$ on $\Omega(Z_1)$ has length 2;*

(ii) *if $u \in \Omega(Z_1) \setminus \Omega_1(w)$, then $u \in \Omega_2^{(4)}(w)$ and $F_{2+}(u) \cap F_{2+}(w) \le C_{G_1}(Y_1)$;*

(iii) *if $u \in \Omega_2^{(4)}(w)$, then:*

    (a) *$|\Omega_1(u) \cap \Omega_1(w)| = 648$;*

    (b) *$(F_{2+}(u) \cap F_{2+}(w))/Y_1 \cong 2_+^{1+20}.U_4(3).2^2$;*

    (c) *except for a 8,064-orbit of $F_{2+}(u) \cap F_{2+}(w)$, every vertex from $\Omega_1(w)$ is at distance at most 2 from u in Ω.*

Proof. These are extractions from lemmas (5.10.11)–(5.10.14) in [Iv99]. $\square$

Lemma 7.5.8 *The following assertions hold:*

(i) *the diameter of Ω is 3;*

(ii) *there is a unique $F_{2+}(w)$-orbit $\Omega_3(w)$ on the set of vertices at distance 3 from w and $|\Omega_3(w)| = 23\,113\,728$;*

(iii) *if $u \in \Omega_3(w)$, then $[\iota(u), \iota(w)] = 1$ and $(F_{2+}(u) \cap F_{2+}(w))/Y_1 \cong F_4(2) \times 2^2$.*

Proof. These are lemmas (5.10.20) and (5.10.21) in [Iv99].   □

Summarizing (7.5.4 (ii)), (7.5.6 (ii)), (7.5.7 (iii) (a)), (7.5.8 (i), (ii)) and the parameters of the commutating graph of the 2-central involutions in $^2E_6(2)$, we conclude that the number vertices in Ω is 13 571 955 000 and hence

$$|F_{2+}/Y_1| = |F_{2+}(w)/Y_1| \cdot |\Omega|$$

is a in (7.5.1 (i)) which completes the proof of that lemma and also serves a definition of the Baby Monster group.

7.6 *p*-locality

In this section we show that under the completion $\varphi : G \to GL(\Pi)$ the members of the further enriched Monster amalgam become a p-local subgroup of the image.

Proposition 7.6.1 *Let H be one of the following subgroups of G*

$$G_1, G_2, G_3, G_5^{(t)}, G_{10}, {}^2G_2, F_{3+}, F_{2+},$$

and let Z be the centre of $O_p(H)$, where $p = 3$ for $H = F_{3+}$ and $p = 2$ for the remaining cases. Then

$$\varphi(H) = N_{\varphi(G)}(\varphi(Z)).$$

Proof. If $H = G_1$, then $Z = Z_1$ and the result follows directly from (6.3.1). If $H = G_2, G_3$ or $G_5^{(t)}$, then $Z = Z_2, Z_3$, or $Z_5^{(t)}$, respectively, in any case Z is an elementary abelian containing Z_1, and H induces on Z the full linear group $GL(Z)$. Therefore, the assertion follows from the equality $H \cap G_1 = N_{G_1}(Z)$. If $H = G_{10}$, then $Z = Z_{10}$; the quadratic form κ on Z_{10} is defined in terms of the classes $2A$- and $2B$-involutions in G. These classes are clearly not fused in $\varphi(G)$ and hence the image of κ is preserved by $N_{\varphi(G)}(\varphi(Z_{10}))$. The subgroups $Z_5^{(l)}$ and $Z_5^{(t)}$ are not conjugate in $\varphi(G)$, since otherwise they would be conjugate in $\varphi(G_1) \cong G_1$, which is obviously not the case. Hence $N_{\varphi(G)}(\varphi(Z_{10}))$ induces on $\varphi(Z_{10})$ the group $\Omega_{10}^+(2)$ (precisely as G_{10} does on Z_{10}) and the assertion follows from the equality $G_1 \cap G_{10} = N_{G_1}(Z_{10})$ (compare (4.3.7)).

The cases when $H = {}^2G_2, F_{3+}$ or F_{2+} are more delicate. By (6.3.1) we know that $\varphi(H)$ contains the centralizer in $N_{\varphi(G)}(\varphi(Z))$ of $\varphi(z)$, which is a 2-central involution in $\varphi(H)/\varphi(Z)$. From this information we can deduce that $\varphi(H)$ is a strongly embedded subgroup in $N_{\varphi(G)}(\varphi(Z))$ and reach a contradiction with Lemma 7.6 in [A94]. Alternatively, we can apply the following characterizations of $^2E_6(2) : 2$, Fi_{24} and BM by their involution centralizers: Theorem 9.1 in [A01], Theorem 35.1 in [A97], and Main Theorem in [Seg91] (after $\varphi(^2G_2)$ is identified with $N_{\varphi(G)}(\varphi(Y_2))$ the centralizer of a non-central involution in $C_{\varphi(G)}(\varphi(Y_1))/\varphi(Y_1)$ is identified, which indeed makes [Seg91] applicable). $\qquad\square$

As immediate direct consequences of the above proposition we have the following.

Lemma 7.6.2 *Let C be a cyclic subgroup in G containing $Z_1 = Z(G_1)$, $X = O_3(F_{3+})$ or $Y_1 = Z(F_{2+})$. Then*

$$N_{\varphi(G)}(\varphi(C)) = N_{\varphi(H)}(\varphi(C)) \cong N_H(C),$$

where $H = G_1$, F_{3+} or F_{2+}, respectively. $\qquad\square$

A specific case of the above lemma is following.

Lemma 7.6.3 *Let C be a cyclic subgroup of order 4 in Q_1. Then $N_{\varphi(G)}(\varphi(C)) \cong 2_+^{1+24}.Co_3$.*

Proof. The image of C under $\chi : Q_1 \to \bar{\Lambda}$ is a 1-subspace, which is non-isotropic with respect to $(,)_{\bar{\Lambda}}$. Therefore (identifying C with its generator) we conclude that $\chi(C) \in \bar{\Lambda}_3$ and the result follows from the basic properties of the Leech lattice. $\qquad\square$

According to [MSh02] every maximal 2-local subgroup of the Monster is a conjugate of G_1, G_2, G_3, $G_5^{(t)}$, G_{10}, $^2G_2$, or F_{2+}. F_{3+} is a maximal 3-local subgroup.

7.7 Thompson group

The classes of 3-elements in the Conway group Co_1 are known (cf. [CCNPW] and Section 51 in [A94]). Considering the action of these elements on the Leech lattice Λ, we can identify the corresponding centralizers in $\bar{\Lambda} = \Lambda/2\Lambda$ and hence in Q_1. These calculations are summarized in the following lemma:

Lemma 7.7.1 *The group $G_1 \cong 2_+^{1+24}.Co_1$ contains four classes of elements of order 3. These classes are closed under inversion and the information about the corresponding normalizers is given in the following table, where $\bar{G}_1 = G_1/Q_1 \cong Co_1$, $\bar{T} = TQ_1/Q_1$, $\chi_\Lambda(T)$ is the character on $\Lambda \otimes \mathbb{R}$ of an element $\hat{t} \in Co_0$ such that $\langle \hat{t} \rangle Z(Co_0)/Z(Co_0) = \bar{T}$, and $d_\Lambda(T) = \dim_{\Lambda \otimes \mathbb{R}} \langle \hat{t} \rangle$, so that $d_\Lambda(T) = \frac{2}{3}\chi_\Lambda(T) + 8$.*

Class	$N_{\bar{G}_1}(\bar{T})$	$\chi_\Lambda(T)$	$d_\Lambda(T)$	$N_{G_1}(T)$
$3A_1$	$3 \cdot Suz : 2$	-12	0	$6 \cdot Suz : 2$
$3B_1$	$3^2 \cdot U_4(3).2^2$	6	12	$2_+^{1+12}.3^2 \cdot U_4(3).2^2$
$3C_1$	$3_+^{1+4}.2.U_4(2) : 2$	-3	6	$2_+^{1+6}.3_+^{1+4} : 2.U_4(2) : 2$
$3D_1$	$A_9 \times S_3$	0	8	$S_3 \times 2_+^{1+8}.A_9$

$\square$

Definition 7.7.2 *The G-conjugates of a subgroup of type $3D_1$ in G_1 are called $3C$-subgroups.*

Recall some basic facts about the classes of elements of order three in $M \cong M_{24}$ (cf. Section 2.13 in [Iv99]). Let S be a sextet and T be a trio, in the Steiner system $(\mathcal{P}, \mathcal{B})$, so that $M(\mathcal{S}) \cong 2^6 : 3 \cdot S_6$ and $M(\mathcal{T}) \cong 2^6 : (S_3 \times L_3(2))$. Let s and t be order 3 elements contained in $O_{2,3}(M(\mathcal{S}))$ and $O_{2,3}(M(\mathcal{T}))$, respectively, and put $S = \langle s \rangle$, $T = \langle t \rangle$.

Lemma 7.7.3 *In the above notations the following assertions hold:*

(i) *M contains two classes of elements of order 3 with representatives s and t;*

(ii) *$N_M(S) \cong 3 \cdot S_6$ is a complement to $O_2(M(\mathcal{S}))$ in $M(\mathcal{S})$;*

(iii) *$N_M(T) \cong S_3 \times L_3(2)$ is a complement to $O_2(M(\mathcal{T}))$ in $M(\mathcal{T})$;*

(iv) *$C_{C_{11}^*}(T)$ is 3-dimensional and its non-zero elements correspond to the seven sextets refining $\mathcal{T}$;*

(v) *T acts on $\mathcal{P}$ fixed-point freely, while S fixes 6 elements;*

(vi) *$N_M(T)$ permutes transitively the eight orbits of T on $\mathcal{P}$ with stabilizer isomorphic to $S_3 \times F_7^3$.* $\square$

Lemma 7.7.4 *Let T be a subgroup of order 3 in G_{12} whose image in $G_{12}/O_2(G_{12}) \cong M_{24}$ is generated by the trio-type element t. Then T is $3D_1$-subgroup in G_1 and hence a $3C$-subgroup in G.*

Proof. Let M be the M_{24}-complement in the monomial subgroup $C_{12}^* : M_{24}$ of Co_0. Then M maps isomorphically onto $G_2^S/O_2(G_2^S)$, and acts on $\Lambda \otimes \mathbb{R}$ as on the permutational module of the natural action on $\mathcal{P}$. Therefore, if $\hat{t}$ is an element of order 3 in Co_0 which maps onto a generator of $TO_{\{2,3\}}(G_2)/O_{\{2,3\}}(G_2)$ and χ is the character on $\Lambda \otimes \mathbb{R}$, then $\chi(\hat{t}) = 0$ by (7.7.3 (v)) and the result follows by comparing this equality with the third column in the table in (7.7.1). $\square$

For the remainder of the section, T denotes a $3C$-subgroup in G as in (7.7.4) with generator t. And D denotes the D_6-subgroup generated by T together with the $2A$-involutions inverting T.

The following result can be easily deduced from (7.7.3 (iii)) by a Frattini argument.

Lemma 7.7.5 $N_{G_2}(T)R_2 = G_2(T)$. $\qquad\qquad\qquad\qquad\qquad\qquad$ $\square$

The subgroup $Z_5^{(t)}$ was defined in Section 4.1 as the preimage in R_2 of the (sextet-pure) subgroup of order 2^3 in $\bar{R}_2 \cong C_{11}^*$ whose non-zero elements correspond to the sextets refining $\mathcal{T}$. Therefore, $Z_5^{(t)}$ is centralized by T.

Lemma 7.7.6 $C_{G_5^{(t)}}(T)/T \cong 2^5 \cdot L_5(2)$ *(the Dempwolff group)*.

Proof. By the paragraph prior the lemma, by the structure of $G_5^{(t)}$, and by a Frattini argument, $C_{G_5^{(t)}}(T)/T$ is an extension of $Z_5^{(t)}$ by $L_5(2)$. The non-splitness of the extension can be deduced from (7.3.2). $\qquad\qquad$ $\square$

Lemma 7.7.7 *Let* $K = {}^2G_2 \cap G_1$. *The intersection of the class of $3C$-subgroups in G with K is a conjugacy class in K. The image of this class in $\bar{K} = K/O_2(K) \cong U_6(2) : S_3$ is generated by a $3G$-element $\bar{t}$ and*

$$N_{\bar{K}}(\bar{t}) \cong S_3 \times L_2(8) : 3.$$

Proof. The subgroup ${}^2G_2 \cap G_1$ is the preimage in G_1 of the stabilizer in $Co_1 \cong G_1/Q_1$ of a 222-triangle (isomorphic to $U_6(2) : S_3$) and the image of ${}^2G_2 \cap G_{12}$ in the M_{24}-factor group is the stabilizer of a 3-subset in $\mathcal{P}$ (isomorphic to $P\Sigma L_3(4) \cong L_3(4) : S_3$). Assuming that the 3-subset is a T-orbit, we deduce that the image of $N_{2G_2 \cap G_{12}}(T)$ in ${}^2G_2 \cap G_{12}/O_2({}^2G_2 \cap G_{12})$ is isomorphic to $F_7^3 \times S_3$. Now the assertion of the lemma can be easily deduced from the fusion patters of the 3-elements of $L_3(4) : S_3$ into $U_6(2) : S_3$, where the former is considered a Levi complement in the stabilizer of a maximal totally isotropic subspace in the natural module. In its turn the fusion pattern can be seen by restricting the rank 22 character of $U_6(2) : S_3$. $\qquad$ $\square$

Lemma 7.7.8 *For* $H = {}^2G_2 \cap G_1$, ${}^2G_2$, *or* F_{3+}, *the class of $3C$-subgroups in G intersected with H forms a conjugacy class in H. If T is a $3C$-subgroup in H and D is the D_6-subgroup generated by T and the triple of $2A$-involutions inverting T, then:*

(i) $C_H(T)/T \cong 2 \times L_2(8) : 3$ *for* $H = {}^2G_2 \cap G_1$;
(ii) $C_H(T)/T \cong {}^3D_4(2) : 3$ *for* $H = {}^2G_2$;
(iii) $C_H(T)/T \cong (3 \times G_2(3)).2$ *for* $H = F_{3+}$.

Proof. The assertion (i) follows from (7.7.7) and the fact that a $3G$-element in $U_6(2) : S_3$ acts fixed-point freely on the exterior cube of the natural module, which is

$$O_2(K)/O_2(^2G_2)Z_1 \cong 2^{20},$$

where $K = {}^2G_2 \cap G_1$. Since the intersection of $N_{2G_2}(T)$ with the maximal parabolic subgroup which maps onto the centralizer of a 2-central involution is known by (7.7.7) (or rather by (i)), the assertion (ii) can be deduced from the known description of the classes of order 3 elements in $^2E_6(2) : S_3$ [GL83]. Finally, (iii) can be deduced from the known data on the classes of order 3 elements in the Fischer group $Fi_{24} \cong F_{3+}/O_3(F_{3+})$ [A97]. In order to achieve the identification it is helpful to apply (7.7.5) and the structure of G_2 to deduce that the normalizer in F_{3+}/D of an elementary abelian subgroup of order 8 is the non-split extension $2^3 \cdot L_3(2)$. Alternatively, we can apply the amalgam characterization of $G_2(3)$ from [HS05]. □

The above achieved information on the normalizers of T in various members of the enriched Monster amalgam is summarized in the table below.

H	Shape of H	$C_H(T)/T$
G_1	$2_+^{1+24}.Co_1$	$2_+^{1+8}.A_9$
G_2	$2^{2+11+22}.(S_3 \times M_{24})$	$2^{2+3+6}.(L_3(2) \times S_3)$
$G_5^{(t)}$	$2^{5+10+20}.(L_5(2) \times S_3)$	$2^5 \cdot L_5(2)$
$^2G_2 \cap G_1$	$2^2.2_+^{1+20}.U_6(2).S_3$	$2 \times L_2(8) : 3$
$^2G_2$	$2^2 \cdot (^2E_6(2)) : S_3$	$^3D_4(2) : 3$
F_{3+}	$3 \cdot Fi_{24}$	$(3 \times G_2(3)).2$

Let T be a $3C$-subgroup in G which is normalized by the central involution of F_{2+}. Then the above collected information on the normalizers of T in the members of the enriched Monster amalgam can be translated in terms of generators and relations, which enables us to apply the main result of [HSW00] to identify F_3 with the sporadic simple group of Thompson.

Proposition 7.7.9 *Let T be a $3C$-subgroup in G inverted by the central involution of F_{2+}. Let F_3 denote the subgroup in F_{2+} generated by the intersections with F_{2+} of the subgroups $N_H(T) \cap F_{2+}$ taken for $H = G_1, G_2, G_5^{(t)}$,*

$^2G_2$, and F_{3+}. Then F_3 is the sporadic simple group of Thompson, whose order is

$$2^{15} \cdot 3^{10} \cdot 5^3 \cdot 7^2 \cdot 13 \cdot 19 \cdot 31.$$

$\square$

Since by (7.6.1 $\varphi(F_{2+})$ is the centralizer in φ of the central involution of $\varphi(F_{2+})$, we also have the following.

Lemma 7.7.10 *In terms of* (7.7.9), *we have*

$$\varphi(F_3) = C_{\varphi(G)}(\varphi(T)).$$

$\square$

7.8 Harada–Norton group

In this section we follow that the strategy similar to that implemented in the previous section to analyse the $5A$-subgroups in G, and to construct the Harada–Norton subgroup in F_{2+}. We start with discussing the classes of order 5 elements in G_1.

Lemma 7.8.1 *The group* $G_1 \cong 2_+^{1+24}.Co_1$ *contains three classes of elements of order 5. Every element is conjugate to all its non-trivial powers (so that all order 5 subgroups are fully normalized in G_1). The information about the corresponding normalizers is given in the following table, where $\bar{G}_1 = G_1/Q_1 \cong Co_1$, $\bar{S} = SQ_1/Q_1$, $\chi_\Lambda(S)$ is the character on $\Lambda \otimes \mathbb{R}$ of an element $\hat{s} \in Co_0$ such that $\langle \hat{s} \rangle Z(Co_0)/Z(Co_0) = \bar{S}$, and $d_\Lambda(S) = \dim_{\Lambda \otimes \mathbb{R}}(\hat{s})$, so that $d_\Lambda(S) = \frac{4}{5}(\chi_\Lambda(S) + 6)$.*

Class	$N_{\bar{G}_1}(\bar{S})$	$\chi_\Lambda(S)$	$d_\Lambda(S)$	$N_{G_1}(S)$
$5A_1$	$(D_{10} \times J_2).2$	-6	0	$(D_{10} \times 2 \cdot J_2).2$
$5B_1$	$(D_{10} \times (A_5 \times A_5).2).2$	4	8	$(D_{10} \times 2_+^{1+8}.(A_5 \times A_5).2).2$
$5C_1$	$5_+^{1+2} : GL_2(5)$	-1	4	$2_+^{1+4}.5_+^{1+2} : GL_2(5)$

Definition 7.8.2 *The G-conjugates of a subgroup of type $5B_1$ in G_1 are called $5A$-subgroups.*

In the following lemma we summarize the well-known and easily provable properties of the 5-subgroups in M_{24}.

Lemma 7.8.3 *The group $M \cong M_{24}$ contains a unique class of order 5 subgroups which are Sylow 5-subgroups. If S is such a subgroup with generator s, then:*

(i) *S is fully normalized in M and*

$$N_M(S) \cong (A_4 \times D_{10}).2 \cong (S_4 \times F_5^4)^+;$$

(ii) *the cyclic type of s on $\mathcal{P}$ is $1^4 5^4$;*

(iii) *the sextet containing the tetrad fixed by S is the unique one stabilized by S;*

(iv) *S stabilizes precisely four octads which are obtained by adjoining a triple of fixed elements to a 5-orbit;*

(v) *both $C_{\mathcal{C}_{11}}(S)$ and $C_{\mathcal{C}_{11}^*}(S)$ are 4-dimensional.* $\square$

Throughout the section, S denotes a Sylow 5-subgroup in G_2 which stabilizes the octad $\mathcal{O}$ such that $G_2(\mathcal{O}) = G_2 \cap G_{10}$ and the sextet $\mathcal{S}$ such that $G_3 \cap G(\mathcal{S}) = G_2(\mathcal{S})$, so that S is contained in $G_3 \cap G_{10}$.

Lemma 7.8.4 *The following assertions hold:*

(i) *$C_{G_3}(S)/S \cong 2^3.2^2.2^6.(3 \times L_3(2))$;*

(ii) *$C_{G_3}(S) \cap R_3$ is elementary abelian of order 2^5 containing $Z_3 \cong 2^3$ and $N_{G_3}(S) \cap Q_3$ induces on it the group generated by the transvections containing Z_3.*

Proof. Since S is a Sylow 5-subgroup in G_3, we assume that S is contained in the hexacode subgroup $Y \cong 3 \cdot S_6$. Every chief factor of Y inside Q_3 is isomorphic, up to duality, either to the 4-dimensional symplectic module for $Y/O_3(Y) \cong Sp_4(2)$ (on which S acts fixed-point freely), or to a hexacode module $\mathcal{H}_6$ (in which S centralizes a 2-subspace). This gives (i), while (ii) follows from the structure of G_3 (compare Section 2.8). $\square$

Lemma 7.8.5 *In the notation adopted in the paragraph prior the lemma the following assertions hold:*

(i) *$C_{Q_{10}}(S)$ is a non-singular 6-dimensional subspace of minus type;*

(ii) *S does not centralize non-identity elements in $Q_{10}/Z_{10} \cong 2^{16}$;*

(iii) *$C_{G_{10}}(S) \cong 2^6.U_4(2)$.*

Proof. The conclusion comes from considering the action of S on the chief factors of $G_2^s(\mathcal{O}) = G_2^s \cap G_{10}$ contained in $O_2(G_2)$ (compare Section 4.3). Every such chief factor is either (a) trivial 1-dimensional; (b) natural 4-dimensional linear; or (c) 6-dimensional orthogonal module for $L_4(2) \cong G_2^s(\mathcal{O})/O_2(G_2^s(\mathcal{O}))$. Such a chief factor contributes 1, 0, or 2 dimensions to the dimension of the centralizer of S in the cases (a), (b), and (c), respectively. Since four 1-dimensional chief factors and a 6-dimensional one are contained in Z_{10}, we obtain (i). On the other hand, every chief factor in Q_{10}/Z_{10} is 4-dimensional, which gives (ii). Finally, (iii) and the non-singularity assertion in (i) follow from the classical description of the classes of 5-elements in $\Omega_{10}^+(2)$ (compare [CCNPW]). $\qquad\square$

If $\{p, q\}$ is a pair in $\mathcal{P}$ fixed by S, then $4p + 4q + 2\Lambda$ is a vector in $\bar{\Lambda}_2$ fixed by S and the χ-preimage of this vector is a $2A$-involution centralized by S. Such an involution can be found in the $L_3(2)$-invariant complement to Z_3 in $C_{R_3}(S)$.

Lemma 7.8.6 *Let Y_1 be an order 2 subgroup in the $L_3(2)$-invariant complement to Z_3 in $C_{R_3}(S)$. For $H = G_1$, G_3, or G_{10}, put*

$$\Phi(H) = (C_H(S) \cap C_H(Y_1))/Y_1.$$

Then:

(i) $\Phi(G_1)/S \cong [2^6].S_5$;
(ii) $\Phi(G_3)/S \cong 4^3 : L_3(2)$;
(iii) $\Phi(G_{10})/S \cong 2^4 : S_5$.

Proof. Since Y_1 is easily seen to be contained in $O_2(C_{G_1}(S))$, (i) follows from the basic properties of extraspecial groups applied to the structure of $C_{G_1}(S)$. The assertion (ii) can be deduced from a description of the structure of G_3 given in Section 2.8. Finally (iii) is immediate from (7.8.5 (iii)), since $U_4(2) \cong \Omega_6^-(2)$ acts on the non-zero vectors of its natural module with two orbits with lengths 27 and 36 formed by the isotropic and non-isotropic vectors, respectively, and with stabilizers isomorphic to $2^4 : A_5$ and S_6, respectively. $\square$

Lemma 7.8.7 *The following assertions hold:*

(i) $C_{F_{3+}}(S)/S \cong 3 \times A_9$;
(ii) $C_{F_{2+}}(S)/S \cong 2 \cdot HS$, *where HS is the Higman–Sims sporadic simple group.*

Proof. The assertion (i) follows from the 5-local structure of the Fischer group F_{24} (cf. [A97]). To prove (ii) we have to show that the subgroup Φ in the Baby Monster group $F_{2+}/Y_1 \cong BM$ generated by the subgroups $\Phi(H)$ as in (7.8.6) for $H = G_1$, G_3, and G_{10} factorized over S is isomorphic to the Higman–Sims group. First of all, considering possible intersections of a normal subgroup in Φ with $\Phi(H)$s, it is easy to show that Φ must be a simple group. Then [JW69] and [GH73] can be applied to characterize HS just by the structure of $\Phi(G_1)/S$ and $\Phi(G_{10})/S$, respectively.　　　　　　　□

I believe that the following is true.

Conjecture 7.8.8 *The Higman–Sims group HS is the universal completion group of the amalgam*

$$\{\Phi(H)/S \mid H = G_1, G_3, G_{10}\}.$$

H	Shape of H	$C_H(S)/S$
G_1	$2_+^{1+24}.Co_1$	$2_+^{1+8}.(A_5 \times A_5).2$
G_3	$2^{3+6+12+18}.(3 \cdot S_6 \times L_3(2))$	$2^{3+2+6}.(3 \times L_3(2))$
G_{10}	$2^{10+16}.\Omega_{10}^+(2)$	$2^6.U_4(2)$
F_{3+}	$3 \cdot Fi_{24}$	$3 \times A_9$
F_{2+}	$2 \cdot BM$	$2 \cdot HS$

The obtained information on the structure of the centralizer of S in various members of the Monster amalgam is gathered in the above table. The first and the last rows in that table allows us to apply the main result of [Seg92] to obtain the following:

Proposition 7.8.9 *Let S be a $5A$-subgroup in G inverted by the central involution of F_{2+}. Let F_5 denote the subgroup in F_{2+} generated by the intersections with F_{2+} of the subgroups $C_H(S) \cap F_{2+}$ taken for $H = G_1, G_3, G_{10},$ and F_{3+}. Then F_5 is the sporadic simple group of Harada and Norton, whose order is*

$$2^{15} \cdot 3^{10} \cdot 5^3 \cdot 7^2 \cdot 13 \cdot 19 \cdot 31.$$

□

The obvious analogue of (7.7.10) holds in the situation discussed in this section.

8

Majorana involutions

8.1 196 883+1=196 884

It was the starting point of the Monstrous Moonshine when J. McKay noticed that the degree of the minimal faithful representation of the Monster (in characteristic zero) plus one is precisely the linear coefficient of the modular invariant $J(q)$. This can be taken as an explanation that calculations in the Griess algebra are easier to perform in the space $\Pi \oplus \Pi_1$ (where Π_1 is a trivial 1-dimensional G-module). This has already be seen in Section 6.4, although we did not adjoin the 1-dimensional subspace to Π in that section since it would spoil the uniqueness aspects and we are adjoining it now. Thus in what follows, $\star_z$ is the algebra on

$$S^2(\Lambda) \oplus \Pi^1_{|\bar{\Lambda}_2|}$$

such that the restriction to $S^2(\Lambda)$ is the Jordan algebra and the restriction to $S^2(\Lambda) \oplus C_{\Pi^1_{|\bar{\Lambda}_2|}}(Z_2)$ is G_2-invariant. Thus the algebra is as described in Section 6.4 with the wording 'projected to the subspace Π^1_{299} of $S^2(\Lambda)$' suppressed. Similarly, $(\, , \,)_z$ is the inner product on $S^2(\Lambda) \oplus \Pi^1_{|\bar{\Lambda}_2|}$, whose restriction to $S^2(\Lambda)$ is the tensor product form on this space and the restriction to the centralizer of Z_2 is G_2-invariant. The inner $(\, , \,)$ and algebra $\star$ products are $(\, , \,)_z$ and $\star_z$, respectively expanded to the whole of $\Pi \oplus \Pi_1$ via G-invariance. Hopefully, the fact that the old notation will be used for the new substance and that the $\star$ will still be called Griess algebra will not cause too much trouble. The following assertion is easy to check:

Lemma 8.1.1 *For the* 196 884-*dimensional Griess algebra* $\star$ *and inner* $(\, , \,)$ *products, the following assertions hold:*

(i) $(\, , \,)$ *is positive definite associative with respect to* $\star$ *in the sense that* $(u, v \star w) = (u \star v, w)$ *for all* $u, v, w \in \Pi \oplus \Pi_1$;

(ii) *the trilinear form* $(u, v, w) \mapsto (u, v \star w)$ *is symmetric.* $\qquad\square$

8.2 Transposition axial vectors

In this section we analyse idempotents in the Griess algebra and show that
the subgroup F_{2+} stabilizes in Π a unique 1-dimensional subspace formed by
the scalar multiples of an idempotent. We start with the following elementary
general statement:

Lemma 8.2.1 *Let* $(V, \circ)$ *be a commutative algebra, let* $x, y \in V$ *satisfy*
$x \circ x = x$ *(so that* x *is an idempotent),* $y \circ y = x$, *and* $x \circ y = y$. *Then*
$\frac{1}{2}(x - y)$ *is an idempotent.*

Proof. The result follows from the equality $(x - y) \circ (x - y) = x \circ x - 2 x \circ y + y \circ y$. $\qquad\square$

Lemma 8.2.2 *For a subset* $\mathcal{S}$ *of* $\mathcal{P}$, *the element* $\frac{1}{2^2} \sum_{p \in \mathcal{S}} p^2$ *is an idempotent*
with respect to the Jordan multiplication on $S^2(\Lambda)$.

Proof. The subspace in $S^2(\Lambda)$ spanned by the elements $p^2 = p \otimes p$ taken
for all $p \in \mathcal{P}$ is the permutation M_{24}-module and the Jordan product on it is
the Hadamard product multiplied by four. $\qquad\square$

Lemma 8.2.3 *If* $\lambda \in \Lambda \otimes \mathbb{R}$, *then*

$$(\lambda \otimes \lambda) \star (\lambda \otimes \lambda) = 2^2 \, (\lambda, \lambda)_\Lambda \, (\lambda \otimes \lambda).$$

Proof. The result is immediate from (6.4.1 (i)). $\qquad\square$

Lemma 8.2.4 *If* $\{p, q\}$ *is a 2-subset in* $\mathcal{P}$, *then the element*

$$a = \frac{1}{2^4}(p^2 + q^2 - (pq)_z - (pq)_x - (pq)_y)$$

is an idempotent of the Griess algebra.

Proof. By (6.4.4) we have

$$(pq)_x = \pi(\vartheta(pq)) + \pi(\vartheta(pq)x), \quad (pq)_y = \pi(\vartheta(pq)) + \pi(\vartheta(pq)y),$$

where $\pi(\vartheta(pq)x) = -\pi(\vartheta(pq)y)$ and $\chi(\pi(\vartheta(pq)) = 4p - 4q + 2\Lambda$,
therefore

$$a = \frac{\lambda \otimes \lambda}{2^8} - \frac{\pi(\vartheta(pq))}{2^3},$$

where $\lambda = 4p - 4q$. Thus the assertion follows from (8.2.3) and (8.2.1) applied
with

$$x = \frac{\lambda \otimes \lambda}{2^7} \quad \text{and} \quad y = \frac{\pi(\vartheta(pq))}{2^2}$$

(compare (6.4.7 (ii)) remembering that the projection to the 299-subspace is suppressed). □

Lemma 8.2.5 *If $Y_1 = \langle \vartheta(pq) \rangle$, then the vector a in (8.2.4) is F_{2+}-invariant.*

Proof. Since $\vartheta(pq)$ is a $2A$-involution contained in Q_1, it generates a G_1-conjugate of Y_1. From the shape

$$a = \frac{\lambda \otimes \lambda}{2^8} - \frac{\pi(\vartheta(pq))}{2^3},$$

given in the proof of (8.2.4), it is clear that a is stable under $C_{G_1}(Y_1)$. On the other hand, the shape of a given in the hypothesis of (8.2.4)

$$a = \frac{1}{2^4}(p^2 + q^2 - (pq)_z - (pq)_x - (pq)_y)$$

shows that it is centralized by an order 3 element in G_2 which commutes with M^δ modulo R_2 and permutes $\{z, x, y\}$. Hence a is also stable under $C_{G_2}(Y_1)$ and the result follows, since $F_{2+} = \langle C_{G_1}(Y_1), C_{G_2}(Y_1) \rangle$. □

Lemma 8.2.6 *Every vector in Π centralized by F_{2+} is the projection of a scalar multiple of a.*

Proof. Let $b \in \Pi$ be a non-zero vector centralized by F_{2+} and assume that $Z(F_{2+}) = \langle \vartheta(pq) \rangle$. Since F_{2+} contains Z_1 and the latter acts fixed-point freely on $[\Pi, Z_1]$, the vector b must be in $C_\Pi(Z_1)$. Since $\chi(Q_1 \cap F_{2+})$ is the orthogonal complement of $\chi(Y_1)$, the only weight subspace of Q_1 in $\Pi^1_{|\bar{\Lambda}_2|}$ centralized by $Q_1 \cap F_{2+}$ is the one containing $\pi(\vartheta(pq))$. Therefore, the projections of a and b onto $\Pi^1_{|\bar{\Lambda}_2|}$ span the same 1-subspace. The action induced by $C_{G_1}(Y_1)$ on Π^1_{299} is isomorphic to $Co2$ and it is easily seen to centralize a single 1-subspace, namely the one spanned by the projection of $(4p - 4q) \otimes (4p - 4q)$. This shows that the projections of a and b onto Π^1_{299} also span the same 1-subspace and also that $C_\Pi(C_{G_1}(Y_1))$ is 2-dimensional. The projection of $\mu \otimes \mu$ onto Π^1_{299} is not stable under $C_{G_2}(Y_1)$, which gives the result. □

8.3 Spectrum

The G-conjugates of the generator $\vartheta(pq)$ of the centre of F_{2+} are the $2A$-involutions also known as *transpositions* and the vector a in (8.2.4) is called the *axial vector* of $\vartheta(pq)$. The axial vector $a(\vartheta)$ of an arbitrary $2A$-involution ϑ in G is the image of a under an element in G which conjugates $\vartheta(pq)$

onto ϑ. We are aiming to show that $a(\vartheta)$ uniquely determines the action of ϑ on the Griess algebra by calculating the spectrum of $\mathrm{ad}(a(\vartheta))$. To achieve this we identify the subalgebras in the Griess algebra generated by pairs of axial vectors corresponding to $2A$-involutions contained in Q_1.

Let ϑ be a $2A$-involution in Q_1. Then by the proof of (8.2.5) and because of the transitivity of G_1 on the set of such involutions we have

$$a(\vartheta) = \frac{\lambda \otimes \lambda}{2^8} - \frac{\pi(\vartheta)}{2^3},$$

where $\lambda \in \Lambda_2$ and $\chi(\vartheta) = \lambda + 2\Lambda$.

Lemma 8.3.1 *Let ϑ and θ be $2A$-involutions in Q_1, let λ and μ be vectors from Λ_2 such that $\chi(\vartheta) = \lambda + 2\Lambda$, $\chi(\theta) = \mu + 2\Lambda$, and assume without loss that $(\lambda, \mu)_\Lambda \le 0$. Then θ is contained in one of five $C_{G_1}(\vartheta)$-orbits specified by the following:*

(i) $\theta = \vartheta$;
(ii) $\theta = \vartheta z$;
(iii) $\vartheta\theta$ *is a $2B$-involution other than z and* $(\lambda, \mu)_\Lambda = 0$;
(iv) $\vartheta\theta$ *is a $2A$-involution,* $(\lambda, \mu) = -16$ *and* $\lambda + \mu \in \Lambda_2$;
(v) $\vartheta\theta$ *is a $4A$-element,* $(\lambda, \mu) = -8$ *and* $\lambda + \mu \in \Lambda_3$.

Proof. Unless $\pi(\vartheta)$ and $\pi(\theta)$ are proportional, there is always an element in Q_1 which centralizes $\pi(\vartheta)$ and negates $\pi(\theta)$. Therefore, the result follows from the well-known information on the action of Co_1 on $\bar{\Lambda}_2$ (cf. Section 4.11 in [Iv99]). $\qquad\square$

We adopt (8.3.1 (v)) as the definition of the class of $4A$-elements in G.

In order to maintain notation invariant under rescaling of the inner product for $u, v \in \Pi \oplus \Pi_1$, put

$$(v|u) = \frac{(u, v)}{(u, u)},$$

which is the cosine of the angle formed by u and v in the most important case when their lengths are equal.

Lemma 8.3.2 *Let ϑ and θ be $2A$-involutions in Q_1. Then*

$$(a(\vartheta)|a(\theta)) = 1, \ \frac{1}{2^3}, \ 0 \ or \ \frac{1}{2^5}$$

depending of whether $\vartheta\theta$ is of type $1A$, $2A$, $2B$, or $4A$.

Proof. We have

$$a(\vartheta) = \frac{\lambda \otimes \lambda}{2^8} - \frac{\pi(\vartheta)}{2^3}, \quad a(\theta) = \frac{\mu \otimes \mu}{2^8} - \frac{\pi(\theta)}{2^3}.$$

If $\pi(\vartheta)$ and $\pi(\theta)$ are not proportional, they are perpendicular and

$$(a(\vartheta), a(\theta)) = \frac{1}{2^{16}}(\lambda \otimes \lambda, \mu \otimes \mu)_{S^2(\Lambda)} = \frac{1}{2^{16}}(\lambda, \mu)^2.$$

If $\pi(\vartheta) = \pm\pi(\theta)$, then $\lambda = \mu$ and calculations are even easier. □

Definition 8.3.3 *Let $u, v \in \Pi \oplus \Pi_1$. Then:*

(i) $\langle u, v \rangle_\star$ *denotes the subalgebra in the Griess algebra generated by u and v;*

(ii) *if $u \star v = \alpha v$, then v is said to be an α-eigenvector of u.*

Lemma 8.3.4 *Let ϑ and θ be $2A$-involutions in Q_1 product to a $2B$-involution. Then:*

(i) $\dim(\langle a(\vartheta), a(\theta) \rangle_\star) = 2$;

(ii) $a(\vartheta) \star a(\theta) = 0.$

Proof. In terms of (8.3.1), the element θ corresponds to (ii) or (iii). In the latter case the product is zero, since λ and μ are orthogonal and so are $\lambda \otimes \lambda$ and $\mu \otimes \mu$. □

Lemma 8.3.5 *Let ϑ and θ be $2A$-involutions in Q_1 product to a $2A$-involution. Then:*

(i) $\dim(\langle a(\vartheta), a(\theta) \rangle_\star) = 3$;

(ii) $a(\vartheta) \star a(\theta) = \frac{1}{2^3}(a(\vartheta) + a(\theta) - a(\vartheta\theta))$;

(iii) $a(\vartheta)$, $a(\theta) - a(\vartheta\theta)$, and $\frac{1}{2^2}a(\vartheta) - a(\theta) - a(\vartheta\theta)$ are 1-$\frac{1}{2^2}$-, and 0-eigenvectors of $a(\vartheta)$, respectively.

Proof. We have

$$\left(\frac{\lambda \otimes \lambda}{2^8} - \frac{\pi(\vartheta)}{2^3}\right) \star \left(\frac{\mu \otimes \mu}{2^8} - \frac{\pi(\theta)}{2^3}\right) =$$

$$\frac{2(\lambda, \mu)}{2^{16}}(\lambda \otimes \mu + \mu \otimes \lambda) - \frac{(\lambda, \mu)^2}{2^{14}}(\pi(\vartheta) + \pi(\theta)) + \frac{\pi(\vartheta\theta)}{2^6}.$$

Since

$$(\lambda + \mu) \otimes (\lambda + \mu) = (\lambda \otimes \lambda) + (\mu \otimes \mu) + (\lambda \otimes \mu + \mu \otimes \lambda),$$

and $(\lambda, \mu) = -16$, (ii) follows. The assertion (iii) follows from (ii) in view of the complete symmetry between ϑ, θ, and $\vartheta\theta$. □

The subgroup $\langle v \rangle$ in (8.3.5) is a cyclic subgroup of order 4 contained in Q_1 and by (7.6.3)

$$N_{\varphi(G)}(\varphi(\langle v \rangle)) \cong 2_+^{1+24}.Co_3.$$

Lemma 8.3.6 *Let υ be an element of type $4A$ in Q_1, let $v \in \Lambda_3$ be such that $\chi(\upsilon) = v + 2\Lambda$, and let*

$$a(\upsilon) = \frac{v \otimes v}{2^6 3}.$$

Then:

(i) *$a(\upsilon)$ is an idempotent in the Griess algebra;*
(ii) *a vector in $\Pi \oplus \Pi_1$ centralized by $N_G(\langle \upsilon \rangle)$ and the vector $a(\upsilon)$ have proportional Π-projections.*

Proof. The assertion (i) is immediate from (8.2.3) since $(v, v)_\Lambda = 48$. Since $N_{G_1}(\langle \upsilon \rangle)$ contains Q_1, it does not centralize vectors outside the centralizer of Q_1 which is $\Pi_{299}^1 \oplus \Pi_1$. The group $N_{G_1}(\langle \upsilon \rangle)/Q_1 \cong Co_3$ acts on $\Lambda \otimes \mathbb{R}$ with two irreducible constituents of dimensions 1 and 23, from which it is immediate that its centralizer in Π_{299}^1 is just 1-dimensional. $\qquad\square$

Lemma 8.3.7 *Let ϑ and θ be $2A$-involutions in Q_1 product to a $4A$-element. Then:*

(i) *$\dim(\langle a(\vartheta), a(\theta) \rangle_\star) = 5$;*
(ii) *$a(\vartheta) \star a(\theta z^\alpha) = \frac{1}{2^6}(3a(\vartheta) + a(\vartheta z) + 3a(\theta z^\alpha) + a(\theta z^{\alpha+1})) - 3a(\vartheta\theta))$;*
(iii) *$a(\vartheta) \star a(\vartheta\theta) = \frac{1}{2^4}(5a(\vartheta) - a(\vartheta z) - 2a(\theta) - 2a(\theta z) + 3a(\vartheta\theta))$;*
(iv) *the vectors $a(\vartheta)$, $a(\theta) - a(\theta z)$, $a(\vartheta z)$,*

$$\sigma = -\frac{1}{2^2}a(\vartheta) + a(\theta) + a(\theta z) + \frac{1}{2}a(\vartheta\theta) \text{ and}$$

$$\delta = -\frac{3}{2^5}a(\vartheta) - \frac{1}{2^4}a(\vartheta z) - \frac{1}{2^3}\sigma + \frac{1}{2^2}a(\vartheta\theta)$$

are 1-, $\frac{1}{2^5}$-,0-, 0-, and $\frac{1}{2^2}$-eigenvectors of $a(\vartheta)$, respectively.

Proof. The assertions (ii) and (iii) can be checked directly using the equalities

$$\lambda \otimes \lambda = 2^7(a(\vartheta) + a(\vartheta z)) \text{ and } (\lambda + \mu) \otimes (\lambda + \mu) = 2^6 3\, a(\vartheta\theta)$$

(compare the proof of (8.3.5).) It is immediate from (ii) that the difference between $a(\theta)$ and $a(\theta z)$ is a $\frac{1}{2^5}$-eigenvector of $a(\vartheta)$. Since ϑ and ϑz product to z, which is a $2B$-element, $a(\vartheta)$ and $a(\vartheta z)$ annihilate each other by (8.3.4). Combining (ii) and (iii) we obtain

$$a(\vartheta) \star (a(\theta) + a(\theta z) + \frac{1}{2}a(\vartheta\theta)) = \frac{1}{2^2}a(\vartheta) = a(\vartheta) \star \frac{1}{2^2}a(\vartheta),$$

which shows that σ is annihilated by $a(\vartheta)$. Next we express the right-hand side of (iii) as a linear combination of $a(\vartheta\theta)$ and the eigenvectors we have found so far. This gives

$$a(\vartheta) \star a(\vartheta\theta) = \frac{9}{2^5}a(\vartheta) - \frac{1}{2^4}a(\vartheta z) - \frac{1}{2^3}\sigma + \frac{1}{2^2}a(\vartheta\theta).$$

Since $a(\vartheta)$ is an idempotent which annihilates both $a(\vartheta z)$ and σ, we conclude that δ is a $\frac{1}{2^2}$-eigenvector of $a(\vartheta)$. $\square$

Proposition 8.3.8 *The spectrum on* $\Pi \oplus \Pi_1$ *of an axial vector of a transposition is* $\left\{1, 0, \frac{1}{2^2}, \frac{1}{2^5}\right\}$.

Proof. The vectors $\lambda \otimes \lambda = 2^7(a(\theta) + a(\theta z))$ taken for all $\lambda \in \Lambda_2$ span $S^2(\Lambda)$, while the vectors $\pi(\theta)$ taken over a transversal of the set

$$\{\{\theta, \theta z\} \mid \theta \in Q_1, \ \chi(\theta) \in \bar{\Lambda}_2\}$$

constitutes a basis of $\Pi^1_{|\bar{\Lambda}_2|}$. This shows that $S^2(\Lambda) \oplus \Pi^1_{|\bar{\Lambda}_2|}$ is spanned by the vectors $\pi(\theta)$ taken for all 2A-involutions θ in Q_1. Hence by (8.3.4), (8.3.5), and (8.3.7), the spectrum of $a(\vartheta)$ on $S^2(\Lambda) \oplus \Pi^1_{|\bar{\Lambda}_2|} = C_{\Pi\oplus\Pi_1}(z)$ is as stated. On the other hand, we know that $a(\vartheta)$ is invariant under an element of order 3 which permutes the elements of $Z_2^\#$. Hence the spectrum on the whole of

$$\Pi \oplus \Pi_1 = \sum_{a\in Z_2^\#} C_{\Pi\oplus\Pi_1}(a)$$

does not become larger. $\square$

8.4 Multiplicities

In this section we calculate the multiplicities of the eigenvalues of the axial vector of a transposition. In the next section we impose some restrictions on where the product of two eigenvectors might end up. These restrictions go under the name of *fusion rules*.

Lemma 8.4.1 *Let* ϑ *be a 2A-element in G, let* $a(\vartheta)$ *be the corresponding axial vector, and let* $\Pi_\alpha^{(\vartheta)}$ *be the subspace in* $\Pi \oplus \Pi_1$ *be the subspace formed by the* α-*eigenvectors of* $a(\vartheta)$. *Then:*

(i) $\Pi \oplus \Pi_1 = \Pi_1^{(\vartheta)} \oplus \Pi_{\frac{1}{2^2}}^{(\vartheta)} \oplus \Pi_0^{(\vartheta)} \oplus \Pi_{\frac{1}{2^5}}^{(\vartheta)}$;

(ii) $\Pi_1^{(\vartheta)}$ *is 1-dimensional spanned by* $a(\vartheta)$;

(iii) $\Pi_1^{(\vartheta)} \oplus \Pi_{\frac{1}{2^2}}^{(\vartheta)} \oplus \Pi_0^{(\vartheta)} = C_{\Pi \oplus \Pi_1}(\vartheta)$, $\Pi_{\frac{1}{2^5}}^{(\vartheta)} = [\Pi \oplus \Pi_1, \vartheta]$;

(iv) *the action of ϑ on $\Pi \oplus \Pi_1$ is uniquely determined by $a(\vartheta)$ and therefore*
$$C_G(a(\vartheta)) = C_G(\varphi(\vartheta)).$$

Proof. The assertion (i) is a reformulation of (8.3.8). Considering the centralizer-commutator decomposition with respect to ϑ of the subalgabras $\langle a(\vartheta), a(\theta) \rangle_*$ in (8.3.4), (8.3.5), and (8.3.7) we deduce (ii) and (iii). Now (iv) follows from (iii). □

Lemma 8.4.2

$$\dim \left(\Pi_{\frac{1}{2^5}}^{(\vartheta)} \right) = 2^6 \cdot 176 + 3 \cdot 24 \cdot 2^{10} = 96\,256.$$

Proof. By (8.3.4), (8.3.5), and (8.3.7) for $\vartheta, \theta \in Q_1$, the subspace $\langle a(\vartheta), a(\theta) \rangle_*$ contains a $\frac{1}{2^2}$-eigenvector of $a(\vartheta)$ if and only if $\vartheta\theta$ has order 4, in which case the eigenvector is

$$a(\theta) - a(\theta z) = -\frac{\pi(\theta)}{2^2}.$$

Since $\pi(\theta)$ spans the weight space $\Pi(\bar\mu)$ (where $\bar\mu = \chi(\vartheta)$), we conclude that the intersection

$$C_{\Pi \oplus \Pi_1}(z) \cap \Pi_{\frac{1}{2^5}}^{(\vartheta)}$$

is the sum of the weight spaces $\Pi(\bar\mu)$ taken over the images in $\bar\Lambda$ of the set

$$\Theta = \{\mu \mid \mu \in \Lambda_2, (\lambda, \mu) = -16\},$$

where $\lambda \in \Lambda_2$ and $\lambda + 2\Lambda = \chi(\vartheta)$. Put

$$\Theta^\varepsilon = \{\mu \mid \mu \in \Theta, \frac{1}{8}(\lambda_4, \mu) = \varepsilon \bmod 2\},$$

where $\lambda = \sum_{p \in \mathcal{P}} 2p$ is the Leech vector such that G_{12} is the stabilizer in G_1 of $\lambda_4 + 2\Lambda$. Elementary calculations in the Leech lattice under the assumption that $\lambda = 4p - 4q$ show that

$$|\Theta^0| = 2^6 \cdot 176 \text{ and } |\Theta^1| = 24 \cdot 2^{10}.$$

If $\mu \in \Theta^\varepsilon$ and $\bar\mu = \mu + 2\Lambda$, then the weight space $\Pi(\bar\mu)$ is contained in $\Pi_{2^6 \cdot 759}^{12}$ or $\Pi_{24 \cdot 2^{11}}^{12}$ depending on whether $\varepsilon = 0$ or 1. Since $\Pi_{2^6 \cdot 759}^{12}$ is stable under G_2 while $\Pi_{24 \cdot 2^{11}}^{12}$ is not, we obtain

$$\dim \left(\Pi_{\frac{1}{2^5}}^{(\vartheta)} \right) = |\Theta^0| + 3 \cdot |\Theta^1| = 2^6 \cdot 176 + 3 \cdot 24 \cdot 2^{10}$$

as claimed. □

Towards calculating the dimension of $\Pi^{(\vartheta)}_{\frac{1}{2^2}}$, we prove the following:

Lemma 8.4.3 *The $\frac{1}{2^2}$-eigenvectors of $a(\vartheta)$ contained in the subalgebras $\langle a(\vartheta), a(\theta) \rangle_\star$ taken for all 2A-elements in Q_1 having order 4 product with ϑ span a 23-dimensional subspace of $S^2(\Lambda)$ spanned by the vectors $(\lambda \otimes v + v \otimes \lambda)$, where $\lambda \in \Lambda_2$ with $\chi(\vartheta) = \lambda + 2\Lambda$ and $v \in \Lambda$ with $(\lambda, v) = 0$.*

Proof. Performing elementary calculations with the expression for δ given in (8.3.7), we obtain

$$2^{11}\delta = \left(\lambda \otimes \left(2\mu + \frac{1}{2}\lambda \right) + \left(2\mu + \frac{1}{2}\lambda \right) \otimes \lambda \right).$$

Since $(\lambda, \lambda) = 32$ and $(\lambda, \mu) = -8$, the result follows. $\qquad\square$

Lemma 8.4.4 $\dim\left(\Pi^{(\vartheta)}_{\frac{1}{2^2}} \right) = 4371.$

Proof. By (8.3.4), (8.3.5), and (8.3.7) for 2A-involutions ϑ and θ in Q_1 a $\frac{1}{2^2}$-eigenvector of $a(\vartheta)$ is contained in $\langle a(\vartheta), a(\theta) \rangle_\star$ if $\vartheta\theta$ is of type 2A or 4A and in each case the multiplicity of the $\frac{1}{2^2}$-eigenvalue is 1. By (8.4.3) the 4A-case contributes 23 to the dimension in question. By (8.3.5) we obtain one eigenvector of $a(\vartheta)$ for each 222-triangle in $\bar{\Lambda}$ containing $\bar{\lambda} = \chi(\vartheta)$. It follows from the parameters of the shortest Leech graph in Section 4.11 in [Iv99] that the number of such triangles is 2300. If $\chi(\vartheta) = \lambda + 2\Lambda$, where $\lambda = 4p - 4q$, then the 2300 pairs $\{\bar{\mu}, \bar{v}\}$ forming 222-triangles together with $\bar{\lambda}$ are divided into three subsets of size 44, 1232, and 1024, depending of whether the union $\Pi(\bar{\mu}) \cup \Pi(\bar{v})$ of the corresponding weight spaces is contained in

$$\Pi^{12}_{2 \cdot 276}, \quad \Pi^{12}_{2^6 \cdot 759}, \quad \text{or} \quad \Pi^{12}_{24 \cdot 2^{11}},$$

respectively. Taking into account the 4A-contribution, we obtain:

$$\dim\left(\Pi^{(\vartheta)}_{\frac{1}{2^2}} \right) = 23 + 44 + 1232 + 3 \cdot 1024 = 4371$$

as stated. $\qquad\square$

As a direct consequence of (8.4.1 (i)), (8.4.2), and (8.4.4) we recalculate the character value of ϑ on Π (compare Section 4.1).

Corollary 8.4.5 $\dim\left(\Pi^{(\vartheta)}_0 \right) = \dim\left(\Pi^{(\vartheta)}_{\frac{1}{2^5}} \right) = 96\,256$ *in particular the character of ϑ on Π is equal to* $\dim\left(\Pi^{(\vartheta)}_{\frac{1}{2^2}} \right) = 4371.$

It follows from (8.4.1 (iv)) that $a(\vartheta)$ is stabilized by $C_G(\varphi(\vartheta))$, in particular it is stabilized by F_{2+}. Since F_{2+} preserves the Griess algebra, it stabilizes $\Pi_\alpha^{(\vartheta)}$ for every $\alpha \in \left\{1, 0, \frac{1}{2^2}, \frac{1}{2^5}\right\}$.

Lemma 8.4.6 *The action of F_{2+} on $\Pi_{\frac{1}{2^2}}^{(\vartheta)}$ is irreducible.*

Proof. It can easily be deduced from the proof of (8.4.4) that the degrees of the irreducible constituents of $(G_1 \cap F_{2+})/Y_1 \cong 2_+^{1+22}.Co_2$ on $\Pi_{\frac{1}{2^2}}^{(\vartheta)}$ are

$$23, \quad 2300 \text{ and } 2048$$

and those of $(G_2 \cap F_{2+})/Y_1 \cong 2^{2+10+20}.(\text{Aut}(M_{22}) \times S_3)$ are

$$1, \quad 3 \cdot 22, \quad 1232, \text{ and } 3 \cdot 1024$$

so that the irreducibility in question already follows from this numerology. In fact, arguing as in Chapter 3, it can be shown that 4371 is the degree of the minimal non-trivial representation of the Baby Monster amalgam. □

Notice that F_{2+} acts irreducibly on $\Pi_{\frac{1}{2^5}}^{(\vartheta)}$, while $\Pi_0^{(\vartheta)}$ is the sum of Π_1 and is an irreducible module.

8.5 Fusion rules

In this section we impose some restrictions on the location of the product of axial vectors of two transpositions. These restrictions go under the name of *fusion rules*. By (8.3.8)

$$S = \left\{1, 0, \frac{1}{2^2}, \frac{1}{2^5}\right\}$$

is the spectrum of $a(\vartheta)$ on $\Pi \oplus \Pi_1$. Let

$$f : S \times S \to 2^S$$

be defined by the following table, where $f(\alpha, \beta)$ is given in the intersection of the αth row and βth column.

	1	0	$\frac{1}{2^2}$	$\frac{1}{2^5}$
1	1	0	$\frac{1}{2^2}$	$\frac{1}{2^5}$
0	0	0	$\frac{1}{2^2}$	$\frac{1}{2^5}$
$\frac{1}{2^2}$	$\frac{1}{2^2}$	$\frac{1}{2^2}$	$1, 0$	$\frac{1}{2^5}$
$\frac{1}{2^5}$	$\frac{1}{2^5}$	$\frac{1}{2^5}$	$\frac{1}{2^5}$	$1, 0, \frac{1}{2^2}$

Lemma 8.5.1 *For ϑ being a 2A-involution in G and $a(\vartheta)$ being the axial vector of ϑ, let $\alpha, \beta \in S = \left\{ 1, 0, \frac{1}{2^2}, \frac{1}{2^5} \right\}$, and let $u_\alpha \in \Pi_\alpha^{(\vartheta)}$, $u_\beta \in \Pi_\beta^{(\vartheta)}$. Then*

$$u_\alpha \star u_\beta \in \bigoplus_{\gamma \in f(\alpha, \beta)} \Pi_\gamma^{(\vartheta)}.$$

Proof. Notice that the function f is symmetric and that the assertion is obvious if at least one of α and β is 1. By (8.4.1 (iii)), ϑ inverts u_α if and only if $\alpha = \frac{1}{2^5}$ and similarly for β. Therefore, ϑ inverts an even number of vectors in the multiset $\{u_\alpha, u_\beta, u_\alpha \star u_\beta\}$, which proves the assertion in all the cases when $\frac{1}{2^5}$ is involved in $\alpha \cup \beta \cup f(\alpha, \beta)$.

Assuming that $\alpha = \beta = \frac{1}{2^2}$, we have to show that the product $u_\alpha \star u_\beta$ has zero projection to $\Pi_{\frac{1}{2^2}}^{(\vartheta)}$. Assuming as above that $\lambda \in \Lambda_2$ satisfies $\chi(\vartheta) = \lambda + 2\Lambda$, denote by L the 1-dimensional subspace in $\Lambda \otimes \mathbb{R}$ spanned by λ and let $L^\perp$ be the orthogonal complement to L in $\Lambda \otimes \mathbb{R}$ with respect to $(\ ,\ )_\Lambda$. By (8.4.3) and the proof of (8.4.4), the set

$$\Xi = \{\alpha(\lambda \otimes \nu + \nu \otimes \lambda) \mid \alpha \in \mathbb{R}, \nu \in L^\perp\}$$

is precisely the intersection of $\Pi_{\frac{1}{2^2}}^{(\vartheta)}$ with $S^2(\Lambda)$. By (6.4.1)

$$\Xi \star \Xi = (L \otimes L) \oplus (L^\perp \otimes L^\perp).$$

Since Ξ, $L \otimes L$ and $L^\perp \otimes L^\perp$ are pairwise orthogonal with respect to the tensor product of $(\ ,\ )_\Lambda$, the projection of $\Xi \star \Xi$ to Ξ is indeed zero. By (8.4.6) the subspace $\Pi_{\frac{1}{2^2}}^{(\vartheta)}$ is F_{2+}-irreducible, therefore by the above $f\left(\frac{1}{2^2}, \frac{1}{2^2}\right) = \{0, 1\}$.

The remaining fusion coefficients $f(0, 0)$ and $f\left(0, \frac{1}{2^2}\right)$ follow from the ones already calculated in view of the associativity of $(\ ,\ )$ with respect to $\star$ (cf. (8.1.1)). $\qquad \Box$

8.6 Main definition

In this section, we axiomatize the established properties of the transpositions and their axial vectors.

Definition 8.6.1 *Let V be a real vector space, let $\circ$ be a commutative algebra product on V, let $\langle\ ,\ \rangle$ be a positive definite symmetric inner product on V associative with $\circ$, and suppose that the Norton inequality*

$$\langle u \circ u, v \circ v \rangle \geq \langle u \circ v, u \circ v \rangle$$

holds for all $u, v \in V$.

*Let μ be an automorphism of $(V, \circ, \langle\ ,\ \rangle)$, and let $a = a(\mu)$ be a vector
in V. Then μ is said to be a* Majorana involution *and a is said to be an* axial
vector *of μ if:*

(i) *$a \circ a = a$, so that a is an idempotent;*

(ii) *V is the sum of s-eigenspaces of*

$$\mathrm{ad}_a : v \mapsto a \circ v$$

   *for s taken from $S = \left\{ 1, 0, \frac{1}{2^2}, \frac{1}{2^5} \right\}$ and the 1-eigenspace is
   one-dimensional spanned by a;*

(iii) *μ inverts every $\frac{1}{2^5}$-eigenvector of a and centralizes the other
   eigenvectors;*

(iv) *if v_s and v_t are s- and t-eigenvectors of a, where $s, t \in S$, then $v_s \circ v_t$ is
   a sum of eigenvectors with eigenvalues $r \in f(s, t)$, where
   $f : S \times S \to 2^S$ is the fusion function given by:*

	1	0	$\frac{1}{2^2}$	$\frac{1}{2^5}$
1	1	0	$\frac{1}{2^2}$	$\frac{1}{2^5}$
0	0	0	$\frac{1}{2^2}$	$\frac{1}{2^5}$
$\frac{1}{2^2}$	$\frac{1}{2^2}$	$\frac{1}{2^2}$	1, 0	$\frac{1}{2^5}$
$\frac{1}{2^5}$	$\frac{1}{2^5}$	$\frac{1}{2^5}$	$\frac{1}{2^5}$	1, 0, $\frac{1}{2^2}$

In what follows we always assume that a Majorana involution possesses a
single axial vector, and that all the Majorana axial vectors are the same length.
Furthermore, unless specified otherwise we assume that $\langle a, a \rangle = 1$ for every
Majorana axial vector a.

Proposition 8.6.2 *Let ϑ be a $2A$-involution in G and $a(\vartheta)$ be the corresponding axial vector. Then the action of ϑ on $\Pi \oplus \Pi_1$ is a Majorana
involution with axial vector $a(\vartheta)$ with respect to the Griess algebra and the
inner product $(\ ,\ )$.* $\square$

Proof. By (8.1.1) in order to show that $(\ ,\ )$ and $\star$ satisfy the requirement,
we only have to check Norton's inequality, since the rest can be viewed as an
axiomatization of the properties established in (8.3.2), (8.4.1 (iii)), and (8.5.1).
For the proof of Norton's inequality, see (8.9.5). $\square$

When 1 is merged with 0 the function f is precisely the fusion rules of the
Virasoro algebra with central charge $\frac{1}{2}$ (cf. (6.18) in [BPZ84] and [DMZ94]).

The algebra describes the *Majorana fermion*, which is equivalent to the Ising model (cf. pp. 104, 117 in [GNT98]). In the area of Vertex Operator Algebras, the notions of *Ising vector* and *Miyamoto-* or τ-involution are now rather standard [Miy95], [Miy03], [Miy04]. The above definition is specifically attached to the finite-dimensional situation.

The next section contains Sakuma's classification of subalgebras generated by pairs of Majorana axial vectors. We say that a pair of Majorana vectors have type $2A$, $2B$, or $4A$ if the subalgebra they generate is isomorphic to that from (8.3.4), (8.3.5), or (8.3.7), respectively. It is a very important and nontrivial property of the Monster group that a maximal $2A$-pure subgroup has order 2^2 (cf. (8.2) in [ASeg92] and (3.6.5) in [GMS89]). Compare this with the following.

Lemma 8.6.3 *Let X be an elementary abelian 2-group all whose non-identity elements are Majorana involutions. Suppose that any two of the corresponding axial vectors generate a subalgebra of type $2A$ whose third axial vector corresponds to an involution in X. Then $|X| \leq 2^2$.*

Proof. It is clearly sufficient to rule out the possibility for X to be of order 2^3. If this is the case, then the involutions in X are points of the Fano plane. Let x_0 be a point and let $\{x_0, x_1, x_2\}$, $\{x_0, x_3, x_4\}$, $\{x_0, x_5, x_6\}$ be the lines passing through x_0. Then by (8.3.5 (iii)), $a(x_1) - a(x_2)$, $a(x_3) - a(x_4)$, and $a(x_5) - a(x_6)$ are $\frac{1}{2^2}$-eigenvectors of $a(x_0)$. Since any two axial vectors generate a subalgebra of type $2A$, by (8.3.5(i)) we have

$$(a(x_1) - a(x_2)) \star (a(x_3) - a(x_4)) = \frac{1}{2^2}(a(x_5) - a(x_6)).$$

This contradicts the fusion rules, since $\frac{1}{2^2} \notin f\left(\frac{1}{2^2}, \frac{1}{2^2}\right)$. $\qquad\square$

In the content of Majorana involutions, we can now prove Conway's bound on the number of axial vectors of transpositions (cf. Section 13 in [C84] and 9.1 in [Miy04]).

Lemma 8.6.4 *Let a and b be distinct axial vectors of Majorana involutions. Then*

$$0 \leq (a|b) \leq \frac{1}{3}.$$

In particular, the total number of such vectors is bounded by a function of the dimension of the underlying vector space.

Proof. The vectors a and b are idempotents of the same length which can be assumed without loss of generality to be 1. By Norton's inequality, we have

$$\langle a^2, b^2 \rangle = \langle a, b \rangle \geq \langle a \star b, a \star b \rangle \geq 0$$

and hence b can be written in the form

$$b = \rho\, a + u,$$

where $0 \leq \rho < 1$ and u is a vector perpendicular to a (so that $\rho = (a|b) = \langle a, b \rangle$, since $\langle a, a \rangle = \langle b, b \rangle = 1$ according to our assumption). In order to prove the assertion, it is sufficient to show that $\rho \leq \frac{1}{3}$. Since the scalar product is associative with respect to $\star$

$$\langle u \star a, a \rangle = \langle u, a \star a \rangle = \langle u, a \rangle = 0$$

and therefore the product $u \star a$ is also perpendicular to a. Since b is an idempotent, we have

$$\rho\, a + u = b = b \star b = (\rho\, a + u) \star (\rho\, a + u) = \rho^2 a + 2\rho\,(a \star u) + u \star u.$$

So that $\rho = \langle a, b \rangle = \rho + \langle a, u \star u \rangle$. On the other hand, $1 = \langle b, b \rangle = \rho^2 + \langle u, u \rangle$ and hence $\langle u, u \rangle = (1 - \rho^2)$. By (8.6.1 (ii)), the eigenvalues of a on the orthogonal complement to a are $0, \frac{1}{2^5}, \frac{1}{2^2}$, and therefore

$$(\rho - \rho^2) = \langle a, u \star u \rangle = \langle a \star u, u \rangle \leq \frac{1}{2^2} \langle u, u \rangle = \frac{1}{2^2}(1 - \rho^2).$$

Hence $(\rho - \rho^2) \leq \frac{1}{2^2}(1 - \rho^2)$, and since $0 \leq \rho < 1$ this gives the required $\rho \leq \frac{1}{3}$. $\square$

Corollary 8.6.5 *The image $\varphi(G)$ of G in $GL(\Pi) \cong GL_{196883}(\mathbb{R})$ is a finite group.*

Proof. By (8.6.2) and (8.6.4), there are only a finite number of axial vectors of transpositions in $\Pi \oplus \Pi_1$. Since these vectors span the whole space and since the image of an axial vector under an element of G is again an axial vector, this gives the result. $\square$

8.7 Sakuma's theorem

It was known since the early stages of investigating the Monster that the $2A$-involutions form a class of 6-transpositions in the sense that the product of any two such involutions has order at most six. Furthermore, the products constitute the union of the following nine conjugacy classes

$$1A, \ 2A, \ 2B, \ 3A, \ 3C, \ 4A, \ 4B, \ 5A, \ \text{and}\ 6A.$$

The orbit of the Monster acting by conjugation on the pairs of $2A$-involutions is uniquely determined by the class containing the product, so that the permutation rank of the Monster of the set of $2A$-involutions in nine.

The subalgebras in (the $196\,884$-dimensional version) of the Monster algebra generated by the pairs of transposition axes as calculated by J.H. Conway and S.P. Norton [C84], [N96] (subject to some rescaling and renaming) are given in the table below.

Type	dim	$(a_0\|a_1)$	Products
1A	1	–	$a_0 \star a_0 = a_0$
2A	3	$\frac{1}{2^3}$	$a_0 \star a_1 = \frac{1}{2^3}(a_0 + a_1 - a_2)$
2B	2	0	$a_0 \star a_1 = 0$
3A	4	$\frac{13}{2^8}$	$a_0 \star a_1 = \frac{1}{2^5}(2a_0 + 2a_1 + a_2) - \frac{3^3 \cdot 5}{2^{11}}u_3,$
			$a_0 \star u_3 = \frac{1}{3^2}(2a_0 - a_1 - a_2) + \frac{5}{2^5}u_3,$
			$u_3 \star u_3 = u_3$
3C	3	$\frac{1}{2^6}$	$a_0 \star a_1 = \frac{1}{2^6}(a_0 + a_1 - a_2)$
4A	5	$\frac{1}{2^5}$	$a_0 \star a_1 = \frac{1}{2^6}(3a_0 + 3a_1 + a_2 + a_3 - 3u_4),$
			$a_0 \star u_4 = \frac{1}{2^4}(5a_0 - 2a_1 - a_2 - 2a_3 + 3u_4),$
			$u_u \star u_4 = u_4$
4B	5	$\frac{1}{2^6}$	$a_0 \star a_1 = \frac{1}{2^6}(a_0 + a_1 - a_2 - a_3 - u_2),$
5A	6	$\frac{3}{2^7}$	$a_0 \star a_1 = \frac{1}{2^7}(3a_0 + 3a_1 - a_2 - a_3 - a_4 + u_5),$
			$a_0 \star a_2 = \frac{1}{2^7}(3a_0 + 3a_2 - a_1 - a_3 - a_4 - u_5),$
			$a_0 \star u_5 = \frac{7}{2^5}(a_1 - a_2 - a_3 + a_4 + u_5),$
			$u_5 \star u_5 = \frac{175}{2^5}(a_0 + a_1 + a_2 + a_3 + a_4)$
6A	8	$\frac{5}{2^8}$	$a_0 \star a_1 = \frac{1}{2^6}(a_0 + a_1 - a_2 - a_3 - a_4 - a_5 + u_2 + u_3),$
			$u_2 \star u_3 = 0,$

The following remarkable theorem proved by S. Sakuma [Sak07] gives strong evidence that Majorana involution is a very efficient tool for studying the Monster.

Theorem 8.7.1 *Let (μ_0, μ_1) be a pair of Majorana involutions and let (a_0, a_1) be the corresponding pair of Majorana axes. Let $D \cong D_{2n}$ be the dihedral group (of order $2n$) generated by μ_0 and μ_1 and let Δ be the subalgebra generated by a_0 and a_1. Then:*

(i) *$n \leq 6$;*

(ii) *$\dim(\Delta) \leq 8$;*

(iii) *Δ is isomorphic to one of the nine 2-generated subalgebras in the Monster algebra.* □

8.8 Majorana calculus

In this section we analyse the 2-generated Majorana algebras with a particular emphasis on the cases where the product of the Majorana generators has order 2, 3, or 5. This is motivated by our intention (which goes beyond the scope of this volume) to classify the Majorana representations of the smallest non-abelian simple group A_5. We pose the following:

Conjecture 8.8.1 *The alternating group A_5 of degree five possesses exactly two Majorana representations whose dimensions are 26 and 21 and which correspond $(2A, 3A, 5A)$- and $(2A, 3C, 5A)$-subgroups in the Monster isomorphic to A_5.*

8.8.1 Elementary properties

In this section we adopt the notation introduced in (8.6.1) and assume that $\langle a, a \rangle = 1$ for every Majorana axial vector a. We start by deducing some consequences of the associativity of the inner product $\langle \, , \, \rangle$ with respect to $\circ$. For a Majorana axial vector $a \in V$, let $V_s^{(a)}$ denote the s-eigenspace of a

$$V_s^{(a)} = \{v \mid v \in V, a \circ v = sv\}$$

for $s \in S$, where $S = \left\{1, 0, \frac{1}{2^2}, \frac{1}{2^5}\right\}$ is the spectrum of (the adjoint action of) a on V. For $v \in V$ and $s \in S$, denote by $(v|a)_s$ the projection of v to $V_s^{(a)}$, put $(v|a)_+$ and $(v|a)_-$ to be the projections of v to $C_V(\mu)$ and $[\mu, V]$, where μ is the Majorana involution associated with a, and denote by v^μ the image of v under μ.

Lemma 8.8.2 *The decomposition*

$$V = V_1^{(a)} \oplus V_0^{(a)} \oplus V_{\frac{1}{2^2}}^{(a)} \oplus V_{\frac{1}{2^5}}^{(a)}$$

is $\langle \, , \, \rangle$-orthogonal.

Proof. Let $u \in V_s^{(a)}$ and $v \in V_t^{(a)}$. If s is zero, while t is not, then

$$\langle u, v \rangle = \frac{1}{t} \langle u, a \circ v \rangle = \frac{1}{t} \langle a \circ u, v \rangle = 0.$$

If both s and t are non-zero, then

$$\frac{1}{s} \langle u, v \rangle = \langle a \circ u, v \rangle = \langle u, a \circ v \rangle = \frac{1}{t} \langle u, v \rangle,$$

so that u and v are perpendicular unless $s = t$. $\qquad\square$

Lemma 8.8.3 *Let $u, v \in V_s^{(a)}$. Then $(u \circ v|a)_1 = 0$ if $s = 0$ and*

$$(u \circ v|a)_1 = \frac{1}{s} \langle u, v \rangle \, a$$

otherwise.

Proof. We have $s \langle u \circ v, a \rangle = s \langle u, a \circ v \rangle = \langle u, v \rangle$ and

$$(u \circ v|a)_1 = \frac{\langle u \circ v, a \rangle}{\langle a, a \rangle} = \langle u \circ v, a \rangle,$$

implying the assertion. $\qquad\square$

Lemma 8.8.4 *The following assertions hold for a being a Majorana axial vector and $v, u \in V$:*

(i) $\langle (u|a)_s, (v|a)_t \rangle = 0$ *whenever $s, t \in S$ with $s \neq t$;*

(ii) $(v|a)_1 = (v|a)a = \frac{\langle v, a \rangle}{\langle a, a \rangle} a$;

(iii) $v = (v|a)_1 + (v|a)_0 + (v|a)_{\frac{1}{2^2}} + (v|a)_{\frac{1}{2^5}}$;

(iv) $(v|a)_- = (v|a)_{\frac{1}{2^5}}$ *and* $v = (v|a)_+ + (v|a)_-$;

(v) $v^\mu = v - 2(v|a)_{\frac{1}{2^5}} = v - \frac{2^9}{7}[3(v|a)_1 + a \circ v - 4a \circ (a \circ v)]$;

(vi) *if* $(v|a)_{\frac{1}{2^5}} = 0$, *then* $a \circ (a \circ v) - \frac{1}{2^2} a \circ v = \frac{3}{2^2}(v|a)_1$.

Proof. (i) is a restatement of (8.8.2), while (ii) and (iii) follow from (8.6.1 (ii)), with (iv) implied by (8.6.1 (iii)). By (ii) we have

$$a \circ v = (v|a)_1 + \frac{1}{2^2}(v|a)_{\frac{1}{2^2}} + \frac{1}{2^5}(v|a)_{\frac{1}{2^5}};$$

$$a \circ (a \circ v) = (v|a)_1 + \frac{1}{2^4}(v|a)_{\frac{1}{2^2}} + \frac{1}{2^{10}}(v|a)_{\frac{1}{2^5}},$$

while by (iv) $\mu : v \mapsto v - 2(v|a)_{\frac{1}{2^5}}$. Now (v) can be deduced via standard algebraic manipulations. In the hypothesis of (vi) we have

$$a \circ v = (v|a)_1 + \frac{1}{2^2}(v|a)_{\frac{1}{2^2}} + \frac{1}{2^5}(v|a)_{\frac{1}{2^5}}$$

and the assertion follows. $\qquad\square$

8.8.2　Types $2A$ and $2B$

Suppose that D is generated by a pair of commuting Majorana involutions μ_0 and μ_1 with different axes a_0 and a_1. We follow the proof of Lemma 6.11 in [Miy96] and put

$$a_1 = \lambda a_0 + \alpha + \beta,$$

where $\lambda = (a_1|a_0)$, $\alpha = (a_1|a_0)_0$, $\beta = (a_1|a_0)_{\frac{1}{2^2}}$. Since a_1 is an idempotent, squaring it and applying the fusion rules, we obtain the following

$$\lambda a_0 + \alpha = (a_1|a_0)_1 + (a_1|a_0)_0 = (a_1^2|a_0)_1 + (a_1^2|a_0)_0 = \lambda^2 a_0 + \alpha^2 + \beta^2;$$

$$\beta = (a_1|a_0)_{\frac{1}{2^2}} = (a_1^2|a_0)_{\frac{1}{2^2}} = \frac{\lambda}{2}\beta + 2\alpha \circ \beta,$$

which gives

$$\beta^2 = (\lambda - \lambda^2)a_0 + \alpha - \alpha^2;$$

$$\alpha \circ \beta = \frac{\beta}{2} - \frac{\lambda}{2^2}\beta.$$

By the above equalities and since $a_0 \circ a_1 = \lambda a_0 + \frac{1}{2^2}\beta$, we obtain

$$a_1 \circ (a_0 \circ a_1) = (\lambda a_0 + \alpha + \beta) \circ \left(\lambda a_0 + \frac{1}{2^2}\beta\right)$$

$$= \left(\lambda^2 + \frac{\lambda - \lambda^2}{2^2}\right)a_0 + \frac{1}{2^2}(\alpha - \alpha^2) + \left(\frac{\lambda}{2^2} + \frac{1}{2^3}\right)\beta.$$

On the other hand, since a_0 and a_1 are axes of commuting Majorana involutions, by (8.8.4) we have

$$a_1 \circ (a_1 \circ a_0) - \frac{1}{2^2}a_0 \circ a_1 = \frac{3}{2^2}\lambda a_1$$

and hence

$$\frac{3}{2^2}\lambda a_1 = \frac{3}{2^2}\lambda^2 a_0 + \frac{3}{2^2}\lambda\alpha + \frac{3}{2^2}\lambda\beta = \left(a_1 \circ (a_0 \circ a_1) - \frac{1}{2^2}a_1 \circ a_0\right)$$

$$= \left(\lambda^2 - \frac{\lambda^2}{2^2}\right)a_0 + \frac{1}{2^2}(\alpha - \alpha^2) + \left(\frac{\lambda}{2^2} + \frac{1}{2^3} - \frac{1}{2^4}\right)\beta.$$

If $\beta = 0$, then $a_0 \circ a_1 = \lambda a_0 = \lambda a_1 = a_1 \circ a_0$ and since $a_0 \neq a_1$ we have $\lambda = 0$, so that $\langle a_0, a_1 \rangle = a_0 \circ a_1 = 0$ and we have the $2B$-algebra.

If $\beta \neq 0$, then

$$\lambda = \frac{1}{2^3}, \alpha^2 = \frac{5}{2^3}\alpha, \alpha \circ \beta = \frac{15}{2^5}\beta, \beta^2 = \frac{7}{2^6}a_0 + \frac{3}{2^3}\alpha,$$

and since $a_0 \circ a_1 = \frac{1}{2^3}a_0 + \frac{1}{2^2}\beta$, by introducing $a_2 = a_1 - 2\beta = \frac{1}{2^3}a_0 + \alpha - \beta$, we obtain another idempotent and the whole $2A$-algebra.

Lemma 8.8.5 *An algebra generated by a pair of Majorana axes corresponding to commuting involutions is isomorphic to $2A$- or $2B$-algebra.* □

8.8.3 Stable vectors

From now on we assume that the dihedral group $D \cong D_{2n}$ generated by the Majorana involutions μ_0 and μ_1 acts faithfully on the subalgebra Δ generated by the corresponding axes a_0 and a_1. In this section we follow [Sak07] to introduce certain vectors in Δ stable under D.

Lemma 8.8.6 *Let $\sigma = a_0 \circ a_1 - \frac{1}{2^5}(a_0 + a_1)$ and let $\lambda = (a_0|a_1)$ Then:*

(i) σ *is fixed by* μ_0 *and* μ_1;
(ii) $\sigma = a_i \circ a_{i+1} - \frac{1}{2^5}(a_i + a_{i+1})$ *for every* $0 \le i \le n - 1$;
(iii) $(\sigma|a_0) = \frac{31\lambda - 1}{2^5}$.

Proof. By (8.8.4) $a_1 = (a_1|a_0)_+ + (a_1|a_0)_-$. The vector $(a_1|a_0)_+$ is fixed by μ_0 and the vector $(a_1|a_0)_-$ is negated by μ_0, while being multiplied by a_0 it becomes 2^5 times shorter. This, together with the obvious symmetry between a_0 and a_1, gives (i). Since μ_0 and μ_1 generate D and since the latter acts transitively on the set of ordered pairs $(i, i + 1)$ taken modulo n, (ii) follows. Now (iii) follows from the equalities $\langle \sigma, a_0 \rangle = \langle a_1 \circ a_0, a_0 \rangle - \frac{1}{2^5}(\langle a_0, a_0 \rangle + \langle a_1, a_0 \rangle)$ and $\langle a_1 \circ a_0, a_0 \rangle = \langle a_1, a_0 \circ a_0 \rangle = \langle a_1, a_0 \rangle$. □

Let

$$a_1 = \lambda a_0 + \alpha + \beta + \gamma,$$

where $\lambda = (a_1|a_0)$, while α, β, and γ are the projections $(a_1|a_0)_s$ of a_1 onto the s-eigenspaces of a_0 for $s = 0$, $\frac{1}{2^2}$, and $\frac{1}{2^5}$, respectively.

Lemma 8.8.7 *In the above notation the following equalities hold:*

(i) $a_{-1} = \lambda a_0 + \alpha + \beta - \gamma$;
(ii) $\alpha = -4\sigma + \left(3\lambda - \frac{1}{2^3}\right)a_0 + \frac{7}{2^4}(a_1 + a_{-1})$;
(iii) $\beta = 4\sigma - \left(4\lambda - \frac{1}{2^3}\right)a_0 + \frac{1}{2^4}(a_1 + a_{-1})$;
(iv) $\gamma = \frac{1}{2}(a_1 - a_{-1})$ *and* $\alpha + \beta + \lambda a_0 = \frac{1}{2}(a_1 + a_{-1})$;

(v) $a_0 \circ \sigma = \frac{7}{2^5}\sigma + \left(\frac{3\lambda}{4} - \frac{25}{2^{10}}\right)a_0 + \frac{7}{2^{11}}(a_1 + a_{-1});$

(vi) $\sigma = \left(\frac{31\lambda - 1}{2^5}\right)a_0 - \frac{1}{2^5}\alpha + \frac{7}{2^5}\beta;$

(vii) $\gamma \circ \sigma = \left(\frac{3\lambda}{4} - \frac{25}{2^{10}}\right)\gamma + \frac{7}{2^{11}}(a_2 - a_{-2}).$

Proof. Since $a_{-1} = a_1^{\mu_0}$, (i) follows from (8.8.4 (iv)). Now (i) and the expression of a_1 as the sum of its projections into the eigenspaces of a_0 give (iv). Since

$$a_0 \circ a_1 = \lambda a_0 + \frac{1}{2^2}\beta + \frac{1}{2^5}\gamma,$$

(iii) can be deduced from (8.8.6). Now (ii) and (vi) follow, while (v) is deduced by expanding the equality $a_0 \circ \alpha = 0$ with α as in (ii). Finally, (vii) follows from (v) and (iv). □

Lemma 8.8.8 *The following assertions hold:*

(i) $(\alpha + \beta) \circ \gamma = \left(\frac{1}{2} - \frac{\lambda}{2^5}\right)\gamma;$

(ii) α^2 *is annihilated by* $a_0;$

(iii) $(\gamma^2|a_0)_1 = (\lambda - \lambda^2)a_0 - (\beta^2|a_0)_1;$

(iv) $(\gamma^2|a_0)_0 = (\alpha - \alpha^2) - (\beta^2|a_0)_0;$

(v) $(\gamma^2|a_0)_{\frac{1}{2^2}} = \left(1 - \frac{\lambda}{2}\right)\beta - 2\alpha \circ \beta.$

Proof. The result is deduced by squaring the right-hand side of

$$a_1 = \lambda a_0 + \alpha + \beta + \gamma,$$

and by equalizing the square with the original expression (a_1 is an idempotent):

$$\lambda a_0 + \alpha + \beta + \gamma = \lambda^2 a_0 + \frac{\lambda}{2}\beta + \frac{\lambda}{2^4}\gamma + \alpha^2 + \beta^2 + \gamma^2 + 2\alpha \circ \beta + 2\alpha \circ \gamma + 2\beta \circ \gamma.$$

By the fusion rule, a term is contained in the eigenspace of a_0 with eigenvalue $\frac{1}{2^5}$ if and only it involves γ exactly once. This gives (i) (which alternatively can be deduced directly from (8.8.6 (ii)) and (8.8.7 (iv))). The assertion (ii) follows directly from the fusion rules. Removing the γ-linear terms from the above equality and rearranging we obtain

$$\gamma^2 = (\lambda - \lambda^2)a_0 + (\alpha - \alpha^2) + \left(1 - \frac{\lambda}{2}\right)\beta - \beta^2 - 2\alpha \circ \beta,$$

which gives (iii), (iv), and (v) because of the fusion rules. □

8.8.4 Types 3*A* and 3*C*

If n is half the order of the dihedral group $D \cong D_{2n}$ generated by the Majorana involutions μ_0 and μ_1, then by (8.8.7) a specific feature of the $n = 3$ case is that γ is a σ-eigenvector. It turns out that when $n = 3$, the subspace Ξ spanned by the projections of a_1 into the eigenspaces of a_0 is closed under the multiplication. Since Ξ is easily seen to contain a_0 and a_1, it is closed under the multiplication if and only if it coincides with the whole of Δ. To prove that the product is closed on Ξ, we follow the proof of Lemma 2.2 in [Miy03] and show that whenever $u, v \in \{a_0, \alpha, \beta, \gamma\}$ the product $u \circ v$ can be written as a linear combination of a_0, α, β and γ. The assertion is obvious for $u = a_0$ and the next lemma takes care of the case $u = \gamma$.

Lemma 8.8.9 *If $n = 3$, then:*

(i) $\gamma \circ \sigma = \left(\frac{3\lambda}{4} - \frac{57}{2^{11}} \right) \gamma$;

(ii) $\gamma \circ \alpha = \frac{93\,(3-16\lambda)}{2^9} \gamma$;

(iii) $\gamma \circ \beta = \frac{23\,(64\lambda-1)}{2^9} \gamma$;

(iv) $\gamma^2 = -\frac{1}{2}\sigma + \frac{15}{2^6}(a_1 + a_{-1}) = \frac{1-\lambda}{2^6}a_0 + \frac{31}{2^6}\alpha + \frac{23}{2^6}\beta$.

Proof. Since 2 and -1 are equal modulo 3, (i) is a specialization of (8.8.7 (vi)). Now (ii) and (iii) follow from (i) in view of (8.8.7 (ii), (iii)). Finally (iv) is by (8.8.7 (iv)) and (8.8.6 (ii)). □

Lemma 8.8.10 *If $n = 3$, then $\alpha \circ \beta = \left(\frac{41}{27} - \frac{\lambda}{2^2} \right) \beta$.*

Proof. By (8.8.9 (iv)), we have $(\gamma^2 | a_0)_{\frac{1}{2^2}} = \frac{23}{2^6}\beta$. Equalizing this with the right-hand side of (8.8.8 (v)) gives the result. □

Lemma 8.8.11 *If $n = 3$, then:*

(i) $(a_0|a_1)_0 = \left(3\lambda^2 - \frac{57\lambda-9}{2^4} \right) a_0 + \left(3\lambda + \frac{7}{2^4} \right) \alpha + \left(3\lambda - \frac{9}{2^4} \right) (\beta + \gamma)$;

(ii) $(a_0|a_1)_{\frac{1}{2^2}} = -\left(4\lambda^2 - \frac{65\lambda-1}{2^4} \right) a_0 - \left(4\lambda - \frac{1}{2^4} \right) (\alpha + \gamma) - \left(4\lambda - \frac{17}{2^4} \right) \beta$;

(iii) $(a_0|a_1)_{\frac{1}{2^5}} = \frac{1-\lambda}{2}a_0 - \frac{1}{2}\alpha - \frac{1}{2}\beta + \frac{1}{2}\gamma$.

Proof. The result can be achieved by acting via the Majorana involution μ_{-1} on the equalities in (8.8.7) and by making use of the invariance of σ established in (8.8.6). Alternatively, we might follow [Miy03] and resolve the following equalities (notice that only the former one requires the condition $n = 3$)

$$a_{-1} = a_1 - 2\gamma = a_0 - 2(a_0|a_1)_{\frac{1}{2^5}};$$

$$\lambda a_0 + \frac{1}{2^2}\beta + \frac{1}{2^5}\gamma = a_0 \circ a_1 = \lambda a_1 + \frac{1}{2^2}(a_0|a_1)_{\frac{1}{2^2}} + \frac{1}{2^5}(a_0|a_1)_{\frac{1}{2^5}};$$

$$a_0 = \lambda(\lambda a_0 + \alpha + \beta + \gamma) + \sum_{s=0,\frac{1}{2^2},\frac{1}{2^5}}(a_0|a_1)_s.$$

$\square$

Lemma 8.8.12 *If $n = 3$, then:*

(i) $\alpha^2 = \frac{3}{2^4}(3 - 2^4\lambda)\alpha$;

(ii) $\alpha \circ \beta = \frac{9}{2^6}(3 - 2^4\lambda)\beta.$

Proof. Write down the equality

$$a_1 \circ (a_0|a_1)_0 = 0$$

in terms of a_0, α, β, and γ, where $a_1 = \lambda a_0 + \alpha + \beta + \gamma$ and (8.8.11 (i)) are taken. Projecting the equality on the 0- and $\frac{1}{2^2}$-eigenspaces of a_0 we get (i) and (ii), respectively. $\square$

By (8.8.9), (8.8.10), and (8.8.12) we know that the set $\{a_0, \alpha, \beta, \gamma\}$ contains a basis of Δ and also has all the relevant structure constants as functions of λ.

Proposition 8.8.13 *An algebra generated by a pair of Majorana axes whose involutions generate D_6 is isomorphic to $3A$- or $3C$-algebra.*

Proof. Comparing (8.8.10) with (8.8.12 (ii)), we conclude that either $\beta = 0$ or $\lambda = \frac{13}{2^8}$. In the latter case, we have the $3A$-algebra, while in the former case, (8.8.9 (iii)) gives $\lambda = \frac{1}{2^6}$ and we obtain the $3C$-algebra.

	$3A$	$3C$
λ	$\frac{13}{2^8}$	$\frac{1}{2^6}$
$a_0 \circ a_1$	$\frac{1}{2^5}(2a_0 + 2a_1 + a_{-1}) - \frac{3^3 \cdot 5}{2^{11}}u_3$	$\frac{1}{2^6}(a_0 + a_1 - a_{-1})$
$a_0 \circ u_3$	$\frac{1}{3^2}(2a_0 - a_1 - a_2) + \frac{5}{2^5}u_3$	$-$
$u_3 \circ u_3$	u_3	$-$
σ	$\frac{1}{2^5}(a_0 + a_1 + a_{-1}) - \frac{3^3 \cdot 5}{2^{11}}u_3$	$-\frac{1}{2^6}(a_0 + a_1 + a_{-1})$
γ	$\frac{1}{2}(a_1 - a_{-1})$	$\frac{1}{2}(a_1 - a_{-1})$
α	$-\frac{25}{2^8}a_0 + \frac{5}{2^4}(a_1 + a_{-1}) + \frac{3^3 \cdot 5}{2^9}u_3$	$-\frac{1}{2^6}a_0 + \frac{1}{2}(a_1 + a_{-1})$
β	$\frac{12}{2^8}a_0 + \frac{3}{2^4}(a_1 + a_{-1}) - \frac{3^3 \cdot 5}{2^9}u_3$	0
$\alpha + \beta$	$-\frac{13}{2^8}a_0 + \frac{1}{2}(a_1 + a_{-1})$	$-\frac{1}{2^6}a_0 + \frac{1}{2}(a_1 + a_{-1})$

$\square$

8.8.5 Less stable vectors

For $0 \leq j < i \leq n - 1$ let $D^{(ji)}$ denote the subgroup in D generated by the Majorana involutions μ_j and μ_i corresponding to the axes a_j and a_i. Then $D^{(ji)}$ is also a dihedral group generated by a pair of Majorana involutions whose order can easily be expressed in terms of n, j and i. If $D^{(ji)}$ is a proper subgroup in D, then the subalgebra generated by a_j and a_i may assumed to be known within an inductive approach, otherwise we might gain an additional symmetry considering different generating sets of D. We start by considering the case $j = 0$ and denote the subgroup $D^{(0i)}$ simply by $D^{(i)}$.

For $1 \leq i \leq n - 1$, put

$$\sigma_i = a_0 \circ a_i - \frac{1}{2^5}(a_0 + a_i), \quad \lambda_i = (a_i | a_0),$$

$$\alpha_i = (a_i | a_0)_0, \quad \beta_i = (a_i | a_0)_{\frac{1}{2^2}}, \quad \gamma_i = (a_i | a_0)_{\frac{1}{2^5}},$$

so that

$$a_i = \lambda a_0 + \alpha_i + \beta_i + \gamma_i$$

is the eigenvector decomposition of a_i with respect to a_0, while σ_1 is the same as σ introduced in Subsection 8.8.3. Notice that $\gamma_i = 0$ precisely when $i = \frac{n}{2}$. The following three lemmas are straightforward generalizations of (8.8.6), (8.8.7), and (8.8.8), respectively.

Lemma 8.8.14 *The following assertions hold:*

(i) σ_i *is fixed by the subgroup* $D^{(i)}$ *of* D *generated by* μ_0 *and* μ_i;
(ii) $\sigma_i = a_j \circ a_{j+i} - \frac{1}{2^5}(a_j + a_{j+i})$ *whenever* a_j *is contained in the* $D^{(i)}$-*orbit of* $\{a_0, a_i\}$;
(iii) $(\sigma_i | a_0) = \frac{31\lambda_i - 1}{2^5}$. $\qquad\qquad \square$

Lemma 8.8.15 *The following equalities hold:*

(i) $a_{-i} = \lambda a_0 + \alpha_i + \beta_i - \gamma_i$;
(ii) $\alpha_i = -4\sigma_i + \left(3\lambda_i - \frac{1}{2^3}\right)a_0 + \frac{7}{2^4}(a_i + a_{-i})$;
(iii) $\beta_i = 4\sigma_i - \left(4\lambda_i - \frac{1}{2^3}\right)a_0 + \frac{1}{2^4}(a_i + a_{-i})$;
(iv) $\gamma_i = \frac{1}{2}(a_i - a_{-i})$ *and* $\alpha_i + \beta_i + \lambda_i a_0 = \frac{1}{2}(a_i + a_{-i})$;
(v) $a_0 \circ \sigma_i = \frac{7}{2^5}\sigma_i + \left(\frac{3\lambda_i}{4} - \frac{25}{2^{10}}\right)a_0 + \frac{7}{2^{11}}(a_i + a_{-i})$;
(vi) $\sigma_i = \left(\frac{31\lambda_i - 1}{2^5}\right)a_0 - \frac{1}{2^5}\alpha_i + \frac{7}{2^5}\beta_i$;
(vii) $\gamma_i \circ \sigma_i = \left(\frac{3\lambda_i}{4} - \frac{25}{2^{10}}\right)\gamma_i + \frac{7}{2^{11}}\gamma_{2i}$. $\qquad\qquad \square$

Lemma 8.8.16 *The following assertions hold:*

(i) $(\alpha_i + \beta_i) \circ \gamma_i = \left(\frac{1}{2} - \frac{\lambda_i}{2^5}\right)\gamma_i$;

(ii) α_i^2 *is annihilated by* a_0;

(iii) $(\gamma_i^2 | a_0)_1 = (\lambda_i - \lambda_i^2) a_0 - (\beta_i^2 | a_0)_1$;

(iv) $(\gamma_i^2 | a_0)_0 = (\alpha_i - \alpha_i^2) - (\beta_i^2 | a_0)_0$;

(v) $(\gamma_i^2 | a_0)_{\frac{1}{2^2}} = (1 - \frac{\lambda_i}{2})\beta_i - 2\alpha_i \circ \beta_i$. $\qquad\qquad\square$

The following lemma is the general version of (8.8.9 (iv)) and (8.8.10).

Lemma 8.8.17 *If n is odd then the following equalities hold:*

(i) $\gamma_1^2 = -\frac{1}{2}\sigma_2 + \frac{15}{2^6}(a_1 + a_{-1}) = \left(\frac{1 + 30\lambda_1 - 31\lambda_2}{2^6}\right)a_0 + \frac{30\alpha_1 + \alpha_2}{2^6} + \frac{30\beta_1 - 7\beta_2}{2^6}$;

(ii) $\alpha_1 \circ \beta_1 = \left(\frac{34}{2^7} - \frac{\lambda_1}{2^2}\right)\beta_1 + \frac{7}{2^7}\beta_2$;

(iii) $\gamma_1\gamma_2 = \sigma_1 - \sigma_3$.

Proof. The first equality is obtained by squaring both sides of $\gamma_1 = \frac{1}{2}(a_1 + a_{-1})$ in (8.8.15 (iv)) and expanding σ_2 by (8.8.15 (vi)). Since $(\gamma_1|a_0)_{\frac{1}{2^2}}$ is precisely the last fraction in the right-hand side sum in (i), the equality (ii) follows from (8.8.16 (v)). Finally, (iii) is by (8.8.15 (iv)). $\qquad\square$

8.8.6 Type $5A$

In this subsection we assume that $n = 5$. Since 5 is a prime number, we have $D^{(1)} = D^{(2)} = D$. The following lemma is due to Sergey Shpectorov.

Lemma 8.8.18 *If $n = 5$, then:*

(i) $\lambda_1 = \lambda_2$;

(ii) Δ *is spanned by* a_i*s taken for* $0 \le i \le 4$ *together with* σ_1 *and* σ_2, *in particular* $\dim(\Delta) \le 7$.

Proof. By the definition of σ_i, we have

$$a_0 \circ a_i = \sigma_i + \frac{1}{32}(a_0 + a_i),$$

while by (8.8.14) $(\sigma_i | a_j) = \frac{31\lambda_i - 1}{2^5}$. Now by the associativity of the inner product $\langle\, , \,\rangle$ with respect to $\circ$, we obtain

$$\lambda_1 + \frac{\lambda_2 - 1}{2^5} = \langle a_1 \circ a_0, a_2 \rangle = \langle a_1, a_0 \circ a_2 \rangle = \lambda_2 + \frac{\lambda_1 - 1}{2^5},$$

which gives (i).

Let Ξ be the subspace of Δ spanned by a_is and σ_js, and let $\xi \in \Xi$. Then by (8.8.14 (ii)), (8.8.15 (v)), and since D stabilizing σ_1 and σ_2 permutes a_is transitively, we have $\xi \circ a_i \in \Xi$ for every $0 \leq i \leq 4$. Therefore, in order to prove (ii) it suffices to show that $\sigma_i \circ \sigma_j \in \Xi$ for all $1 \leq i, j \leq 2$. Suppose that $\sigma_i \circ \sigma_j \notin \Xi$. Then by (8.8.15 (ii))

$$\sigma_i \circ \sigma_j = \frac{1}{2^4}\alpha_i \circ \alpha_j \bmod \Xi,$$

while by the fusion rules $\alpha_i \circ \alpha_j$ is a 0-eigenvector of a_0 and hence $a_0 \circ (\sigma_i \circ \sigma_j) = 0 \bmod \Xi$. On the other hand, by (8.8.15 (ii), (iii))

$$\sigma_i \circ \sigma_j = -\frac{1}{2^4}\alpha_i \circ \beta_j \bmod \Xi,$$

and by the fusion rules, $\alpha_i \circ \beta_j$ is a $\frac{1}{2^2}$-eigenvector of a_0, so that $a_0 \circ (\sigma_i \circ \sigma_j) = \frac{1}{2^2}\sigma_i\sigma_j \bmod \Xi$, which contradicts what we have seen above. $\quad\square$

The above lemma enables us to drop the subscripts of λ_1 and λ_2;

Lemma 8.8.19 *Let $n = 5$ and $i \in \{1, 2\}$. Then:*

(i) $\gamma_i^2 = \left(\frac{1-\lambda}{2^6}\right) a_0 + \frac{30\alpha_i + \alpha_{3-i}}{2^6} + \frac{30\beta_i - 7\beta_{3-i}}{2^6};$

(ii) $\gamma_1\gamma_2 = -\frac{1}{2^6}(\alpha_1 - \alpha_2) + \frac{7}{2^6}(\beta_1 - \beta_2);$

(iii) $\alpha_i \circ \beta_i = \left(\frac{34}{2^7} - \frac{\lambda}{2^2}\right)\beta_i + \frac{7}{2^7}\beta_{3-i};$

(iv) $(a_0|a_i)_0 = \left(3\lambda^2 - \frac{57\lambda - 9}{2^4}\right) a_0 + 3\lambda\alpha_i + (3\lambda - 1)\beta_i + (3\lambda - \frac{1}{2^3})\gamma_i + \frac{7}{2^4}(\alpha_{3-i} + \beta_{3-i} + \gamma_{3-i});$

(v) $(a_0|a_i)_{\frac{1}{2^2}} = -\left(4\lambda^2 - \frac{65\lambda - 1}{2^4}\right) a_0 - 4\lambda\alpha_i - (4\lambda + \frac{1}{2^3})\beta_i - (4\lambda - \frac{1}{2^3})\gamma_i + \frac{1}{2^4}(\alpha_{3-i} + \beta_{3-i} + \gamma_{3-i}).$

Proof. If $i = 1$, then (i) is just (8.8.17 (i)), while for $i = 2$ the equality can be obtained by squaring the both sides of $\gamma_2 = \frac{1}{2}(a_2 - a_{-2})$ (notice that $a_2a_{-2} = \sigma_1 + \frac{1}{2^5}(a_2 + a_{-2})$ since $n = 5$). Since in the considered situation σ_3 is the same as σ_2, the equality (ii) follows from (8.8.17 (iii)). Now (iii) follows from (i) and (8.8.16 (v)). In order to obtain (iv), we first apply μ_{-2} to (8.8.15 (ii)) to get

$$(a_0|a_i)_0 = -4\sigma_1 + \left(3\lambda - \frac{1}{2^3}\right)a_i + \frac{7}{2^4}(a_0 + a_2)$$

and then expand the right-hand side of the above using (8.8.15). In a similar way we obtain (v). $\quad\square$

	$5A$
λ	$\frac{3}{2^7}$
$a_0 \circ a_1$	$\frac{1}{2^7}(3a_0 + 3a_1 - a_{-1} - a_2 - a_{-2} + u_5)$
$a_0 \circ a_2$	$\frac{1}{2^7}(3a_0 + 3a_2 - a_1 - a_{-1} - a_{-2} - u_5)$
$a_0 \circ u_5$	$\frac{7}{2^5}(a_1 + a_{-1} - a_2 - a_{-2} + u_5)$
$u_5 \circ u_5$	$\frac{175}{2^5}(a_0 + a_1 + a_{-1} + a_2 + a_{-2})$
σ_1	$\frac{1}{2^7}(-a_0 - a_1 - a_{-1} - a_2 - a_{-2} + u_5)$
σ_2	$\frac{1}{2^7}(-a_0 - a_1 - a_{-1} - a_2 - a_{-2} - u_5)$
γ_1	$\frac{1}{2}(a_1 - a_{-1})$
γ_2	$\frac{1}{2}(a_2 - a_{-2})$
α_1	$-\frac{3}{2^7}a_0 + \frac{15}{2^5}(a_1 + a_{-1}) + \frac{1}{2^5}(a_2 + a_{-2}) - \frac{1}{2^5}u_5$
α_2	$-\frac{3}{2^7}a_0 + \frac{1}{2^5}(a_1 + a_{-1}) + \frac{15}{2^5}(a_2 + a_{-2}) - \frac{1}{2^5}u_5$
$\beta_1 = -\beta_2$	$\frac{1}{2^5}(a_1 + a_{-1} - a_2 - a_{-2} + u_5)$

8.9 Associators

Let V be a real vector space endowed with a non-zero symmetric positive definite inner product $\langle \,,\, \rangle$ and a commutative algebra product $\circ$. Suppose that $\circ$ is associative with respect to $\langle \,,\, \rangle$ in the sense that

$$\langle u \circ v, w \rangle = \langle u, v \circ w \rangle$$

for all $u, v, w \in V$, and let H be an automorphism group of $(V, \langle \,,\, \rangle, \circ)$. Recall that *Norton's inequality* requires that

$$\langle u^2, v^2 \rangle \geq \langle u \circ v, u \circ v \rangle$$

for all $u, v \in V$.

Lemma 8.9.1 *The validity of Norton's inequality is not effected by independent rescaling of* $\langle \,,\, \rangle$ *and* $\circ$.

Proof. The claim follows from the observation that both sides of the Norton inequality are linear in $\langle \,,\, \rangle$ and quadratic in $\circ$. $\qquad \square$

Lemma 8.9.2 *Suppose that Norton's inequality holds. Then:*

(i) *all the nilpotent elements of* $(V, \circ)$ *are contained in the radical;*
(ii) *the angle between any two idempotents is sharp;*
(iii) *if* V *contain a non-zero idempotent and* H *is finite, then* H *fixes a 1-subspace in* V.

Proof. If u is non-radical nilpotent element in V, then for a suitable $v \in V$ we have

$$0 < \langle u \circ v, u \circ v \rangle \le \langle u^2, v^2 \rangle = \langle 0, v^2 \rangle = 0,$$

which is impossible and (i) follows. If u and v are non-zero idempotents, then

$$0 < \langle u \circ v, u \circ v \rangle \le \langle u^2, v^2 \rangle = \langle u, v \rangle,$$

since $\langle \, , \, \rangle$ is positive definite, which gives (ii). Finally, let u be a non-zero idempotent. Then by (ii) the vector $\sum_{h \in H} h(u)$, which is clearly H-stable is non-zero, and (iii) follows. $\qquad \square$

For $u, v, w \in V$, define the *associator* of the ordered triple (w, u, v) to be

$$a(w, u, v) = (w \circ u) \circ v - (w \circ v) \circ u.$$

The following lemma is an immediate consequence of the bilinearity of the product $\circ$:

Lemma 8.9.3 *The following assertions hold:*

(i) *for any given* $u, v \in V$, *the mapping*

$$a(\bullet, u, v) : w \mapsto a(w, u, v)$$

is linear on V;
(ii) *the mapping* $V \times V \to \operatorname{End}(V)$ *defined via*

$$(u, v) \mapsto a(\bullet, u, v)$$

is bilinear and antisymmetric. $\qquad \square$

Let $A(V)$ be the subspace in $\operatorname{End}(V)$ spanned by the *associator operators* $a(\bullet, u, v)$ taken for all $u, v \in V$. By (8.9.3 (ii)), $A(V)$ is a quotient of the exterior square of $A(V)$. Let β_A and q_A be the forms on $V(A)$ defined via

$$\beta_A(a(\bullet, u, v), a(\bullet, x, y)) = \langle a(u, x, y), v \rangle,$$

$$q_A(a(\bullet, u, v)) = \beta_A(a(\bullet, u, v), a(\bullet, u, v)) = \langle a(u, u, v), v \rangle.$$

Lemma 8.9.4 *The following assertions hold:*

(i) β_A *is a symmetric bilinear form on $A(V)$;*
(ii) q_A *is a quadratic form on $A(V)$, whose associated bilinear form is β_A;*
(iii) *Norton's inequality holds if and only if q_A is non-negative definite.*

Proof. Applying the associativity of $\langle\ ,\ \rangle$ with respect to $\circ$ we obtain

$$\beta_A(a(\bullet, u, v), a(\bullet, x, y)) = \langle (u \circ x) \circ y, v \rangle - \langle (u \circ y) \circ x, v \rangle =$$
$$\langle (x \circ u) \circ v, y \rangle - \langle (x \circ v) \circ u, y \rangle = \beta_a(a(\bullet, x, y), a(\bullet, u, v)),$$

so that β_A in indeed symmetric, while it's bilinearity follows from that of $\langle\ ,\ \rangle$ and $\circ$.

For $u, v \in V$, the Norton inequality requires

$$\langle u^2, v^2 \rangle - \langle u \circ v, u \circ v \rangle \geq 0,$$

which is equivalent to

$$\langle u, u \circ (v \circ v) - v \circ (u \circ v) \rangle \geq 0,$$

which is precisely

$$\langle u, a(v, v, u) \rangle \geq 0.$$

and (iii) follows. $\qquad\qquad\square$

The following assertion was established in Sections 16 and 17 of [C84].

Proposition 8.9.5 *Norton's inequality holds for the $196\,884$-dimensional Griess algebra on $\Pi \oplus \Pi_1$.*

Proof. If $V = \Pi \oplus \Pi_1$, then, calculating with the characters, we conclude that the exterior square $\bigwedge^2 V$ is the sum of three distinct irreducibles $V^{(1)}$, $V^{(2)}$, and $V^{(3)}$ of degree

$$196\,883, \quad 21\,296\,876, \quad \text{and } 1\,936\,002\,527,$$

respectively. It is claimed that $V^{(1)} + V^{(3)}$ is the kernel of the homomorphism $\rho : \bigwedge^2 V \to A(V)$. For the case of $V^{(1)}$, the assertion is easy since $V^{(1)}$ is spanned by the vectors $\pi_1 \wedge \pi$ taken for all $\pi \in \Pi$, where $0 \neq \pi_1 \in \Pi_1$. Since π_1 associates with every vector in the algebra, $V^{(1)} \leq \ker(\rho)$. For the case of $V^{(3)}$, the argument the runs as follows. Let $\lambda, \mu \in \Lambda_2$ and $(\lambda, \mu)_\Lambda = 0$. Then the weight of subspaces $\Pi_{\bar\lambda}$ and $\Pi_{\bar\mu}$ are spanned by Majorana axial vectors $a(\vartheta)$ and $a(\theta)$ corresponding to a pair of $2A$-involutions ϑ and θ, whose product is a $2B$-involution. By (8.3.4), $a(\vartheta)$ and $a(\theta)$ annihilate each other, in particular they associate and hence $a(\vartheta) \wedge a(\theta)$ is in the kernel of ρ. Counting the number of ('linearly independent') choices for λ and μ we get

$$\frac{1}{2}|\Lambda_2| \cdot 46\,575 > 196\,883 + 21\,296\,876$$

and hence $A(V) = V^{(2)}$ (notice that $A(V)$ is non-trivial since the algebra is non-associative). Since $A(V)$ is irreducible, in order to apply (8.9.4 (iii)) it is sufficient to check that q_A takes on some positive values, which can easily be done. $\qquad\square$

An alternative proof of (8.9.5) was given in [Miy96] in the context of Vertex Operator Algebras. The proof is based on the existence of a positive definite form on the whole Fock space together with the fact that the space of level 3 operators in the Moonshine module is isomorphic to

$$\Pi_1 \oplus \Pi \oplus A(\Pi).$$

9

The Monster graph

In Chapter 7 we accomplished the construction of the Monster as the image $\varphi(G)$ of the minimal non-trivial (196 883-dimensional) representation of the universal completion G of the Monster amalgam $\mathcal{M}$. We have seen (cf. Sections 6.6 and 7.5) that G is a finite simple group possessing

$$\varphi(G_1) \cong G_1 \cong 2_+^{1+24}.Co_1 \text{ and } \varphi(F_{2+}) \cong F_{2+} \cong 2 \cdot BM$$

as involution centralizers. In this chapter we finish the uniqueness part of the project by showing that $\varphi(G) = G$. We achieve this by establishing the simple connectedness of coset geometry associated with the embedding of the Monster amalgam $\mathcal{M}$ into $\varphi(G)$.

9.1 Collinearity graph

At first glance it appears easier to deal with the collinearity graph of the tilde geometry associated with a faithful completion of $\mathcal{M}$ since its local structure possesses a natural description in terms of the Leech lattice. Let us start by recalling some basic properties of the collinearity graph.

Let $\psi : \mathcal{M} \to H$ be a faithful generating completion of the Monster amalgam. Then H is also a completion group of the enriched Monster amalgam, and the coset geometry associated with the embedding into H of the amalgam

$$\{G_1, G_2, G_3, G_4, G_5^{(t)}\}$$

is a rank 5 tilde geometry $\mathcal{G}(H)$ with the diagram

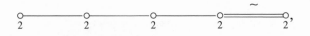

on which H acts as a flag-transitive automorphism group. Let us call the elements corresponding to the leftmost and the second left nodes on the diagram *points* and *lines* of $\mathcal{G}(H)$, respectively. Let Θ be the *collinearity graph* of $\mathcal{G}(H)$ which is the graph on the point-set in which two points are adjacent whenever they are incident to a common line. It can be seen from the diagram of $\mathcal{G}(H)$ that every line is formed by a triple of points and hence such a line corresponds to a triangle in Θ. The line-triangles passing through a given vertex x correspond to the points of the residue of x in $\mathcal{G}(H)$: the rank 4 tilde geometry $\mathcal{G}(Co_1)$ of the first Conway group (cf. Section 4.12 in [Iv99]).

The points of $\mathcal{G}(Co_1)$ are the elements of $\bar{\Lambda}_4$ and the lines are the images under Co_1 of the triple

$$\left\{ 8p + 2\lambda, \sum_{q \in T} 4q + 2\lambda, -8p + \sum_{q \in T} 4q + 2\lambda \right\},$$

where $p \in T \subset \mathcal{P}$ and T is a tetrad. The following result can be deduced directly from the structure of the parabolics G_1 and G_2 (compare Lemma 5.3.1 in [Iv99]).

Lemma 9.1.1 *Let $\psi : \mathcal{M} \to H$ be a faithful generating completion of the Monster amalgam $\mathcal{M}$. Let $\Theta = \Theta(H)$ be the collinearity graph of the tilde geometry $\mathcal{G}(H)$ associated with the embedding $\mathcal{M}$ in H. Let x be a vertex of Θ stabilized by $\psi(G_1) \cong G_1$ in the natural action of H on Θ and let $\Theta(x)$ be the set of neighbours of x in Θ. Then there is a $\psi(G_1)$-mapping η of $\Theta(x)$ onto the point-set of the residual tilde geometry $\mathcal{G}(Co_1)$ of the Conway group such that:*

(i) *for $y, z \in \Theta(x)$ with $y \neq z$, the equality $\eta(y) = \eta(z)$ holds if and only if $\{x, y, z\}$ is a line of $\mathcal{G}(H)$ (equivalently if and only if $\{y, z\}$ is a $O_2(\psi(G_1))$-orbit);*

(ii) *two distinct vertices $y, z \in \Theta(x)$ are adjacent in Θ if and only if either $\eta(y) = \eta(z)$, or $\eta(y)$ and $\eta(z)$ are adjacent in $\mathcal{G}(Co_1)$.* □

By the above lemma, the local structure of $\Theta(H)$ is independent of the particular choice of the faithful generating completion H. Therefore, the graph covering $\zeta : \Theta(G) \to \Theta(H)$ induced by the homomorphism $G \to H$ of completion groups is a *local isomorphism* in the sense that ζ is an isomorphism when restricted to the subgraph induced by $\{x\} \cup \Theta(x)$ for a vertex x of $\Theta(G)$.

Proposition 9.1.2 *The graph $\Theta(\varphi(G))$ is simply connected in the sense that its triangles generate the fundamental group.*

From what we have seen above, (9.1.2) implies that $\zeta : \Theta(G) \to \Theta(\varphi(G))$ is an isomorphism and hence $\varphi(G) = G$ is the universal completion of $\mathcal{M}$. Since $\varphi(G)$ is simple, it must be the only non-trivial completion.

9.2 Transposition graph

The *transposition graph* Γ of the Monster (or simply the *Monster graph*) is a graph on the conjugacy class of $2A$-involutions in the Monster (the centralizer of such an involution is $F_{2+} \cong 2 \cdot BM$) in which two vertex-involutions are adjacent if and only if their product is again a $2A$-involution. The transposition graph has played an essential role in the uniqueness paper [N82]; where the principal steps in the proof of the following theorem were shown.

Theorem 9.2.1 *Any group where the centralizers of elements of prime order are isomorphic to those in the Fischer–Griess Monster has a* 196 883-*dimensional representation.* □

This theorem justifies the most crucial assumption in Thompson's uniqueness paper [Th79a].

It has been already mentioned in Section 8.7 that the permutation rank of the Monster acting by conjugation on the set of $2A$-involutions is nine and that the orbit containing a given pair of such involutions is uniquely determined by the conjugacy class containing their product. The nine classes and the corresponding 2-point stabilizers are given in the following tables.

K	$1A$	$2A$	$2B$	$3A$	$3C$
Stabilizer	F_{2+}	$2^2.(^2E_6(2))$	$2.2_+^{1+22}Co_2$	Fi_{23}	F_3

K	$4A$	$4B$	$5A$	$6A$
Stabilizer	$2_+^{1+22}.McL$	$2.F_4(2)$	F_5	$2.Fi_{22}$

The Monster graph Γ has diameter 3. If $\Gamma_i(a)$ denotes the set of vertices at distance i in Γ from a given $a \in 2A$ and $\Gamma_K(a) = \{b \in 2A \mid ab \in K\}$ for a conjugacy class K, then

$$\Gamma_0(a) = \{a\} = \Gamma_{1A}(a), \quad \Gamma_1(a) = \Gamma_{2A}(a);$$
$$\Gamma_2(a) = \Gamma_{2B}(a) \cup \Gamma_{3A}(a) \cup \Gamma_{4A}(a)\Gamma_{4B}(a) \cup \Gamma_{6A}(a);$$
$$\Gamma_3(a) = \Gamma_{3C}(a) \cup \Gamma_{5A}(a).$$

The intersection parameters p_{KL} are defined as

$$p_{KL} = \{c \mid c \in 2A, \ ac \in 2A, \ bc \in K\},$$

where $a, b \in 2A$, $ab \in L$ and they are commonly arranged into a matrix. The intersection matrix of the Monster graph is given in Table 4 in [N82]. This matrix is too massive to be reproduced here and the calibre of the entries can be grasped from the number of vertices and the valency of Γ which are

$$97\,239\,461\,142\,009\,186\,000 \text{ and } 27\,143\,910\,000,$$

respectively. The former number multiplied by the order of F_{2+} gives the order of the Monster.

The general theory of association schemes and multiplicity-free permutation representations [BI84] supplies explicit formulas for degrees of primitive idempotents of Bose–Mesner algebras. The degrees are precisely the ranks of the irreducible constituents in the corresponding multiplicty-free permutational character. This is the way it was shown in [N82] that one of the degrees is the required number 196 883.

Projecting the elements of the standard basis of the permutation module to the 196 883 irreducible constituent (compare Section 5.3), we obtain the principal component of the Majorana axial vectors. The inner products ρ_K of these projections depend only on the class K containing the product. The inner products are computable from the eigenvalues of the intersection matrix, and are given in the following table. We can notice that they fit perfectly the inner products allowed by Sakuma's theorem. In a sense the purpose of increasing the dimension by one is to validate Norton's inequality.

K		$1A$	$2A$	$2B$	$3A$	$3C$	$4A$	$4B$	$5A$	$6A$
ρ_K		752	80	-16	23	-4	8	-4	2	1
$\lambda_K = \frac{\rho_K + 16}{768}$		1	$\frac{1}{2^3}$	0	$\frac{13}{2^8}$	$\frac{1}{2^6}$	$\frac{1}{2^5}$	$\frac{1}{2^6}$	$\frac{3}{2^7}$	$\frac{5}{2^8}$

It should be emphasized that the inner products ρ_K were obtained in [N82] as a result of extensive calculations with the intersection matrix which has to be calculated first, while Majorana's inner products are universal and relatively easy to calculate. Anyone experienced in calculating with intersection matrices of association schemes and their eigenvalues knows that any partial information on the intersection numbers does not allow to evaluate the inner product.

In [N82] S. Norton (who credited B. Fischer and J. Thompson for earlier partial data) also describes the orbits of the Monster on the triples (a, b, c) of $2A$-involutions such that a and b are adjacent in the Monster graph (which means that $ab \in 2A$). Being inspired by Sakuma's theorem we might ask whether the relevant 37 three-generated Majorana algebras exhaust the set of all such algebras containing a two-generated subalgebra of type $2A$.

9.3 Simple connectedness

My proof of the simple connectedness result (9.1.2) for the tilde geometry of the Monster as presented at the July 1990 Durham conference, comprised of the following three ingredients:

Ingredient I. It was shown (cf. [Iv92] and Section 5.12 in [Iv99]) that the transposition graph Γ can be recovered from the collinearity graph of the tilde geometry and that the completion homomorphism $G \to \varphi(G)$ induces a local isomorphism $\Gamma(G) \to \Gamma(\varphi(G))$ of the corresponding 'transposition' graphs. Thus, in order to settle the simple connectedness question it is suffices to prove the following statement:

Proposition 9.3.1 *The Monster graph $\Gamma(\varphi(G))$ is simply connected in the sense that its triangles generate the fundamental group.* $\square$

If H is a faithful generating completion of the enriched Monster amalgam, then $\Gamma(H)$ can be defined as a graph on the set of cosets in H of the image of $F_{2+} \cong 2 \cdot BM$ in which two coset-vertices are adjacent whenever they intersect a common coset of the image of $^2G_2 \cong 2^2.(^2E_6(2)) : S_3$.

Ingredient II. By comparing the intersection parameters of the Monster graph against the 37 triple stabilizers in [N82], we can easily deduce the following:

Proposition 9.3.2 *Let Σ be a cycle in the transposition graph of the Monster such that the distances between any two vertices from Σ calculated in Σ and in the whole of Γ coincide. Then there is an involution in the Monster which fixes Σ vertex-wise.* $\square$

It is immediate from (9.3.2) that the fundamental group of the Monster graph is generated by the cycles fixed by involutions. A $2A$-involution a is a vertex of the Monster graph Γ and the vertices it fixes are precisely those in

$$\{a\} \cup \Gamma_{2A}(a) \cup \Gamma_{2B}(a).$$

The generator z of the centre of G_1 is a $2B$-involution and G_1 acts on the set of vertices fixed by z with two orbits, which correspond to the $2A$-involutions contained in Q_1 and in $G_1 \setminus Q_1$, respectively. This can be seen from the following assertion established in [MSh02] under the assumption that G_1 is the centralizer of an involution in a finite simple group.

Proposition 9.3.3 *Adopt the bar convention for the images in $\bar{G}_1 = G_1/Q_1 \cong Co_1$ of elements and subgroups of G_1. Then G_1 contains precisely seven conjugacy classes of involutions with representatives e_i, $1 \leq i \leq 7$, such that:*

(i) *$e_1, e_2, e_3 \in Q_i$ with e_1 being the generator of Z_1, $\chi(e_2) \in \bar{\Lambda}_2$, and $\chi(e_3) \in \bar{\Lambda}_4$;*
(ii) *$\bar{e}_4, \bar{e}_5$, and $\bar{e}_6$ are 2-central in $\bar{G}_1$ with $C_{\bar{G}_1}(\bar{e}_i) \cong 2^{1+8}_+.\Omega^+_8(2)$ for $4 \leq i \leq 6$, while $C_{\bar{G}_1}(\bar{e}_7) \cong 2^{11}.\mathrm{Aut}(M_{12})$;*
(iii) *$e_5 = e_1 e_4$ and*

$$C_{G_1}(e_4) \cong C_{G_1}(e_5) \cong 2.2^{16}.2^{1+8}_+.\Omega^+_8(2);$$

(iv) *e_2 and e_4 are $2A$-involutions, while the other five representatives are $2B$-involutions.* □

Ingredient III. The final ingredient is not hard to imagine.

Proposition 9.3.4 *The local isomorphism $\Gamma(G) \to \Gamma(\varphi(G))$ induced by the completion isomorphism $G \to \varphi(G)$ is an isomorphism when restricted to a connected component of the subgraph induced by the vertices fixed by z or by a (where the latter element generates the centre of F_{2+}).* □

The isomorphisms $F_{2+} \cong \varphi(F_{2+})$ and $G_1 \cong \varphi(G_1)$ immediately show that the restrictions in (9.3.4) are bijective but some further efforts are required to show that they are actually graph isomorphisms.

9.4 Uniqueness systems

When in June 1990 I met Geoffrey Mason at the Free University of Berlin and told him about my simple connectedness results for the large sporadic simple groups, he immediately suggested that these brought me just one step away from the uniqueness proof for those groups. One month later at the Durham conference I learned about the *uniqueness systems* by M. Aschbacher and Y. Segev which is a machinery for deducing uniqueness of groups from the simple connectedness of specific geometries. Within the uniqueness system approach, (9.3.1) was proved in [ASeg92] extracting the necessary information on the transposition graph from [GMS89]. Thus for the proof of (9.1.2) already the first ingredient (amounting to (9.3.2)) becomes sufficient.

It might be desirable to deduce the whole structure of the transposition graph directly from the Monster amalgam (this way the simple connectedness would be achieved as a byproduct). This project was fully implemented in [Iv04] for the fourth Janko group with the details being so complicated that I would have difficulty naming a living person who read it all through. In the Monster case the local structure of the Monster graph was deduced directly from the Monster amalgam in Section 5.12 in [Iv99]. Having recovered F_3 and F_5 in Sections 7.7 and 7.8 we can easily reconstruct the right suborbits at distance 3 from a given vertex. With some extra work (considering the subgraphs induced by the vertices fixed by involutions or otherwise) it can be shown that there are no further suborbits at this or higher distances. I believe that by developing the Majorana theory we might expect a significant simplification of the construction.

Fischer's story

At the April 2008 Oberwolfach 'Groups and Geometry' conference, the news that John Thompson and Jacques Tits had been awarded the Abel prize was still fresh and certainly extremely welcome. A special evening session took place, where Bernd Fischer and Richard Weiss shared their professional and personal enthusiasm for our prize-winning colleagues (who were not present at the conference). A particularly memorable part of the evening was Fischer's account of his collaboration with John Thompson on the early stages of studying the Monster. After the session a common opinion was expressed by Martin Liebeck, one of the organizers, that these recollections must be recorded and passed on to the next generation of finite group theorists. Bernd kindly gave me a more detailed account of his story about how the Monster was discovered. Below is my written version of this.

As a warm-up, Bernd browsed through a draft of this volume. Everything appeared to be very familiar to him except for the newly introduced terms like *anti-permutation module*, *trident group*, and a few others which I had to explain. Then he mentioned two further stories involving the new Abel prize winners. The first one concerned a time when Jacques Tits visited Bernd in Bielefeld during the first half of 1970s and was shown Fischer's generators for the Monster satisfying the relations as on the famous picture given below. He wondered whether the non-identity central element of the D_6-subgroup generated by the elements in the top left corner is equal to the generator in the right bottom corner of the hexagon. Bernd confirmed that indeed these elements are equal in the Monster. Tits conjectured that these properties would then be sufficient to define the monster.[1]

---

[1] After the proof of the Y-conjecture (cf. Chapter 8 in [Iv99] and references therein) it became certain that the corresponding relation together with the Coxeter relations provide a presentation for the Monster.

The second story goes back to 1978 when John Thompson gave his lecture at Princeton on his uniqueness proof for the Monster published in [Th79a]. Immediately after the lecture Bernd asked Thompson about the possibility of turning the uniqueness proof into a construction, but Thompson responded that this would be far too complicated.[2]

Then Fischer turned to the central topic: the discovery of the Monster group. Fischer said that his knowledge at that time (1973) concerned the Baby Monster and its 'essential' subgroups generated by the natural involutions (which are $\{3, 4\}$-transpositions) including $Fi_{23} < BM$. Before showing that Fi_{23} is contained in the Baby Monster, Fischer had established the embedding

$$Fi_{22} <\, ^2E_6(2)$$

which was a true sensation. In 1970 Fischer gave a series of six lectures on this topic at Bowden College in Maine. Walter Feit had noticed that in the longer induced sequence

$$\Omega_7(3) < Fi_{22} <\, ^2E_6(2)$$

every group has Sylow 3-subgroups of order 3^9. Therefore, the well-known fact that $^2E_6(2)$ has a non-trivial 3-part in the Schur multiplier would (by Gaschütsz theorem) be a non-trivial 3-part in the Schur multiplier of $\Omega_7(3)$. Thus the possibility of the embedding $Fi_{22} <\, ^2E_6(2)$ was questioned because of an unpublished result by Robert Steinberg claiming that the Schur multiplier of $\Omega_7(3)$ was classical (therefore with the 3-part being trivial). Steinberg looked

---

[2] We know that Bob Griess did this some two years later [Gri82a]. The construction presented in this volume probably bears even stronger resemblance to Thompson's uniqueness proof.

back at his notes and found that this particular case (3-part of the Schur multiplier of $\Omega_7(3)$) had not been covered in the written proof and hence Fischer went on with his lectures.

Here is a brief review of Fischer's construction of Fi_{22} inside $^2E_6(2)$. The conjugacy class D of 2-central involutions in $^2E_6(2)$ is a class of $\{3,4\}^+$-transpositions. This means that for distinct $a, b \in D$ the product ab has order 2, 3, or 4 and if it is 4, then $(ab)^2 \in D$. If $d \in D$, then the set D_d of transpositions from D distinct from d and commuting with d generate the centralizer $C_{2E_6(2)}(d)$ of the form $2_+^{1+20} : U_6(2)$. Take $x \in D$ such that $(dx)^3 = 1$ and put $K = \langle D_d \cap D_x \rangle$ so that $K \cong U_6(2)$ is a Levi complement in the centralizer of d. Modulo $\langle d \rangle$ centralizer contains three further classes of complements and a complement K_1 can be chosen so that $K_1 \cap K \cong \Omega_6^-(3).2$ (here K_1 is a subgroup of the centralizer of of d, such that $d \in K_1$, $K_1/\langle d \rangle \cong U_6(2)$, and K_1 does not split over $\langle d \rangle$). It is shown that

$$\langle K_1, x \rangle \cong Fi_{22}.$$

This is achieved in a geometrical way through considering the following subset of D:

$$E = \{d\} \bigcup (K_1 \cap D_d) \bigcup x^{K_1}$$

(with the terms in the union having size 1, 693, and 1408, respectively). It was shown via geometrical considerations that E is closed under conjugation by its elements and therefore it forms a conjugacy class (of 3-transpositions) in the subgroup it generates.

Timmesfeld worked on $\{3,4\}^+$-transpositions and Fischer turned to the general classification problem of $\{3,4\}$-transposition groups. If G is such a group, D is a class of $\{3,4\}$-transpositions, and if $d \in D$, then it is common to consider the partition

$$D = \{d\} \cup D_d \cup A_d \cup V_d,$$

where D_d are the transpositions distinct from d and commuting with d, while A_d and V_d are the transpositions commuting with d products of order 3 and 4, respectively. The set D_d is further subdivided into two subsets, say X_d and Y_d so that $D_d = X_d \cup Y_d$, $X_d \in O_2(C_G(d))$ in Timmesfeld's case $X_d = \{d^v | v \in V_d\}$. The case when Y_d is empty is covered by the $\{3,4\}_+$-transposition groups completely classified by Franz Timmesfeld, a former student of Fischer.

Aiming to accomplish the generic case we should first build up a reliable list $\mathcal{L}$ of examples to be as complete as possible (so as not to run into a completely new example in the middle of the proof). The classification itself is divided into two cases: (1) where $C_G(d)$ has a large normal 2-subgroup, and (2) where

$C_G(d)$ is 'essentially' simple. The latter case is where new examples were most likely to appear and one of the subcases to consider was where the simple group involved in $C_G(d)$ is $^2E_6(2)$ (this is precisely where the Baby Monster emerged).

It has been already mentioned that $^2E_6(2)$ contains a class of 2-central involutions with centralizer $2^{1+20}_+.U_6(2)$ which is a class of $\{3, 4\}_+$-transpositions. Another class of $\{3, 4\}$-transpositions (which does not satisfy the $+$ condition) can be found in the extension of $^2E_6(2)$ by an outer automorphism of order 2. The centralizer in $^2E_6(2)$ of such an outer $\{3, 4\}$-transposition is $F_4(2)$. For the construction to be successful both the inner and outer classes of $\{3, 4\}$-transpositions must be taken to form the set D_d.

The action of $C_G(d) \cong 2 \cdot {}^2E_6(2) : 2$ on A_d must be similar to its action on the cosets of a Fi_{22}-subgroup (that is why it was important that this subgroup was revealed in the first place). The set X_d is empty and the squares of the order 4 products of transpositions form a class of 2-central involutions with the centralizer of the form

$$2^{1+22}_+.Co_2.$$

Finally, the centralizer in $C_G(d)$ of a transposition from $V_d = Y_d$ is of the form $2^{1+20}_+.U_4(3).2^2$. This makes it possible to calculate the size of D as the sum of five indices

$$|D| = 1 + [{}^2E_6(2) : 2^{1+20}_+.U_6(2)] + [{}^2E_6(2) : F_4(2)]$$
$$+ [{}^2E_6(2) : Fi_{22}] + [{}^2E_6(2) : 2^{1+20}_+.U_4(3)].$$

Fischer mentioned that his wife helped him to sum up the five numbers to obtain

$$|D| = 13\,571\,955\,000$$

which gives for the group order the number

$$|2 \cdot {}^2E_6(2) : 2| \cdot |D| = 2^{41} \cdot 3^{13} \cdot 5^6 \cdot 7^2 \cdot 11 \cdot 13 \cdot 17 \cdot 19 \cdot 23 \cdot 31 \cdot 41$$

(the order of the Baby Monster group).

After that (in 1973) Fischer tried to perform another extension but soon realized that it is impossible to accomplish the extension staying in the class of $\{3, 4\}$-transposition groups. It turned out that such a extension can be constructed in the class of $\{3, 4, 5, 6\}$-transposition groups and this is precisely the Monster group.

In fact Fischer traced back the first shadow of the Monster to the Oxford conference in 1968 when, as a joke, Conway noticed that Fischer's newly constructed sporadic simple group group Fi'_{24} was larger than his group Co_1. He suggested a proper way to resolve this tension is to construct a larger group which would contain both Co_1 and Fi'_{24}. Each of Co_1 and Fi_{24} contains an extension of 2^{11} by M_{24} where 2^{11} is the irreducible Golay code module in Co_1 and the irreducible Todd module in Fi'_{24}. So we might start thinking what a 2-local containing both the extensions might look like. In 1968, this did not go very far and nothing happened till five years later. The stronger evidence for the existence of the Monster started emerging rapidly during a group theory conference in Bielefeld in November 1973, which was attended by a number of mathematicians from Cambridge, including John Thompson, John Conway, Koichiro Harada, John Walter and Simon Norton.[3] Bernd remembered that when he arrived at the lecture room, Harada was reporting on his recent non-existence result for a particular involution centralizer problem. It is noticeable that the configuration Harada was illuminating turned out to be rather similar to the one which is realized in the Harada–Norton group discovered a few weeks later. Already at that time Fischer had some evidence that such a configuration might exist. In fact, a group having the centralizer of an order 2 subgroup (modulo this subgroup itself) isomorphic to the Baby Monster must have involution centralizer of the form $2^{1+24}_+.Co_1$ (since in the Baby Monster it is $2^{1+22}_+.Co_2$). Considering further the centralizer of an element of order 5 we arrive at the Harada–Norton involution centralizer configuration.[4]

After Fischer had revealed the order of the Baby Monster at the Bielefeld 1973 conference he was asked to carry on speaking. Soon after that, Fischer himself went to Cambridge and, as he put it, 'during the next fortnight everything happened'. By this he presumably meant that properties of the Monster and related new sporadic simple groups were discovered.

The advantage of working in Cambridge was that they had computed the character table of the Conway group Co_1, and in particular they knew the classes of the elements in this group. Making use of this information, it became possible to calculate the centralizers in

$$C(2B) \cong 2^{1+24}_+.Co_1$$

---

[3] When asked whether Simon was rather young that time Fischer responded 'Yes, he was much younger, but he always was the same person'.

[4] Compare Section 7.8.

of various odd order elements. It was known by Cambridge people that Co_1 contains an element of order 3 centralized by A_9, whose centralizer in 2_+^{1+23} is an extraspecial group.[5] This gives

$$(C(2B) \cap C(3C))/3C \cong 2_+^{1+8}.A_9.$$

Starting with these data, Thompson had deduced that $C(3C)/3C$ must be a new finite simple group now known as the Thompson group F_3. During this fortnight when Fischer was in Cambridge, Thompson found the order of F_3 as well as the degree 248 of its minimal representation. He made extensive use of his exclusive knowledge of the group $G_2(3)$ and a special version of a very natural $G_2(2)$-subgroup in it.[6]

In a similar way, working with a suitable 5-element in Co_5 we obtain

$$(C(2B) \cap C(5A))/5A \cong 2_+^{1+8}.(A_5 \times A_5).2,$$

which is the involution centralizer of the Harada–Norton group F_5 also discovered that fortnight. This situation has an additional advantage, since the element which squares the generator of the 5-subgroup induces an outer automorphism of F_5 with centralizer isomorphic to S_{10}. Fischer had proved earlier that a group whose commutator subgroup is simple and which contains involutions centralized by S_{10} must contain S_{12} as a subgroup.

Within the revealing of the subgroup structure of the Monster were a few crucial things, one of them being the structure of the $3A$-normalizer. It became definite rather soon that modulo the normal subgroup of order 3 the normalizer is the largest Fischer 3-transposition group Fi_{24}. Also, because of the way it intersects some other subgroups already identified, the extension must not split. On the other hand, at that time there was a belief [7] that the Schur multiplier of F_{24}^i is trivial. Were this the case, the whole existence of the Monster would be jeopardized. Next morning Thompson resolved the situation in favour of the non-trivial Schur multiplier.

Next Fischer turned to the naming of the newly discovered sporadic groups. When he discovered his sporadic 3-transposition groups, he chose names which reflected the analogy with the series of large Mathieu groups which start with a classical group

$$L_3(4) = M_{21}, \ M_{22}, \ M_{23}, \ M_{24}.$$

[5] Compare Section 7.7.

[6] A progress report on construction and uniqueness proof for F_3 can be found in [Th76]. In particular the step 'Hope for the best' in the outline of his program.

[7] Compare p. 355 in [Gri73b] and pp. 445–446 in [Gri74].

Recall that the next group in this sequence is a transitive extension of the previous one. Fischer gave to his 3-transposition groups the names

$$U_6(2) = M(21), \ M(22), \ M(23), \ M(24).$$

Here also the next group is an extension of the previous one, but in the sense of involution centralizers. The relevance to the Mathieu groups lies in the property that a maximal set of pairwise commuting transpositions in $M(21+i)$ has size $21+i$ and its stabilizer is an extension of an elementary abelian group of order 2^{9+i} by M_{21+i} (the extension splits when $i \leq 1$ and does not split when $i = 2$ or 3). Fischer suggested Conway follow a similar strategy when naming his groups related to the leech Lattice, but obviously Conway went along with his own naming.

In the case of the involution centralizer extensions of $^2E_6(2)$, the following notation was suggested first

$$^2E_6(2) = M^{21}, \ M^{22}, \ M^{23}, \ M^{24}.$$

This is justified by the analogy between $^2E_6(2)$ and $U_6(2)$ (the latter is the simple factor group of the centralizer of a central involution in the former one). But on the one hand $M(22)$ is contained in $^2E_6(2)$ and $M(23)$ is contained in the extensions of $^2E_6(2)$. This suggests the following shifted list of names

$$^2E_6(2) = M^{22}, \ M^{23}, \ M^{24}.$$

Thus the naming involved some ambiguity, since $^2E_6(2)$ can be called M^{21} or M^{22} according to which system of names is accepted. In order to avoid this ambiguity in the internal communications, John Conway suggested calling the extensions of $^2E_6(2)$ the *Baby Monster*, the double extension the *Middle Monster*, and the triple extension the *Super Monster*. Fischer was not in favour of this terminology being used externally. When it was shown that $^2E_6(2)$ can only be extended twice and therefore that the Super Monster does not exist, the prefix *Middle* was dropped and the name *Monster*, as we know it, emerged. After Fischer's visit to Cambridge in 1973, John Conway went to the USA to give a series of lectures where he freely used the terms Monster and Baby Monster. This way the names became official.

Later, at a mathematical congress during the early 1980s (it must have been in Helsinki), Bob Griess approached Fischer and suggested renaming the Monster as the *Friendly Giant*. Apparently, there was an encrypted double meaning in the suggested name. In any event, for one reason or another, as we know, this attempt was not successful.

Next, Fischer discussed his role in the computer construction of the Baby Monster. He said it was probably in 1976 when he visited Canberra for five

weeks and at that time Charles Sims was there on a sabbatical for the whole
year. They discussed the possibility of constructing the Baby Monster in the
way the Lyons group had been constructed not much earlier. For such a construction to be successful, a good identification is required of the 3-point
stabilizers and their centralizers. In a situation where the permutation action is
on the set of involutions these subgroups are generated by some special triples
of transpositions. There is an additional advantage in this case: whenever a set
is closed under the 'internal' conjugations, it is a conjugacy class in the subgroup it generates. Since the set of transpositions in the Baby Monster is very
large, we cannot perform usual coset enumerations and Sims said that in order
to perform the calculations he must be able to solve the word problem in a
practical way. Thus he should be able to recognize a permutation by a small
fragment (compare the notion of *base* of a permutation group). Therefore, they
were looking for a subgroup which would be (1) small enough to allow efficient multiplication and recognition procedures and also (2) large enough not
to have too many orbits on the underlying set. Sims believed the subgroup
Fi_{22} in the Baby Monster to be ideal for this purpose. An additional reason in
favour of this subgroup was Conway's description of Fi_{22} in terms of *hexads*,
the latter being a collection of six pairwise commuting transposition products
to the identity element. Sims thought that the permutation action of Fi_{22} on the
hexads was optimal for his word problem. He needed a description of all the
triple centralizers inside the Baby Monster group in terms of the centralizer
of a given triple. There appeared to be about one hundred cases to consider
and some of them were really difficult. Fischer mentioned that particularly bad
things happened inside a Sylow 3-subgroup. This time Fischer also tried to
figure out what it would take to move forward from the Baby Monster to the
Monster. Apparently, the number of cases to be considered increased from one
hundred to a couple of thousands (in the Monster case we should probably take
a Fi_{23}-subgroup instead).

Sims took notes of the information they deduced during Fischer's stay in
Canberra. Some time later Fischer suddenly got the three-page paper on the
construction of the Baby Monster.[8] Hardly any details were included in the
paper, but Fischer presumed Leon and Sims had performed the construction
in the way Sims had originally wanted to do. At the end of the computer construction story, Fischer mentioned that at the 1968 Oxford conference Sims
talked about construction of smaller sporadic groups. Fischer said that during
the lecture he was sitting next to Donald Higman and that the description was

---

[8] J.S. Leon and C.C. Sims, 'The existence and the uniqueness of a simple group generated by
3, 4-transpositions', *Bull. Amer. Math. Soc.* **83** (1977), 1039–1040.

so good that they could check the calculations without a computer. But even compared to the case of the Lyons group, the Baby Monster case was much more elaborate.

The final topic covered by Fischer was about the representations and characters of the Monster. It was realized rather early in 1973 in Cambridge that the Monster has exactly two classes of involutions and that the centralizer of a central involution has the shape

$$C(2B) \sim 2_+^{1+24}.Co_1.$$

The exact isomorphism type of this group took a little while to identify. From the representation-theoretical view point, the difference between the isomorphism types can be seen in the degrees of the minimal representations. By Blichfeldt's theorem (from the Curtis–Reiner textbook), every faithful irreducible representation of the group of that shape has degree equal to 2^{12} times the degree of a projective representation of Co_1. On the other hand, although the Baby Monster itself possesses a rather small representation (of degree 4371), a lower bound for the minimal degree of $C(2A) \cong 2 \cdot BM$ (which is a non-split extension) is about 95,000. This way it was concluded that $C(2B)$ could not possess a 2^{12}-dimensional representation and thus the smallest degree is $2^{12} \times 24$. After that, making use of the fusion, the lower bound 196 883 for the minimal representation of the Monster was deduced.

In 1974 when Fischer came to Birmingham his intention was to compute the character table of the Baby Monster and he thought he had a strong starting point which is the minimal non-trivial character degree, but Donald Livingston suggested calculating the character table of the Monster instead (he believed that the calculations would be easier). The counter argument of Fischer was that they did not even know the degree of the minimal representation, since the 196 883 representation was only a conjecture at that time. Then Livingstone suggested to start calculation assuming that this representation existed. If at the end they were to complete the character table, and it would turn out that there is no such representation, then they would have a table which looked like a character table, meaning that it satisfied all the necessary conditions, but actually not a character table. Livingstone's argument was that such a table might be even more valuable than the character table of the Monster. Thus being joined by Mike Thorne, Fischer and Livingstone launched the project, which was tremendous on any scale and took more than a year to complete.

The method was to get the character tables of the subgroup $2_+^{1+24}.Co_1$ and $2^{10+16}.\Omega_{10}^+(2)$ first using a new method to make Clifford's theorem applicable for groups of this size. Fischer said it took five weeks to do the corresponding Clifford matrices. He said that each one constituted a whole book and invited me to Bielefeld one day to come and see them. The character tables of the sub-groups themselves formed an enormous project to calculate. With computer capacity at that time, it once took six hours just to transfer the data from one computer to another. They determined the fusion pattern of the Baby Monster elements into the Monster, although Fischer mentioned that he had never finished the work on the Clifford matrices in the general setting, and that he was still planning to accomplish this.

At the very end of his story, Fischer said that he had enjoyed all his life but this was a particularly great time.[9]

---

[9] The Oberwolfach archive keeps Fischer's abstract of the January 1974 meeting with the orders of the four new sporadic simple groups: F_1, F_2, F_3, and F_5. This is probably the first official record of the Monster era.

References

[A86] M. Aschbacher, *Finite Group Theory*, Cambridge Univ. Press, Cambridge, 1986.

[A94] M. Aschbacher, *Sporadic Groups*, Cambridge Univ. Press, Cambridge, 1994.

[A97] M. Aschbacher, *3-Transposition Groups*, Cambridge Univ. Press, Cambridge, 1997.

[A01] M. Aschbacher, A characterization of $^2E_6(2)$, in *Advanced Studies in Pure Math.* **32**, pp. 209–244, MSJ, Tokyo, 2001.

[ASeg92] M. Aschbacher and Y. Segev, Extending morphisms of groups and graphs, *Ann. Math.* **135** (1992), 297–324.

[BI84] E. Bannai and T. Ito, *Algebraic Combinatorics I. Association Schemes*, Benjamin, Menlo Park, California, 1984.

[BPZ84] A.A. Belavin, A.M. Polyakov, and A.B. Zamolodchikov, Infinite conformal symmetry in two-dimensional quantum field theory, *Nuclear Physics* **B241** (1984), 333–380.

[Be78] G. Bell, On the cohomology of finite special linear groups I and II, *J. Algebra* **54** (1978), 216–238 and 239–259.

[Bi93] N. Biggs, *Algebraic Graph Theory* (2nd edn.) Cambidge Univ. Press, Cambridge, 1993.

[Bo85] R.E. Borcherds, Vertex algebras, Kac–Moody algebras, and the Monster, *Proc. Nat. Acad. Sci. USA* **83** (1986), 3068–3071.

[CGS78] P.J. Cameron, J.-M. Goethals, and J.J. Seidel, The Krein condition, spherical designs, Norton algebras and permutation groups. *Nederl. Akad. Wetensch. Indag. Math.* **40** (1978), 196–206.

[Car02] R.W. Carter, *Representations of the Monster*, Dept. Mat., Univ. de Coimbra **33**, Portugal, 2002.

[Con71] J.H. Conway, Three lectures on exceptional groups, in *Finite Simple Groups* (M.B. Powell and G.Higman, eds.), pp. 215–247, Acad. Press, New York, 1971.

[C84] J.H. Conway, A simple construction for the Fischer–Griess monster group, *Invent. Math.* **79** (1984), 513–540.

[CCNPW] J.H. Conway, R.T. Curtis, S.P. Norton, R.A. Parker, and R.A. Wilson, *Atlas of Finite Groups*, Clarendon Press, Oxford, 1985.

[CN79]   J.H. Conway and S.P. Norton, Monstrous moonshine, *Bull. London Math. Soc.* **11** (1979), 308–339.

[CNS88]   J.H. Conway, S.P. Norton, and L.H. Soicher, The bimonster, the group Y_{555} and the projective plane of order 3, in *Computers in Algebra*, pp. 27–50, Dekker, New York, 1988.

[CS88]   J.H. Conway and N.J.A. Sloane, *Sphere Packing, Lattices and Groups*, Springer-Verlag, New York 1988.

[CR62]   C.W. Curtis and I. Reiner, *Representation Theory of Finite Groups and Associative Algebras*, John Wiley & Sons, New York, 1962.

[DGM89]   L. Dolan, P. Goddard, and P. Montague, Conformal field theory of twisted vertex operators, *Nuclear Physics* **B338** (1990), 529–601.

[DMZ94]   C. Dong, G. Mason, and Y. Zhu, Discrete series of the Virasoro algebra and the moonshine module, in *Algebraic Groups and their Generalizations: Quantum and Infinite-dimensional Methods*, pp. 295–316, *Proc. Sympos. Pure Math.*, **56** Part 2, AMS Providence, RI, 1994.

[DG98]   C. Dong and R.L. Griess, Rank one lattice type vertex operator algebras and their automorphism groups. *J. Algebra* **208** (1998), 262–275.

[F69]   B. Fischer, *Finite Groups Generated by 3-transpositions*, Univ. of Warwick Lecture Notes, 1969.

[F71]   B. Fischer, Finite groups generated by 3-transpositions, *Invent. Math.* **13** (1971), 232–246.

[FLM84]   I.B. Frenkel, J. Lepowsky, and A. Meurman, A natural representation of the Fischer–Griess Monster with the modular function J as character, *Proc. Nat. Acad. Sci. USA* **81** (1984), 3256–3260.

[FLM88]   I. Frenkel, J. Lepowsky, and A. Meurman, *Vertex Operator Algebras and the Monster*, Acad. Press, Boston, 1988

[Gan06a]   T. Gannon, Monstrous moonshine: the first twenty-five years, *Bull. London Math. Soc.* **38** (2006), 1–33.

[Gan06b]   T. Gannan, *Moonshine Beyond the Monster*, Cambridge Univ. Press, Cambridge, 2006

[GNT98]   A.O. Gogolin, A.A. Nersesyan, and A.M. Tsvelik, *Bosonozation and Strongly Correlated Systems*, Cambridge Univ. Press, Cambridge 1998.

[Gol80]   D. Goldschmidt, Automorphisms of trivalent graphs, *Annals Math.* **111** (1980), 377–406.

[G68]   D. Gorenstein, *Finite Groups*, Harper & Row, New York, 1968.

[GH73]   D. Gorenstein and M.E. Harris, A characterization of the Higman–Sims group, *J. Algebra* (1973), 565–590.

[GL83]   D. Gorenstein and R. Lyons, The local structure of finite groups of characteristic 2 type, *Memoirs AMS* **42** (1983)

[Gri73a]   R.L. Griess, Automorphisms of extra special groups and nonvanishing degree 2 cohomology, *Pacific. J. Math.* **48** (1973), 403–422.

[Gri73b]   R.L. Griess, Schur multipliers of finite simple groups of Lie type, *Trans. Amer. Math. Soc.* **183** (1973), 355–421.

[Gri74]   R.L. Griess, Schur multipliers of some sporadic simple groups. *J. Algebra* **32** (1974), 445–466.

[Gri76]   R.L. Griess, The structure of the 'monster' simple group, in *Proceedings of the Conference on Finite Groups* (Univ. Utah, Park City, Utah, 1975), pp. 113–118, Academic Press, New York, 1976.

[Gri82a]   R.L. Griess, The friendly giant, *Invent. Math.* **69** (1982), 1–102.

[Gri82b]   R.L. Griess, The Monster and its nonassociative algebra, in *Finite Simple Groups, Coming of Age* (J. McKay, ed.), Contemp. Math., **45**, pp. 121–158, AMS, Providence, RI, 1982.

[Gri87]   R.L. Griees, Sporadic groups, code loops and nonvanishing cohomology, *J. Pure and Appl. Algebra* **44** (1987), 191–214.

[Gri88]   R.L. Griess, Code loops and a large finite group containing triality for D_4, in *Group Theory and Combinatorial Geometry, Rend. Circ. Mat. Palermo (2) Suppl.* **19** (1988), 79–98.

[Gri98]   R.L. Griess, *Twelve Sporadic Groups*, Springer-Verlag, Berlin, 1998.

[GMS89]   R.L. Griess, U. Meierfrankenfeld, and Y. Segev, A uniqueness proof for the Monster, *Ann. of Math. (2)* **130** (1989), 567–602.

[H59]   M. Hall, Jr., *The Theory of Groups*, Macmillan Co., New York, 1959.

[HSW00]   G. Havas, L.H. Soicher, and R.A. Wilson, A presentation for the Thompson sporadic simple group, in *Groups and Computation III* (W.M. Kantor and A. Seress, eds.), Ohio State Univ. Math. Res. Inst. Publ. 8, pp. 193–200, de Gruyter, Berlin, 2001.

[HS05]   C. Hoffman and S. Shpectorov, New geometric presentation for Aut $G_2(3)$ and $G_2(3)$, *Bull. Belg. Math. Soc.* **12** (2005), 813–816.

[Iv91]   A.A. Ivanov, A geometric approach to the uniqueness problem for sporadic simple groups, *Dokl. Akad. Nauk SSSR* **316** (1991), 1043–1046. [In Russian]

[Iv92]   A.A. Ivanov, A geometric characterization of the Monster, in *Groups, Combinatorics and Geometry*, Durham 1990, M. Liebeck and J. Saxl eds., LMS Lecture Notes **165**, Cambridge Univ. Press, Cambridge 1992, pp. 46–62.

[Iv93]   A.A. Ivanov, Constructing the Monster via its Y-presentation, in *Paul Erdös is Eighty*, pp. 253–269, Keszthely, Budapest, 1993.

[Iv94]   A.A. Ivanov, Presenting the Baby Monster, *J. Algebra* **163** (1994), 88–108.

[Iv95]   A.A. Ivanov, On geometries of Fischer groups, *Europ. J. Combin.* **16** (1995), 163–183.

[Iv99]   A.A. Ivanov, *Geometry of Sporadic Groups I. Petersen and Tilde Geometries*, Cambridge Univ. Press, Cambridge, 1999.

[Iv04]   A.A. Ivanov, J_4, Oxford Univ. Press, Oxford, 2004.

[Iv05]   A.A. Ivanov, Constructing the Monster amalgam, *J. Algebra* **300** (2005), 571–589.

[ILLSS]   A.A. Ivanov, S.A. Linton, K. Lux, J. Saxl, and L.H. Soicher, Distance-transitive representation of the sporadic groups *Comm. Algebra* **23(9)** (1995), 3379–3427.

[IPS01]   A.A. Ivanov, D.V. Pasechnik, and S.V. Shpectorov, Extended F_4-building and the baby Monster, *Invent. Math.* **144** (2001), 399–433.

[ISh02]   A.A. Ivanov and S.V. Shpectorov, *Geometry of Sporadic Groups II. Representations and Amalgams*, Cambridge Univ. Press, Cambridge, 2002.

[ISh05]   A.A. Ivanov and S.V. Shpectorov, Tri-extraspecial groups, *J. Group Theory*
           8 (2005), 395–413.

[Jack80]   D.J. Jackson, Ph.D. Thesis, Cambridge, circa 1980.

[JW69]   Z. Janko and S.K. Wong, A characterization of the Higman–Sims group, *J.
           Algebra* **13** (1969), 517–534.

[JP76]   W. Jones and B. Parshall, On the 1-cohomology of finite groups of Lie type,
           in *Proc. of the Conference on Finite Groups* (W.R. Scott and R. Gross eds.),
           Academic Press, New York, 1976, pp. 313–327.

[Kac80]   V.G. Kac, A remark on Conways–Norton conjecture about the 'Monster'
           simple group, *Proc. Nat. Acad. Sci. USA* **77** (1980), 5048–5049.

[Ki88]   M. Kitazume, Code loops and even codes over $\mathbb{F}_4$, *J. Algebra* **118** (1988),
           140–149.

[Lie87]   M.W. Liebeck, The affine permutation groups of rank three, *Proc. London
           Math. Soc.* **54** (1987), 477–516.

[M1860]   E. Mathieu, Mémoire sur le nombre de valeurs que peut acquérir une fonc-
           tion quand on y permut ses variables de toutes les manières possibles, *J. de
           Math. Pure et App.* **5** (1860), 9–42.

[M1861]   E. Mathieu, Mémoire sur l'étude des fonctions de plusières quantités, sur
           la manière des formes et sur les substitutions qui les laissent invariables, *J.
           de Math. et App.* **6** (1861), 241–323.

[Ma01]   A. Matsuo, Norton trace formulae for the Griess algebra of a Vertex
           Operator Algebra with large symmetry, *Comm. Math. Phys.* **224** (2001),
           565–591.

[Mei91]   T. Meiner, Some polar towers, *Europ. J. Combin.* **12** (1991), 397–417.

[MSh02]   U. Meirfrankenfeld and S.V. Shpectorov, Maximal 2-local subgroups of the
           Monster and Baby Monster, Preprint 2002.

[MN93]   W. Meyer and W. Neutsch, Associative subalgabras of the Griess algebra,
           J. Algebra **158** (1993), 1–17.

[Miy95]   M. Miyamoto, 21 involutions acting on the moonshine module, *J. Algebra*
           175 (1995), 941–965.

[Miy96]   M. Miyamoto, Griess algebras and conformal vectors in vertex operator
           algebras, *J. Algebra* **179** (1996), 523–548.

[Miy03]   M. Miyamoto, Vertex operator algebras generated by two conformal
           vectors whose τ-involutions generate S_3, *J. Algebra* **268** (2003), 653–671.

[Miy04]   M. Miyamoto, A new construction of the moonshine vertex operator
           algebra over the real number field, *Ann. Math.* **159** (2004), 535–596.

[N82]   S.P. Norton, The uniqueness of the monster, in *Finite Simple Groups,
           Coming of Age* (J. McKay ed.), *Contemp. Math.*, **45**, pp. 271–285, AMS,
           Providence, RI, 1982.

[N96]   S.P. Norton, The Monster algebra: some new formulae, in *Moonshine, the
           Monster and Related Topics*, Contemp. Math. **193**, pp. 297–306, AMS,
           Providence, RI, 1996.

[N98]   S.P. Norton, Anatomy of the Monster I, in *The Atlas of Finite Groups: Ten
           Years On*, LMS Lect. Notes Ser. 249, pp. 198–214, Cambridge Univ. Press,
           Cambridge, 1998.

[NW94]   S.P. Norton and R.A. Wilson, Anatomy of the Monster. II, *Proc. London
           Math. Soc.* **84** (2002), 581–598.

[PW04a]   C.W. Parker and C.B. Wiedorn, A 5-local identification of the monster, *Arch. Math.* **83** (2004), 404–415.

[PW04b]   C.W. Parker and C.B. Wiedorn, 5-local identifications of the Harada Norton group and of the Baby Monster, Preprint 2004.

[Pa94]   A. Pasini, *Diagram Geometries*, Oxford Univ. Press, Oxford 1994.

[R95]   T.M. Richardson, Local subgroups of the Monster and odd code loops, *Trans. Amer. Math. Soc.* **347** (1995), 1453–1531.

[Ron06]   M. Ronan, *Symmetry and the Monster*, Oxford Univ. Press, Oxford 2006.

[RSm80]   M.A. Ronan and S. Smith, 2-Local geometries for some sporadic groups, in *Proc. Symp. Pure Math.* No. 37 (B. Cooperstein and G. Mason eds.), AMS, Providence, R.I., 1980, pp. 283–289.

[Sak07]   S. Sakuma, 6-transposition property of τ-involutions of vertex operator algebras, *International Math. Research Notes* **2007** article rnm030, 19 pages.

[Sc77]   L.L. Scott, Some properties of character products, *J. Algebra* **45** (1977), 253–265.

[Seg91]   Y. Segev, On the uniqueness of Fischer's Baby Monster, *Proc. London Math. Soc.* **62** (1991), 509–536.

[Seg92]   Y. Segev, On the uniqueness of the Harada–Norton group, *J. Algebra* **151** (1992), 261–303.

[Serr77]   J.-P. Serre, *Arbres, amalgams, SL_2*, Astérisque, **46**, 1977.

[Shi07]   H. Shimakura, Lifts of automorphisms of vertex operator algebras in simple current extensions, *Math. Z.* **256** (2007), 491–508.

[Sh88]   S.V. Shpectorov, On geometries with diagram P^n, preprint, 1988 [In Russian]

[Sm77]   S.D. Smith, Nonassociative commutative algebras for triple covers of 3-transposition groups, *Michigan Math. J.* **24** (1977), 273–287.

[Sm79]   S.D. Smith, Large extraspecial subgroups of widths 4 and 6, *J. Algebra* **58** (1979), 251–281.

[Soi89]   L.H. Soicher, From the Monster to the Bimonster, *J. Algebra* **121** (1989), 275–280.

[Soi91]   L.H. Soicher, More on the group Y_{555} and the projective plane of order 3, *J. Algebra* (1991), 168–174.

[Tay92]   D.E. Taylor, *The Geometry of the Classical Groups*. Heldermann Verlag, Berlin, 1992.

[Th76]   J.G. Thompson, A simple subgroup of $E_8(3)$, in *Finite Groups, Sapporo and Kyoto 1974* (N. Iwahori ed.), pp.113–116, JSPS, Tokyo, Japan, 1976.

[Th79a]   J.G. Thompson, Uniqueness of the Fischer–Griess Monster, *Bull. London Math. Soc.* **11** (1979), 340–346.

[Th79b]   J.G. Thompson, Some numerology between the Fischer–Griess Monster and the elliptic modular function, *Bull. London Math. Soc.* **11** (1979), 352–353.

[Th81]   J.G. Thompson, Finite-dimensional representations of free products with an amalgamated subgroup, *J. Algebra* **69** (1981), 146–149.

[Tit74]   J. Tits, *Buildings of Spherical Type and Finite BN-pairs*, Lect. Notes Math. **386** Springer-Verlag, Berlin, 1974.

[Tit82a]   J. Tits, A local approach to buildings, in *The Geometric Vein (Coxeter–Festschrift)*, Springer Verlag, Berlin, 1982, pp. 519–547.

[Tit82b] J. Tits, Remarks on Griess' construction of the Griess–Fischer sporadic group, I, II, III, IV, Mimeographed letters, 1982–1983.

[Tit83] J. Tits, Le Monstre [d'après R.Griess, B.Fischer *et al.*] *Seminar Bourbaki, 36-e année,* 1983/1984 No. 620, pp. 1–18.

[Tit84] J. Tits, On Griess' 'Friendly giant', *Invent. Math.* **78** (1984), 491–499.

[Ve98] A. Venkatesh, Graph coverings and group liftings, Unpublished manuscript, 1998.

[Vi94] J.W. Vick, *Homology Theory: An Introduction to Algebraic Topology,* Springer-Verlag, New York, 1994.

[W38] E. Witt, Über Steinersche Systeme, *Abh. Math. Sem. Univ. Hamburg* **12** (1938), 265–275.

Index

amalgam
    method, xi
associator, 225
automorphism
    central, 66
    standard, 52
    twisted, 52
automorphism group
    code, 23

Baby Monster
    amalgam, 185
branch
    core, 130
bunch of roots, 130

code
    doubly even, 2
    even, 2
completion, 42
    faithful, 42
    generating, 42
    group, 42
    map, 42
    universal, 42
Conway
    group, 1

dent, 170
dodecad, 3

extension
    extraspecial, 23

form
    trace, 149

fusion rules, 205, 208

geometry
    tilde, 111
Golay
    code, 2
    set, 2
graph
    collinearity, 229
    short vector, 165
    transposition, 230
Griess
    algebra, xiii, 123, 148
group
    Baby Monster, xii
    extraspecial, 13
    hexacode, 32, 63
    Higman–Sims, 197
    Monster, 1
    trident, 171

heptad, 4
hexacode, 31
holomorph, 14
    standard, 15
    twisted, 15

Ising
    vector, 211

Jordan
    algebra, 156

Krein
    algebra, 126

Leech
  frame, 26
  lattice, 25
local isomorphism, 229

Majorana
  axial vector, xiii, 210
  fermion, 211
  involution, xiii, 210
map
  intersection, 10
Mathieu
  group, 1, 4
Miyamoto
  involution, 211
module
  anti-heart, 11
  anti-permutation, 10
  elementary induced, 129
  Golay code, 1, 5
    irreducible, 5
  hexacode, 32
  multiplicity-free, 126
Monster
  amalgam, xi, 41, 76
  graph, xii, 230
  group, 42

Norton
  inequality, 209, 224
  pair, 122

octad, 3
  sublattice, 115

pair, 3, 91
Parker
  loop, 16, 59

quasi-complement, 66

reper, 130
representation
  monomial, 128

semidirect product
  partial, 8
sextet, 3, 31
subgroup
  monomial, 27
  singular, 111
  three bases, 37, 38

tetrad, 3, 31
Todd
  module, 1, 3, 5
transformation
  adjoint, 149
triple, 3

uniqueness systems, 234

vector
  axial, 201